Terminal Signs

Approaches to Semiotics
90

Mouton de Gruyter
Berlin · New York

Terminal Signs

Computers and Social Change in Africa

by

Bennetta Jules-Rosette

Mouton de Gruyter
Berlin · New York 1990

Mouton de Gruyter (formerly Mouton, The Hague)
is a Division of Walter de Gruyter & Co., Berlin.

⊚ Printed on acid-free paper.
(ageing resistant — pH: 7, neutral)

Library of Congress Cataloging in Publication Data

Jules-Rosette, Bennetta.
Terminal signs : computers and social change in Africa / by Bennetta Jules-Rosette.
p. cm. — (Approaches to semiotics ; 90)
Includes bibliographical references.
ISBN 0-89925-616-3 (alk. paper)
1. Computers—Social aspects—Africa. I. Title. II. Series.
QA76.9.C66J85 1989
303.48′34′096—dc20 89-13751
CIP

Deutsche Bibliothek Cataloging in Publication Data

Jules-Rosette, Bennetta:
Terminal signs : computers and social change in Africa / by Bennetta Jules-Rosette. — Berlin ; New York : Mouton de Gruyter, 1990
(Approaches to semiotics ; 90)
ISBN 3-11-012221-9
NE: GT

Typesetter: Asian Research Service, Hong Kong. — Printing: Ratzlow-Druck, Berlin. — Binding: Dieter Mikolai, Berlin. — Printed in Germany

For the Kenyan and Ivorian computer professionals,
whose courage will shape the future

Preface

Technology's impact on social change in two African settings is the subject of this study. The preparation of this book has involved field research over a four-year period in Kenya and Ivory Coast. A longitudinal study has permitted me to trace the enormous changes that have taken place in the availability of microcomputers in both settings and the growth of firms and educational institutions dedicated to computer use. Part I of this book establishes the theoretical and ethnographic background for this comparative study. Part II presents ethnographic data on computer adoption drawn from work and educational settings in Kenya and Ivory Coast. Part III concludes the book with a semiotic analysis of computer adoption and post-modernity in Africa. Many institutions and individuals have supported me in this exploration of Africa and the computer. They include intellectual dialogants, research organizations, granting agencies, and, above all, the computer professionals of Kenya and Ivory Coast, whose inspiration helped me to conceive this study.

In 1984, the first phase of this study was funded by Grant #Int. 84-07272 from the International Division of the National Science Foundation. Mr. Robert Bell, then head of the Subsaharan Africa Division of the Science in Developing Countries Program at the National Science Foundation, was instrumental in assisting me to develop the research design for the pilot study. He provided me with invaluable contacts in Kenya and monitored the initiation of the research project in Nairobi. The computer experiment described in Chapter Five of this book was a byproduct of my discussions with Mr. Bell. I would like to thank Mrs. C.A. Mwango of the Office of the President in Kenya for her attention to my research projects over the years and her personal concern with my progress. Mr. Aleke-Dondo of the National Council for Science and Technology in Nairobi was helpful in guiding me in the design of the early stages of this project.

The second phase of this project was supported by a fellowship from the John Simon Guggenheim Memorial Foundation in 1985 and 1986. This fellowship enabled me to conduct follow-up research on small firms adopting computers in Kenya and to spend an academic term as a research fellow at the Development Centre of the Organization for Economic Cooperation and Development (OECD) in Paris. In particular, I appreciate the assistance of Jean Bonvin, Director of Coordination for the Development Centre during my fellowship period, and the encouragement of my OECD research colleagues, Carliene Brenner and Winifred Weekes-Vagliani, who taught me much about the interrelationship of technology and economy.

During the 1985 research period in Kenya, I was assisted in the collection of interview data by Mary Omosa, a graduate student in the Department of Sociology at the University of Nairobi, and David Holzman, then a doctoral candidate in the Department of Sociology at the University of California, San Diego. Both Mary and David added pertinent insights to this study and alerted me to important trends in Nairobi's business world. Willie Mushioki, a member of the Nairobi Handicraft Cooperative, assisted with local interviewing and participated in the computer experiment at the cooperative. To Willie and to the management and staff of the Nairobi Handicraft Cooperative, I am more than grateful for their time, support, and eagerness to help with the computer experiment. They ventured into unknown territory with me and shared a mutual learning experience. Kevin Crews, then a Peace Corps volunteer in Nairobi, enthusiastically contributed his advice and expertise by monitoring the computer experiment in 1984 and 1985. The firm managers, computer professionals, and computer vendors in Nairobi kindly consented to share their time and expertise with me. Their interviews are the empirical foundation of this study.

The faculty members of several educational institutions in Nairobi have provided me with research support, critical feedback, and intellectual stimulation. Dr. Erasato Muga, former chairperson of the Department of Sociology at the University of Nairobi, helped me to initiate this project. As acting chairperson of the Department of Sociology, Dr. Roberta Mutisyo was instrumental in the research clearance process and offered helpful comments on this study. Dr. Joshua A'Konga of the University of Nairobi has long been supportive of this project and of my previous field research efforts in Kenya. I owe him special thanks.

Professor Robert J.P. Scott, former head of the Institute of Computer Science at the University of Nairobi, has reviewed draft research proposals, offered critical comments, and shared with me his expertise and extensive publications on computer education in Kenya. I hope that he will find my interpretations of his research of particular interest. John P. Nordin, formerly a member of the faculty at the Institute of Computer Science, directed me toward firms to study and was an important sounding board for my research questions.

As computer instructors, David Wilson and Glyn Turner of the Starehe Boys Centre along with Brian Wray of the Aga Khan Academy were helpful in introducing me to computer education in Nairobi's secondary schools. Special thanks are due to Mr. Oyomno, head of the Mathematics, Statistics, and Computing Department of Kenya Polytechnic; Mrs. N.G. Fazal, cofounder and director of the Computer Training Centre in Nairobi; and Mr. Nishit Chauhan, head of the Computer Department at Graffins College in Nairobi. To the computer students and graduates of these institutions, I extend my admiration and encouragement in their quest for a bright future as computer professionals.

The 1986 phase of my research in Kenya was funded by a stipend from the Transaction/Wiley Matching Gift Program and a research grant from the Academic Senate of the University of California, San Diego. I would like to thank Irving Louis Horowitz of Rutgers University and Transaction Publishers for his support and special interest in my research. Mary Friedland, former editor-in-chief of *East African Computer News*, provided helpful research suggestions and offered me a complete set of back issues of the journal for 1985 and 1986. Excerpts from articles in the 1985 and 1986 issues of *East African Computer News* have been reproduced with the permission of the journal.

For unflagging personal support in Kenya, I would like to thank the family of Dr. Curtis Powell without whom my 1986 research period in Kenya would have been impossible. Mr. Noah N. Gathenge of Nairobi and Gatundu, Kenya saw to it that my interview schedule was maintained in the face of extraordinary obstacles. All of these individuals contributed, in their own ways, to making my research in Kenya come to fruition.

The Ivory Coast research in 1987 was funded by Grant #4863 from the Wenner-Gren Foundation for Anthropological Research and a research grant from the Academic Senate of the University of California, San Diego. I would like to thank the Wenner-Gren Foundation for its support and helpful comments on this research project and on my previous field studies in Africa. In 1987, Dominique

Glaub, a doctoral candidate at the U.E.R. de Sciences Sociales of the Université de Paris V, accompanied me on interviews in Abidjan, tirelessly sought new research contacts, and attuned me to crucial nuances of the Franco-Ivorian business world.

I especially appreciate the interest taken in my work by the Université Nationale de Côte d'Ivoire and the Secrétariat Général à l'Informatique of Ivory Coast. I could not have proceeded without the kind of support and careful assistance of the Secrétariat's staff, in particular Rose Balou. As a young research engineer, Rose deepened my understanding of the role of women computer professionals in Ivory Coast and on the international scene.

Mr. Fofana Zémogo, Secretary General of the Association Professionnelle des Informaticiens de Côte d'Ivoire, generously gave of his time to interpret the complexities of the computer profession in Ivory Coast to me. Mr. Seri-Bi N'Guessan, head of the Central Informatics Service at the Institut National Supérieur de l'Enseignement Technique (INSET) in Yamoussoukro, Ivory Coast presented an extensive discussion of the institute's organization and made special efforts to make me feel welcome there. Blagué Valet and Raymond Kragby of SIEPAT-Ordinateur, a computer college in Abidjan, graciously opened their institute of learning to me. To these individuals and many other computer professionals and educators in Ivory Coast, I owe a special debt of gratitude.

Scholars on both sides of the Atlantic have offered helpful comments and suggestions on this manuscript. The inspiration of the groundbreaking semiotic studies of Thomas A. Sebeok and Algirdas Julien Greimas should be evident throughout this volume. I received incisive comments on earlier drafts of this manuscript from Paolo Fabbri and Manar Hammad of the Groupe de Recherches Sémiolinguistiques in Paris and Serge A. Tornay of the Université de Paris X. Nelson H. Graburn, John U. Ogbu, and Neil J. Smelser of the University of California, Berkeley and Gerald M. Platt of the University of Massachusetts, Amherst provided valuable suggestions as the research project unfolded. My University of California colleagues Jeffrey C. Alexander, Teshome H. Gabriel, Dean MacCannell, and Juliet Flower MacCannell challenged me with provocative debates about postmodernity. Rae Lesser Blumberg, Alain J.-J. Cohen, Hugh B. Mehan, and Allan Mitchell of the University of California, San Diego voiced caveats and encouraging comments. Julie Chappell of the University of Washington, Seattle and Tom Myles of TLM Datamatics in Berkeley, California offered useful remarks on several draft chapters. Members of my 1985 and 1986 advanced seminars in the sociology of knowledge at the University of California, San Diego also engaged in lively discussions of these chapters.

The Humanities Research Institute of the University of California, Irvine was a stimulating experimental environment in which to complete the final draft of this manuscript. I appreciate the interest of the Institute's founding director, Murray Krieger, in the implications of my research as a case study in the applied humanities. Productive discussions with fellow researchers and the support of the Institute's staff saw me through the final stages of writing this book.

Suzanne Grant Lewis of Stanford University has been a fellow traveler in the journey toward understanding new technologies in Africa. My colleague, Pierce Julius Flynn, has read many versions of this manuscript and collaborated on development of the framework for analyzing audiovisual transcripts used in Chapter Seven. No words can express my thanks to him.

Over the years, several people have been involved in the transcribing of interviews and the preparation of this manuscript. They include Berta Alexander, Paula Demong, Theodora Garr, Sylvia Moeller, and Muriel Vitaglione. Sallie Bayless has energetically and valiantly seen the manuscript preparation on this project through to the end. I owe special thanks to the staff of the Office of Learning Resources at the University of California, San Diego for their attention to detail in the reproduction of tables and diagrams for this book.

Finally, I would like to thank my parents, Walter and Bennetta Washington, for their confidence in my work, financial support, and willingness to share the daily trials of research and writing with me. I appreciate the patience and moral support of Violaine and Fred Thompson. It is my hope that the young computer professionals of Kenya and Ivory Coast will find a genuine source of encouragement in this book.

Figure 1-5 in Chapter One has been reproduced with permission from *La Micro-électronique, la Robotique, et l'Emploi* (1982:25) published by the Organization for Economic Cooperation and Development in Paris. The depiction of computer specialists in Figure 2-1 of Chapter Two has been reproduced with the permission of *East African Computer News*. In Chapter Four, Figure 4-1 has been translated and reproduced with permission from Ivory Coast's *Plan Informatique, 1981-1985* (1981:20 bis), and Appendix A has been reprinted with permission from the *Plan Informatique National, 1986-1990, Annexe II* (1986:68). Information contained in Appendix B has been reprinted with permission from *Computers Kenya* (1982: 41, 42, 44, 46, and 47) by Robert J.P. Scott of the Institute of Computer Science, University of Nairobi. Two Kamba folktales have been excerpted and quoted with permission from John Mbiti's *Akamba Stories* (Claredon Press, 1966:64, 134-135). I would like to thank these publishers for their cooperation.

La Jolla, California *Bennetta Jules-Rosette*

Contents

Part I: Computers in an Alien Environment 1

Introduction. 3

Introductory Remarks 3
Plan of the Book 5
The Methodological Challenge 9
Illustrations of Technological Change 10
The Broken Promise of Modernity in Africa 13
The Simulation of Postmodernity 16
Computers and Office Work 17
The New Computer Professionals 21
Conclusions 25

Chapter One
The Symbolic and the Social:
Computer Use in Two African Settings 27

Structuring Interaction with Computers 27
Two Models of Narrative and Performance 28
A Narrative Program for Computer Adoption in Africa 30
Social and Economic Environments of Computer Use in
Kenya and Ivory Coast 41
Focus on Small and Medium – Sized Enterprises 43
External Computing and Internal Change 45
New Technologies as Representations of Development 48

Chapter Two
Computer-Related Successes and Excuses:
The Discourse of Confrontation 51

Introduction 51
The Discourse of Human-Machine Confrontation 53
Computer-Related Successes and Excuses 57

New Technologies as Bureaucratic Justifications 71
Conclusions 75

Chapter Three
New Technicians of the Sacred:
Technology, Belief Systems, and Social Control 77

Introduction 77
New Technicians of the Sacred: Technology and
Sectarian Processes 78
External Control of Technology and Religion 85
Computers as Symbols of Change 98
Technology and Religion as Modeling Systems 100
Conclusions 103

Part II: New Technologies, Work Organization, and the Administrative Revolution 105

Chapter Four
The Last Train of the Twentieth Century:
The Computer Revolution in Ivory Coast 107

Abidjan On-Line 107
Interpreting Computer Adoption in Ivory Coast 110
The Ivorian Informatics Plans 113
INSET: The University of the Future 118
The Role of the Commercial Schools 123
The Computer Consulting Firms (SSII) 129
Overview of the Computer Consulting Firms 145
Computer Vendors 148
Conclusion: The Ivorian Picture in Comparative Perspective 152
Appendix A 155

Chapter Five
Display, Domination, and Mastery:
Computers in the Kenyan Setting 157

Decentralization vs. Centralization: The Background for
Computer Adoption 157
Application of the Narrative Program to the Kenyan Setting 158

Six Computer Case Studies 160
Implications of the Kenyan Data 183
The Computer Experiment 191
The Educational Ceiling 198
Conclusions 200
Appendix B 203

Chapter Six
Socializing Workers into the New Mechanical
Solidarity 205

Philosophies of Computer Education in Kenya and Ivory Coast . . 205
Computer Training and Employment in Comparative Perspective 212
Public Training Institutions 221
Secondary Schools 226
Commercial Colleges 229
Profiles of Starehe Graduates 238
The Conversion Experience in Computer Education 244

Part III: Simulating Postmodernity 247

Chapter Seven
Indigenizing the Computer:
Social and Interpretive Practices Surrounding New Technologies . . 249

The Computer as a Tutor 249
Computers and the Semiotics of Invention. 253
Analysis of the Image-Discourse Sets 256
Computers and the Discourse of Domination 269
Learning to Labor with Computers 275
The Politics of Signification in Human-Machine Interaction 277
Appendix C 280

Chapter Eight
The Computer Contract:
A Sociosemiotic Analysis of Computer Adoption 283

Introduction 283
Contractual Structures 284
Contractual Exchanges in Computer Adoption 288

Text and Paratext in Computer Use 310
The Simulacrum and the Model 312
Information and Ideology. 317

Chapter Nine
Terminal Signs 321

Introduction. 321
Signs of Postmodernity. 323
Postmodern Narrativity African-Style. 335
Africa and the Computer 342
Africa in the Postmodern Era. 344
Conclusions 345

Notes 349

Glossary 371

References 379

Index 395

Part I
Computers in an Alien Environment

Introduction

> The increasing skill, speed, and scope of popularizing and disseminating knowledge enhances the prestige of the trial-and-error method at the expense of the old-fashioned inculcation of sacred or quasi-sacred texts (Karl Mannheim, 1950:250).

Introductory Remarks

Prometheus, the mythological son of a Greek Titan, was condemned to perpetual torture for introducing fire to mortals. The Promethean myth is a cultural narrative that reveals a long and deep-seated fear of human mastery through technology. Yet, advances in technology are at the core of contemporary society. New Technologies have radically changed the business world and the international economy. These technologies have the potential for changing the face of Africa. Africa, however, has been ignored in the equation of technological change.

By Africa, I mean not only a continent with a specific cultural and political history but also a symbol inherent in modern thought. This invented Africa has been the nostalgic symbol of tradition and, more recently, of unfulfilled struggles. In this book, I present the results of a comparative study of computer adoption in Kenya and Ivory Coast. The conclusions, however, are not limited to these African settings. Computer adoption has invariant features associated with technological learning in any cultural context. In situations where new technologies introduce the potential for abrupt change, computer adoption is a qualifying test and a challenge.

Technology's relationship to culture is much debated and often misunderstood. The signs and symbols that comprise our cultural understandings form the backdrop for receptivity to new technologies. Therefore, semiotics, the study of processes of signification, offers effective conceptual and methodological tools for examining the computer as a cultural object. Programmatic studies assessing the impact of new technologies for purposes of policy and prediction begin by enumerating machines and measuring changes in work productivity (cf. Wad and Radnor, 1984:18-20). The human element involved in learning to use these technologies is often overlooked.

Interpretations of technology emerge through talk and human interactions with technological objects. Therefore, much of this book will be devoted to discourse about computers at various stages of their use. The semiotic study of narrative will be employed as a framework for analysis.

Several assumptions support a narrative approach to technology, culture, and society. Each culture embodies narratives about the transformation of human potential through the mastery of technology and the domination of nature. The Promethean myth still lives. Technological mastery promises a utopian world in which the forces of nature can be conquered. In discussing the computerization of society, Daniel Bell (1981:xi) refers to the "magic wand" of technological change through which communications, the economy, and lifestyles are transformed. Expectations that such transformations will, indeed must, occur are embedded in cultural narratives that are played out in contrasting ways under different political, economic, and social constraints.

In this book, I apply the program for the study of narrative developed by the semiotician A.J. Greimas to the analysis of cultural understandings and social processes involved in computer adoption.[1] My version of the narrative program has four stages: (1) confrontation of the human subject with new technologies; (2) mastery of technology; (3) attribution of meaning to the technologies mastered; and (4) final narrative resolution through a process of terminal signification. In Kenya and Ivory Coast, the two cases that are my primary foci, the narrative program unfolds in contrasting ways. In Kenya, a fragile computer contract is established in which new technologies are both promoted and feared. This ambivalent response to computers and related technologies denotes a reaction against foreign intervention counterbalanced by a dream of progress. The ambivalence toward technology runs deep in Kenyan society. It is reflected in traditional folktales, everyday discourse, media reports, and government policies. In contrast, Ivory Coast more openly embraces new technologies. Through a progressive computer contract, the Ivorian government attempts to promote and regulate the use of computers and new telecommunications systems as tools for economic development and sources of cultural capital.

The meaning of discourse is shaped by the cultural narratives that ground it. Greimas (1987:64-65) asserts:

> [T]he generation of meaning does not first take the form of the production of utterances and their combination in discourse; it is relayed, in the course of its trajectory, by narrative structures, and it is these that produce meaningful discourse articulated in utterances.

As I interpret it, this statement does not mean that all utterances in everyday discourse constitute a narrative about the subject's actions. Rather, a narrative program that comprises the goals and motivations of acting subjects underscores discourse and offers the structural basis for its interpretation. This assertion is pertinent to analyses of discourse about technology in which the human subject is both the sender and receiver of messages and is transformed by interaction with the technological object. A narrative analysis is relevant to studies of culture and technology because narratives contain presuppositions that emerge in talk and documents about new technologies.

Plan of the Book

This book is divided into three parts that parallel the confrontation, domination, and attribution stages of the canonical narrative program. Part I applies the narrative program to an analysis of the initial stage of confrontation with computers in two African settings. Chapter One describes the social and symbolic contexts into which computers are introduced in Kenya and Ivory Coast. In Chapter Two, I analyze the discourse of confrontation found in computer-related success and disaster stories from Kenyan and Ivorian firms. These micronarratives exhibit similar patterns of utopian discourse, oriented toward final outcomes rather than procedures of computer use. The discourse about the benefits and shortcomings of computers is framed in the subjunctive mode and addresses the unrealized technological potential of computers.

Chapter Three examines the emerging roles of computer professionals – the new technicians of the sacred – as they carve out a domain of expertise in changing societies. Mastery of technology is compared with the acquisition of secret religious knowledge, and the role of technology in African societies is examined as a new arena of knowledge control. The discourse of confrontation, which has utopian religious characteristics, is the linguistic tool of the new technicians.

In Part II, I present case studies of computer adoption in Ivory

Coast and Kenya. Chapter Four contains an analysis of Ivory Coast's progressive computer contract. Following the example of the 1978 Nora/Minc report in France, Ivory Coast sponsored the first Conference for Integration of African Informatics in 1979 and has produced two national plans for the regulation of computer-related technologies. These plans cover areas ranging from education to computer contracts and sales. Their objective is to bring Ivory Coast into the twenty-first century as an important contender in technological development on the world scene. I use these plans to analyze the discourse of public policy in Ivory Coast and the relationship of public discourse to computer adoption. Focus is placed on the computer and engineering consulting companies (SSII) that are mushrooming in Ivory Coast. Their competition with major international vendors introduces fascinating complications into the narrative of computer adoption. The government attempts to regulate this competition by favoring certain vendors and computer consulting companies in their bids for government sales and contracts. In spite of the growing market for computers in Ivory Coast, the government remains the largest purchaser of computer equipment. In Chapter Four, I analyze the consequences of computer diffusion as it is depicted in interviews with computer managers and vendors. In addition, I examine the relationship of the market for computer sales and services to the educational system in a brief profile of INSET, Ivory Coast's University of the future.

The narrative of technological domination and mastery proceeds with the Kenyan case in Chapter Five. The government's ambivalence toward the importation of computers by private companies is evidenced by high import duties and taxes that restrict the availability of computers. The results are importation of outmoded equipment and attempts to circumvent government regulations. In the absence of a formal plan coordinating computer policies, the Kenyan situation superficially appears to be one of unbridled free market competition.

In fact, conditions for the sale of computers in Kenya have been so restrictive that major international vendors have withdrawn from the country. The economic impact of the withdrawal of these multinational firms affects all sectors of the business world and not merely the computer industry. Computer-related success stories appear in firms where the management and computer personnel have made special efforts to obtain new technologies. Case studies of successful computerization in Kenyan government agencies exhibit a pattern of reliance on foreign consultants, who have been quick to publicize their efforts. The lasting effects of these projects, however, are subject to much speculation. In the domain of education, the consequences of weak technological planning are evident in the rise of commercial computer colleges. Although these colleges respond to a felt need in the Kenyan economy, their quality and standards are mixed, and the criteria for their accreditation are unstable.

On the level of discourse, the fragile computer contract is manifested in considerable optimism at the moment of initial confrontation with new technologies. Frustration emerges during the implementation process and is evidenced in complaints about incompatible and outdated computer systems. Computer managers and systems analysts whom I interviewed in Kenya used the ambiguity of this situation to improve their job positions and increase control. They have employed technology as a source of professional entitlement. This process entails what Harold Garfinkel (1967:200-201) has referred to as the "entitlement to read" organizational texts in a privileged manner. Entitled reading enables computer experts to recode and reinforce their positions as new technicians of the sacred.

In Chapter Six, I conclude the comparative ethnographic study of the Kenyan and Ivorian settings with a focus on computer education. I examine the process of socializing workers into the new mechanical solidarity by providing an overview of public and private computer training institutions in Kenya and Ivory Coast. Using interviews with computer educators and employers, I contrast the effects of the fragile and progressive computer contracts in terms of educational content, evaluations, and outcomes. This chapter concludes with profiles of two young Kenyan computer professionals. Their cases illustrate individual successes within a restrictive technological environment.

Part III begins the shift from the domination stage of the technological narrative to final attribution. In Chapter Seven, I continue the focus on computer education with microsociological and sociosemiotic analyses of learning to labor with computers. I examine student programming sessions at a secondary school in Nairobi and emphasize the similarity between human-computer interactions and classroom instruction. A moment-by-moment analysis of videotaped learning sessions reveals the students' tendency to focus on acquisition of technical programming skills rather than on the cultural content of their lessons. The students' technical grasp of the computer's functions is more profound than their understanding of the questions that the computer program asks about Western culture. Classroom instruction is elliptical, ritualized, and mechanical, following the pattern of the student-computer interactions. I compare the students' emergent mastery of computer skills with a computer operator's confident control of his work situation at a Kenyan car dealership. In this analysis, I probe the ritualized ways in which the computer is manipulated as part of a code of empowerment in the business world.

In Chapter Eight, I analyze the contractual exchanges that emerge at the final stage of the narrative program for computer adoption. By explaining how contracts are interpreted in everyday discourse, I demonstrate that the concept of contract is at the core of all business transactions involving computers. Contractual exchanges between the government and firms and between firms and clients constitute the fragile and progressive computer contracts. I contrast the ways in which these contractual exchanges occur in Kenya and Ivory Coast with reference to the narrative program of computer adoption. This chapter combines semiotic theories of narrative with ethnographic and textual data.

The book concludes with an analysis of terminal signs. These signs are the final narrative utterances that draw the technological scenario to a close. Terminal signs contain an ultimate attribution of meaning. These signs reverse the narrative program. Formal attribution permits the reinterpretation of all previous successes and struggles. While terminal signs bring the narrative to a close, they simultaneously deconstruct it. And a new scenario unfolds. Terminal signs index the reflexive status of the narrative program. Rather than being a method for the assessment of new technologies, these signs link technology to culture. They indicate how the meaning of technological objects is incorporated into discourse, gestures, and social actions. Four terminal signs are identified: (1) modeling devices, (2) representational signs, (3) totalizing signs, and (4) analeptic signs. I use these signs to develop a cultural projection for Africa in the postmodern era. At the end of the book, I provide a glossary of technical and popular computer-related terms used in Kenya and Ivory Coast.

The Methodological Challenge

Many studies of technology are unsatisfying to scholars interested in cultural analysis. The meaning of technologies for their users is either ignored or hopelessly stereotyped. The major methodological challenge that I have faced during the course of this study has involved confronting how the users of new technologies in Africa view themselves and their technological objects. What cultural messages do they send and receive via new technologies? In responding to this question, I have employed semiotics as a methodological resource. Semiotic theory, however, does not automatically translate into the social applications of semiotics, or a sociosemiotics (cf. Greimas and Courtés, 1979:355-358; Jones, 1985:235-252). This uncharted region is fertile ground for transdisciplinary collaboration among social scientists and semioticians. In this book, I shall attempt to steer a course through these waters by combining development studies, cross-cultural research, and ethnography with the methodological tools provided by the semiotic analysis of narrative and discourse. This combination requires the intersection of diverse and mutually enriching literatures. The ways in which the signs and symbols of new technologies are manipulated in Africa increase our knowledge of processes of signification. Conversely, an understanding of the cultural narratives that produce the receptivity to technological innovation is essential to a grasp of the challenges faced in Africa as the world enters the post-industrial era. Postmodern culture is characterized by the breakdown of the narratives of certainty and progress that define modernity. New technologies challenge the meaning structures of modernity wherever they appear.

My methodological task demands imparting ethnographic and historical facts about the emergence of postmodernity in Africa. Although ethnography's death and its replacement by the "writing of culture" has been predicted (cf. Clifford, 1986:6-8), I shall argue that our ability to understand cultural narratives depends precisely upon resuscitating ethnography. Part II of this book emphasizes the essential place of ethnography in cultural interpretation. Nevertheless, for analyses of cultural change, ethnographic studies alone are not sufficient. Examination of the fragile and progressive computer contracts in Kenya and Ivory Coast is supplemented by both a comparative method and a semiotic approach.

Discourse is linked to narrative structures by more than a cumulative process. Emanuel Schegloff (1987:229) argues: "Efforts to link the level of culture and society in the search for variation are unassured of success and uncertain in motive."[2] He claims that the analysis of speech exchange systems automatically contains a delineation of the contexts in which this speech occurs. Instead, I am utilizing the structure of cultural narratives as a resource for interpreting discourse. This approach permits analysis of spoken discourse, interaction, gestures, and written texts as part of the narrative program. Consequently, the distinction between the microanalysis of discourse and the structural analysis of cultural meaning is muted. Discourse takes on its significance as it unfolds with in the narrative program. In agreement with Schegloff, I contend that the specific work, educational, and policy-making settings in which this discourse occurs are less important than the structures of exchange that they exhibit. However, I locate these structures with reference to the progressive stages of the narrative program for computer adoption. This narrative program is relevant as an object for sociosemiotic study only as it is thematized and displayed in interactions and talk about technology. Analysis of the links between social interaction and cultural narratives requires ethnographic studies.

Illustrations of Technological Change

Three assumptions are basic to my approach. (1) *Computers are actors in a technological narrative.* (2) *Computers are agents of social change.* (3) *Computers are interactants in the discourse and practices of the workplace.* Computers have been a force in African development over the past two decades. Only recently, however, have computers contributed to large-scale transformations within the workplace. Mainframe computers used in government planning for African nations have been recordkeepers. They have served primarily as memory devices storing outdated information.

A senior programmer at a Kenya wholesale outlet attempts to manage the transition from a labor-intensive NCR keypunch computer to a new electronic multi-user system while increasing the number of employees for whom he is responsible. If he increases the number of employees, although their individual work tasks have diminished, he will obtain the title of data processing manager and double his salary. Government administrators in Ivory Coast strive to develop a computer policy that will limit the activities of multinational computer vendors such as IBM, BULL, and UNISYS, while simultaneously encouraging these companies to invest in the country. The Kenyan government organizes the computerization of key government ministries but imposes stringent restrictions on public access to computers.

These cases are not merely management conundrums. They share a common theme. Each case illustrates an effort to manipulate the narratives of public discourse in order to delimit everyday practices that constitute the adoption of new technologies. Computer policies project a specific representation of development and change. In some African cases, such as Mali and Sudan, new technologies figure only dimly on the agenda for development planning. In other cases, such as Ivory Coast and Gabon, computers have become the technological means for entering the postindustrial era. A comprehensive analysis of the impact of computers on the African continent would encompass all of these examples.

In this study, I have selected Kenya and Ivory Coast as cases in which computer use is well underway. Yet, strikingly different technological orientations have been taken in the two countries. Adoption of computers in Kenya has been slow and unregulated. In Ivory Coast, the approach to computerization has been utopian and progressive. As of late 1986, the Office of the President in Kenya had appointed representatives from several ministries along with foreign consultants as members of a commission charged with developing a national computer policy. Computerization of government ministries, district offices, and educational institutions was to be a primary concern of the commission. In the end, this commission did not manage to devise a national computer policy. On the other hand, Ivory Coast's national computer agenda emphasizes the positive consequences of regulating computerization and automation. The goal of the national informatics plans is to establish Ivory Coast as a major regional leader in engineering research and development.

Émile Durkheim's classic society under mechanical solidarity is one in which the division of labor is simple, and social control is direct and repressive (cf. Durkheim, 1949:61-69). New technologies leave behind the evolving functional division of labor that Durkheim describes as organic solidarity. Postmodern culture has been heralded as "haunted by the . . . representation of a lost 'organic' society" (cf. Lyotard, 1984:15). According to this view, automation and electronic technologies eradicate old forms of work specialization and increase the purview of state control. They introduce a new mechanical solidarity. Some authors argue that alternative forms of societal analysis must be developed to examine the postindustrial era (cf. Bell, 1973; Habermas, 1975; Gouldner, 1979). Daniel Bell (1973:20) has described scientific progress, theoretical knowledge, and the control of technology as key principles motivating policy formation in post-industrial society. The attention of sociologists and popular critics concerned with postmodern discourses and cultural forms has focused almost exclusively on the United States, Western Europe, and Japan. African nations have been included in this scenario only in order to illustrate their losing political and economic positions in a competitive technological game. My research reveals that postmodernity has arrived in Africa. Its advent is exhibited by the spread of new technologies and changes in public discourse. New technologies have become the hope for a great leap into a future-oriented Africa.

Computers are a concrete manifestation of changes in class structure and indigenous control in postcolonial Africa. They provide a point of entrée for studies of the new professional class. Computers offer a way of seeing the modeling of postmodernity in microcosm. The channels through which computers enter a Third World nation reflect existing patterns of domination and state control. Areas of indigenous entrepreneurship are those into which microcomputers have the potential to be used in the most innovative ways.

By examining the computer as a symbol of change, it is possible to juxtapose old and new work patterns, aspirations, and organizational styles among office workers. The innovations of these workers resemble those encountered in the informal worlds of trade and commerce in African cities. My previous studies of social change in Kenya focused on the emergence of a new commercial form – the grass-roots factory.[3] I gradually became aware that my ethnographic research also afforded me a glimpse of a much larger social and economic picture. I sought a way to utilize ethnography to uncover the formal world of African business, which I assumed was, on some level, driven by local innovations. In *Symbols of Change* (cf. Jules-Rosette, 1981:12), I argued for an analysis of how "interpretations of the past are recreated as new experiences arise." In this study, I have continued to explore the ways in which these new beliefs crystallize as symbols of modernity and postmodernity.

The open investment climates in Kenya and Ivory Coast led me to believe that computers were important concerns for businesses dealing in foreign capital. The 1976-80 development plan for Ivory Coast was designed to encourage foreign investment in all economic sectors. During the 1970s, Masini *et al.* (1979:9) found that direct mergers of capital between major foreign investors and local firms in Ivory Coast were still "relatively rare." Hence, I wished to investigate whether computer use would be more widespread in private sector enterprises involving substantial outside investment than it is in the government and small local enterprises. As a result of the growing urban and rural insider elite in Kenya, I had expected to find a larger number of computers in use among smaller, local-level enterprises. The differences in official attitudes toward technology in both countries, however, created a situation quite different from the one that I had anticipated. A positive computer policy stimulated the spread of computerization across economic sectors in Ivory Coast whereas lack of regulation produced mixed results in Kenya.

The Broken Promise of Modernity in Africa

Self-determination in many African nations brought with it the bright promises of opportunity and modernity. Pre-independence movements throughout the continent fueled a restratification of local societies. Mining, the railroads, and the heavy industrial interests of the colonial era forged a new African proletariat. In countries with an agrarian base such as Kenya and Ivory Coast, a rural proletariat emerged. Integration of more and more individuals into the wage labor system expanded the government's control over community life. However, contrary to many Marxist theories of development, state control does not determine all forms of cultural and political adaptation to social change. At one dialectical moment, economy influences society. More generally, subtle patterns of cultural innovation condition visible changes in social and economic organization. This assertion emphasizes the importance of a theory of symbols of change for interpreting the driving forces of African modernity. The broken promise of modernity refers to the sense of disillusionment generated by the social and economic obstacles to development and change in newly independent African nations. The dream of postmodernity involves the wish to fulfill and move beyond the broken promise.

With independence, the new elite and the rising bourgeoisie influenced the course of development chosen by African nations (cf. Langdon, 1981:3). Both the proletariat and the *compradore* bourgeoisie were products of the colonial political and economic experience. Colonialism introduced modern bureaucracy as a form of administrative control. Bureaucracy brought with it the civil service and the structure and atmosphere of the modern workplace. Computers have introduced a new form of organizational hierarchy based on technological expertise. This hierarchy is not dependent upon social structures inherited from the colonial past.

The newly bureaucratized workplace embodied its own hope of progress, advancement, and opportunity. A rising group of professionals – civil service administrators and office clerks – began to differentiate themselves from the proletariat and the bourgeois business interests. Research concerning the socioeconomic status of the white-collar service class in the United States and Europe is well developed (cf. Goldthorpe, 1982:1962-185; Abercrombie and Urry, 1983:122-130). This group is distinct from the working class by virtue of its level of bureaucratic responsibility, the nature of the work contract for professionals, and its economic interests. Although white-collar employment conditions may not mold a class interest group in the African context, they suggest an expanded sphere of knowledge, influence, and privilege for office workers.

As early as 1940 in the Belgian Congo, a new group of *évolués* developed. René Lemarchand (1964:94) characterizes the *évolués* as "a new class . . . composed of clerks, medical assistants, traders, artisans, and so on." Journalists, political leaders, and religious figures emerged from this group in spite of administrative efforts to woo and co-opt them. *Évolués* are an interesting case because of their foothold in the lower levels of the civil service and their indoctrination into the world of office work. They are the predecessors of today's computer professionals.

Lemarchand (1964:96) quotes a 1945 description of the *évolués* by Antoine-Roger Bolamba, a member of their ranks and editor of *La Voix du Congolais*.

> The *évolués* are well dressed and wear fasionable clothes. They meet in front of bookshops and look for new books. They read Georges Duhamel, Henri Bordeaux, Roland Dorgelès, and all the modern writers. They discuss, rightly or wrongly, current affairs. You meet them in cafés exchanging ideas on current problems with a rare exuberance.

These Congolese *évolués*, who had their counterparts in other African nations, were part of the emerging bureaucratic elite and subelite. For example, Abner Cohen (1981:45-48) points out the importance of civil service training and administrative positions for the development and domination of Sierra Leone's Creole population. The Creoles were a colonial civil service elite whose controlling position has been weakened in the contemporary context. Armed with limited education and resources, both Creoles and *évolués* represented a cultural rather than an economic challenge to the state. What has become the *évolués* today? Justin-Daniel Gandoulou (1984) suggests that they have not disappeared. In Brazzaville, they have resurfaced as the new group of *sapeurs*. Leaving secondary education without diplomas or specialized certificates, these young people form social clubs that emphasize appearance, dressing well (*se saper*), speaking well, and traveling. They spend disproportionate sums on the latest technological gadgets (a telephone at the very least), new clothes, and their initiatory trip to Europe from which they return with more gadgets and more clothes. The *sapeurs* have replaced the broken dream of modernity by the simulation of postmodernity. Marginalized by modern life, they choose to benefit from its appearances and the allure of new technological and social forms.

The *sapeurs* and the urban subelite constitute an emerging social class in African cities. This group is young and predominantly male, notwithstanding the studies of women's clerical employment in the Zambian case.[4] The subelite has limited access to formal education and little economic control because of dependence on wage labor in the lower rungs of the white-collar world. What the subelite lacks in economic capital, however, it makes up for in terms of cultural capital or, more precisely, the desire for cultural capital and an unflagging belief in prosperity. Often school leavers, the members of the subelite are not given preference in high-level bureaucratic posts. Their credentials are vocational or job-related certificates that qualify them well for the socioeconomic niche that they occupy, but their aspirations are for wealth, status, and reputation.[5] Discontent with the broken promise of modernity, the urban subelite aspires toward a new world of social opportunity. Born neither of industrial exploitation nor economic advantage, members of the new subelite seek their rightful place in the world of urban work. Computer training provides an arena of mobility for them.

The Simulation of Postmodernity

Automated offices constitute a social microcosm in which the contemporary world is simulated and reflected. According to Jean Baudrillard (1981:10), simulation is not the creation of a referential reality, a system of signs that stand on behalf of real objects. Simulation involves reference to an infinite order of signs. The computer creates a simulation of aspects of human thought. In an applied context such as the office, the computer is part of a referential system that orders all forms of reportage, documentation, and fiscal planning. It is a work master whose systems control and organize those who use it. The computer bears the brunt of organizational and technical breakdowns. The computer-related successes and excuses discussed in Chapter Two mask a host of bureaucratic rearrangements in the workplace. In businesses, argues Rosabeth Moss Kanter (1977: 54): "Computers take over many complicated forecasting and analysis functions." In so doing, computers provide a rationale for reducing the uncertainty of the enterprise's future and the workers' roles. In the Kenyan and Ivorian firms that I have studied, computers are seldom used to make business projections and forecasts. Instead, they are harnessed to narrow economic and organizational tasks – constructing payrolls, ledgers, and stocks. In these cases, the computer's restricted use increases the workers' sense of uncertainty while expanding their repertoire of skills.

Carrying the implications of Baudrillard's argument further, the computer's simulation of social reality may be seen as a vision of the way things ought to, and will, be. The computer involves reclaiming of the broken promise of modernity. Young trainee programmers such as those whom I interviewed in Kenyan data bureaus and Ivorian computer consulting companies advance rapidly to senior programming and administrative positions and triple their initial salaries. This ever-present hope of unlimited achievement through technological competence is part of the computer's simulation of the postmodern dream. *Sapeurs* don new clothes in order to feel cosmopolitan and sophisticated. Technocrats learn new computer programs in order to gain a sense of social and bureaucratic control.

When access to the computer is manipulated skillfully, it may become the source of technological empowerment, bureaucratic control, and organizational restratification. The computer is a self-referential sign, referring back to its own presence in the workplace and to a host of changes that it instigates in the microcosm of office work. Moreover, it is a symbolic representation of the context of postmodernity in which it appears. Not only do terminal signs conclude the narrative of the computer adoption, they are also the signs of postmodernity. They emphasize the ways in which technology influences and models human conduct.

Computers and Office Work

Computer use in Kenya and Ivory Coast has characteristics not shared by Western countries. In Kenya, import duties and restrictions limit the availability of computer equipment. The likelihood of purchasing inappropriate equipment is high. Computer managers and consultants in Kenya complain about lack of vendor support for new computer systems and limitations of on-the-job training programs and vendor assistance. These complaints are less frequent in Ivory Coast where the computer market is more vigorous. In Ivory Coast, however, competition among vendors leads to the possible reduction of after sales support. There is a tendency to distribute readily available equipment rather than waiting for higher quality technologies to arrive.

In Part II, I analyze new technologies, work organization, and the administrative revolution. The rise of office work is a distinctive feature of modern industrial and postindustrial society. Michel Crozier (1971:10) describes this development as the "administrative revolution," which he believes is equal in scope and impact to the industrial revolution of the nineteenth century. The automated workplace is the seedbed of a revolution within a revolution. Bowles and Gintis (1976:204) point out that the percentage of salaried managers and professionals in the U.S. economy has increased over sevenfold since the turn of the century. Office and commercial employees more than doubled between 1940 and 1960 in the United States.[6] Between 1966 and 1973, 92% of all new jobs created in the United States were in the finance, insurance, government, and service sectors employing white-collar workers (cf. Norman, 1980:30). Although these workers number over nine million in the United States, and are on the rise, the entire investment, banking, and insurance sector in Kenya contained only 38,547 workers as of 1979.[7] There is a time lag between the occurrence of the so-called administrative revolution in

Africa and the West. With the 1976-78 coffee boom in Kenya, employment in the public administrative sector expanded. This period coincides with the widespread introduction of mainframe and batch computing into the University of Nairobi, several government agencies, and a number of private concerns.

The contrast in the size and growth of the Kenyan finance and banking sectors and those in the West is yet another indicator of the problem of economic development in the global economy and of what Colin Norman (1980:51) terms the influence of "microprocessors on political processes." These observations suggest that computer workers, and office workers generally, are part of a small segment of privileged employees in Kenya. Complaints are common throughout government, industry, and small private concerns that there are simply not enough experienced computer professionals to meet work needs. The grounds for this complaint will be discussed at more length as part of the analysis of the Kenyan computer training and the credentialing processes in Chapters Five and Six. Now it is sufficient to note that the nature of the computer training process and the crisis over computer credentials limit the available pool of workers and their level of expertise. Many computer posts in government are vacated by reskilled computer experts who leave for more lucrative jobs in private industry.

Cross-cultural literature on the effects of computers in the workplace contains mixed findings. Some studies of new technologies in the United States and Britain argue that computerization and automation centralize power and reinforce the status of managers and computer professionals at the expense of computer operators and the clerical workers (cf. Attewell and Rule, 1984:1188; Perrolle, 1987:134-135). Contrasting studies (Markus, 1984:55) note the uneasy balance of power between bureaucratic managers and computer professionals. Results concerning the increased productivity of computerized organizations vary (cf. Attewell and Rule, 1984:1184-1185). In their report on the adoption of computers in Kenya's Ministry of Agriculture, Pinckney, Cohen, and Leonard (1987:67-93) found that the introduction of microcomputers increased the ministry's efficiency and facilitated the production of accurate and timely budgetary estimates and reports. This project, which is described in Chapter Three, relied heavily on foreign consultants. Due to conflicts between bureaucratic managers and the computer personnel, the implementation of a stable local staff to handle the computerized budget was difficult.

Pinckney, Cohen, and Leonard (1987:88) conclude their evaluation of this project by stating:

> Retention of staff is a serious problem. Microcomputer use is expanding in the developing countries, and operators will be needed in the private sector. At present, there is no government mechanism for promoting skilled microcomputer operators in Kenya.

This problem re-emphasizes the strong relationship between government policy and work conditions in the automated office. In Kenya, weak national policy, fluctuating titles, and ill-defined work tasks in the computer world are evidence of a fragile computer contract in which technological empowerment is shaped by an ambivalence toward the presence of computers. Debates concerning the deskilling and reskilling of workers are less relevant than understanding how jobs are transformed and symbolically recoded by the presence of new technologies in the workplace. Ivory Coast has confronted the problem of job entitlement in its informatics plans. Educational prerequisites are outlined for computer employees, and their job titles are specified as part of the plan. This policy filters down to the level of office work. Company managers describe entry-level qualifications and advancement potential of engineers and computer programmers with reference to the guidelines specified in the informatics plans. These guidelines do not remove negotiation and conflict between computer professionals and general management. Nevertheless, they establish a framework within which new employment patterns can evolve.

Part of the colonial bureaucratic legacy in Kenya and Ivory Coast is the overbureaucratized workplace. In both government and the private sector, overstaffed offices that operate on a strict line authority model with little decision-making autonomy and initiative at middle administrative and lower levels characterize this type of highly stratified work environment. Workers and middle managers wait long hours for directives and information from the top. Because of backlogged filing and reporting systems, the information necessary for processing reports, ledgers, orders, and sales is often not forthcoming or must be sought from an outside agency. Customers' needs are seldom filled quickly. In Kenya, the introduction of batch computing and mainframe computers in government and private businesses has partially remedied this situation by mechanized processing. This cumbersome system, which involves coding and preprocessing of data and potentially long waits for returns from data processing bureaus, is characterized by information analepsis. The backlog of data may simply be hand-processed and placed on the computer after the fact in a card-punched or disk-logged format.

Within many public and private offices in Kenya, mainframe computers predominated until 1983. Under this system, data processing managers initially became the analogs of general office managers, often challenging their authority and control in the workplace while placing an equally complex set of hierarchical rules and requirements on computer operators and clerical workers. Senior programmers, systems analysts, and operations managers filled out the ever-growing office hierarchy and created a new bureaucratic layer through which information had to be filtered. The old lines of authority remained rigidly fixed with the mainframe computer, and bureaucratic control moved into the hands of foreign computer experts and outside consultants. The key word "expert" has become synonymous with new manifestations of bureaucratic control.

Microcomputers threaten this overbureaucratized system of office hierarchy. Users are able to teach themselves after brief introductory lessons. Information, ideally at least, comes more directly under the control of individuals at their work stations, and recycling of computer employees into areas related to their original work tasks is a real possibility (cf. Weekes-Vagliani with Pralong, 1983:9-11). Sociologists in the West have not viewed this type of development with unguarded optimism (cf. Braverman, 1974:342-346; Bowles and Gintis, 1976: 203-219; Crompton and Jones, 1984:52-54). They have argued that the spread of microcomputers and in-house computing have led to a deskilling of workers and a proletarianization of white-collar employees. The decline in skills and remuneration accompanying work tasks creates a fall in prestige of the job.

In Kenya, however, it is more difficult to speak of the proletarianization of a sector of employees who are a privileged minority. Other factors intervene. In some instances, as illustrated by the computerization of Kenya's government ministries, simpler computer programs that could be carried out by reskilled workers have been superseded in a short period by more complex programs requiring systems analysts with more computer training and higher organizational standing. After the initial introduction of a relatively simple microcomputer, more complex systems are installed that rebureaucratize offices following the old format of line authority. Senior programmers replace general office managers as technocratic leaders but continue to play the strategic organizational role of controlling the knowledge, work tasks, and advancement of office workers.[8] Thus, a continuum might be drawn between the old, overbureaucratized workplace and the new, more open office of the future. The tension, however, between old and new formats persists even in small enterprises. Consequently, strategies arise to rebuild old hierarchical forms when they are challenged by the workers' increasing familiarity with new technologies.

The New Computer Professionals

Regarding the social experience of working with computers as measurable in terms of economic productivity flaws much of the literature on new technologies. This approach overlooks the symbolic process of recoding work tasks and job titles. One means of controlling this recoding process is the establishment of professional computer associations, unions, and syndicates. In Kenya, such associations have failed to materialize despite computer professionals' abortive attempts to organize during the early 1970s.[9] This situation is not merely a byproduct of low technological density, or the scarcity of computers in the country. Tanzania, for example, had only 419 institutionally owned microcomputers as of 1986 according to a survey by Lewis and Sheya (1987:9). Ivory Coast had approximately 1,221 microcomputers in 1985 as reported by the second national informatics plan (1986:26). In both cases, professional associations were formed to protect and promote the interests of computer specialists.

The Association Professionelle des Informaticiens de Côte d'Ivoire (APICI) was founded in 1973 in order to promote the interests of computer experts, regulate standards within the profession, and foster outlets for computer education. Members of the association include individual computer experts, vendors, firms, and educational institutions.[10] Their activities are carried out through seminars, monthly meetings, lobbies, and publications in the association's journal. Some computer professionals that I interviewed saw APICI as an arm of the state and a vehicle for the implementation of a uniform computer policy. However, the association's existence predates the national informatics plans. The APICI has been influential in lobbying for the implementation of policies oriented toward technological expansion in the private sector.

GIMI (Groupement Interprofessionel pour le Management et l'Informatique) is a syndicate that represents private sector companies. This organization promotes the interests of computer managers and company heads and is strongly supported by Ivory Coast's computer consulting companies (SSII). Many computer professionals are individual members of the APICI and members of GIMI on behalf of their firms. GIMI occasionally takes an adversarial stand with regard to state regulation of competition among computer consulting firms and vendors. Both APICI and GIMI reinforce the status and recognition of computer professionals. They operate as boundary maintaining organizations that separate neophyte computer users from certified professionals. These professional associations help to crystallize the position of computer specialists as the new technicians of the sacred with a secret code and a set of criteria for displaying their elite status. The APICI even has a Committee on Deontology (Commission Déontologie) designed to establish and propagate ethical and legal standards in the profession. In its maintenance of professional exclusivity, the deontology process has sectarian overtones.

According to Lewis (1987:3), interest in a national computer policy among computer professionals in Tanzania dates back to 1969. In 1986, the Computer Association of Tanzania (COMPAT) was formed as a reaction against restrictions on computer imports imposed by Tanzania's National Computer Advisory Committee. In the manner of the APICI, COMPAT attempts to balance the interests of computer professionals against national technological goals as perceived by the profession. Regardless of their on-the-job experience with computers, COMPAT members are required to hold formal credentials exhibiting their computer training. In a survey of these computer professionals, Lewis (1987:10) found that approximately 60% had overseas training. COMPAT brings together vendors, computer consultants, and managers in a single association and special interest lobby.

While the model for professionalization in the APICI derives from French professional syndicalism, the COMPAT model has been influenced by the formation of associations in the accounting profession (cf. Lewis, 1988:283). This pattern is particularly interesting because it resembles the history of professionalization in bookkeeping and clerical work in the United States and Britain (cf. Strom, 1987: 63-81). This fact emphasizes that Tanzania, like Kenya, follows an Anglo-Saxon pattern of technological change and deregulation. Lewis (1988:268) considers the Tanzanian computer association an "elite within an elite." This association creates an alliance between computer professionals in the public and private sectors, fosters professional exclusivity, and is a source for controlling knowledge about the profession. Although I agree with this interpretation, I would further argue that computer professionals must be situated within the new postcolonial class structures emerging across Africa. Their professional and economic standing represents a break from colonial class structures. Moreover, these new professionals are part of a larger urban elite and subelite who challenge the bureaucratic and cultural hegemony of the older bourgeoisie. Considering the new computer experts as members of an emerging class as well as a profession permits an analysis of their shared cultural beliefs and narratives about the opportunity structures of contemporary society. Abner Cohen (1981:8) suggests that the cult of power and dramaturgy surrounding it be treated as a critical feature of class formation in Africa. Such an analysis includes the lifestyles, conspicuous displays, and aspirations of cultural groups as dimensions of social stratification.

The literature on class formation in Africa is fraught with debates and ambiguities. Much emphasis has been placed on the dual commitments of the emerging working class to the wage labor force and to self-sufficiency in the rural areas. Peasants who relocate to urban areas to sell their labor attempt to retain an independent agricultural base. Variations on this pattern can be found among African office workers who return to rural areas to oversee their farms on weekends. Janet Bujra (1979:48) points out that the subtleties of these processes of migration, social change, and cultural combination have been buried in the discourse of modernization theory by references to the dichotomy between traditional and modern lifestyles. Although the topic of emerging lifestyles will be considered in this study, a counterbalancing issue involves the type of collective experience that develops as a result of employment in the bureaucratic office as an idealized sociocultural microcosm.

A compelling argument concerning the influence of the new computer professionals may be developed if we examine their situation from the perspective of cultural analysis. Computer users share personal aspirations and a sense of cultural identity. The historical precedent for this sense of collective identity may be traced to the *évolué* civil service and clerical workers of the 1940s and the subelite office workers of the 1960s, prior to the recent advent of automation in the African workplace. These groups are unified by their discontent with the broken promise of African modernity and efforts to realize a dream of prospertity through occupational advancement and the conspicuous symbolic display of status. Some computer professionals resemble the older *évolué* class by virtue of the limitations of their educational and economic possibilities. There, however, the similarity ends. *Évolués* saw political independence as the key to a better life. Computer professionals begin within an independent political context that has left their personal expectations unfulfilled.

Workers who routinely use and come into contact with computers may be divided into five categories in Kenya and Ivory Coast: (1) clerical workers; (2) programmers; (3) engineers, systems analysts, and data management experts; (4) computer vendors and their support staff; and (5) computer educators. In most Kenyan enterprises, word processing is still at a minimum. The computer is used for inventories, stocks, ledgers, budget management, and financial planning. At one large data bureau in Nairobi, over thirty people are engaged exclusively in data entry through keypunch operations. Their skills are highly developed and they are carefully screened. As is typical of the Kenyan clerical workplace, these keypunch operators are predominantly male. The feminization of the clerical work force dealing with computers has not yet occurred in Kenya and is just beginning in Ivory Coast.[11]

Scholars studying the feminization of the workplace in the West, in particular Britain, argue that the deskilling of clerical work, along with its decline in prestige and remuneration, has affected women more significantly than it has men (cf. Cromptom and Jones, 1984: 2-3). They claim that men, who are often presumed to be more geographically mobile than women, are able to progress more easily from clerical to managerial and professional positions through their participation in an internal labor market that rewards a combination of technical and managerial skills. In the Kenyan case, this internal labor market is the basis for the mobility aspirations of many young male computer professionals and trainees. In contrast, women in private sector offices are chained to typewriters and are rarely found among the programmer trainees and higher level programmers. Word processing systems, which could easily be added to multi-user computer networks, are considered too expensive for these offices. Paradoxically, however, women appear to outnumber men in rapidly growing Kenyan commercial colleges that prepare workers for careers in word processing and computer use. The restricted role of women in the clerical and computer professions is rooted in colonial history. Administrative and higher level clerical positions are still occupied by men in many African nations (cf. Little, 1973:26-28).

Senior programmers and systems analysts are key figures in the new office hierarchy. They establish the technological redefinition and power base for new office work. Symbolically, they are the image leaders of the computer field. It is their positions toward which computer users aspire. These positions allow them to fulfill the dreams of a new technocratic subelite in terms of salary and with respect to relative autonomy in the workplace. Vendors and educators facilitate and feed off of the rising computer industry. The vendors provide the machines and basic instruction in their use. The educators, who comprise a varied group, earn their livelihood through the training and placement of new computer users.

Conclusions

Local controversies over computer policies in African nations have global consequences. In 1980 at the twenty-first session of the UNESCO General Conference, a resolution was passed to reduce the information imbalance between developing countries and the rest of the world.[12] The resolution called for improved technologies of information transfer and communication in the developing world and "the sincere will of developed countries to help them attain these objectives." A year later, the "Berg Report" on accelerated development in subsaharan Africa submitted to the World Bank suggested that problems of economic development could be resolved by the implantation of effective administrative policies by local governments.[13] These governments were viewed as standing in the way of their own progress due to poor planning, fiscal mismanagement, and lack of attention to effective development strategies. The combined effect of the Belgrade resolutions and the "Berg Report" has created an implicit mandate for international consultants to exploit possibilities for technological innovation in Africa. More importantly, these reports issue a call to action for African nations. New technologies provide an unprecedented economic resource for Africa. These technologies, however, involve both risks and responsibilities. Timothy Luke and Stephen K. White (1985:29) underscore this technological challenge when they state:

> Although the expansion of informational capital extends in new ways the loss of freedom and the loss of meaning in modern life, it nevertheless has an ambivalent character, at least in relationship to some of its new technologies. These, under radically altered conditions, could become supportive of greater human autonomy. . . . The moral-practical sphere could expand in the sense of greater participation and control of average citizens of economic and political decisions.

Problems of meaning in the moral-practical sphere of action are part of a cultural narrative. They exhibit the relationship of technological change to the political and social structures that limit human potential and autonomy. The analysis of computer adoption in Africa is a small, often overlooked aspect of the process of global technological change in the postmodern era. The computer is a self-referential sign of postmodernity's simulation and promise of progress.[14] Through the interactions surrounding it, the computer reshapes life in the postindustrial workplace and aspirations for life beyond work. As a simulation, it presents a symbolic forecast of events to come and is a modeling device for the course of individual and organizational behavior. The computer provides a convenient technological excuse for the breakdown of the new social and mechanical forms that it engenders. Human endeavors and lapses may be repaired by reference to technological errors. The computer's presence, thus, implies its own schema for the modification and repair of social behaviors in transition. The computer in Africa is a symbol of change and a rallying point for the emergence of a new profession and a new social class.

As a consequence of their access to computers, the subelite of office workers may evolve into a new technocratic class – the new technicians of the sacred who manipulate the ideologies and dreams of postmodernity in Africa. This social class will be very different from other class and interest groups molded out of the old colonial infrastructure. To the extent that this group masters and indigenizes the computer, it will not be dependent on outside sources for credentials, influence, or status. Yet, as with the older subelites of Africa, this group may remain marginal, or it may splinter into a new white-collar proletariat, controlled and enslaved by the very machines that promise to free it.

Chapter One
The Symbolic and the Social: Computer Use in Two African Settings

> When adequately domesticated and decolonized, the computer in Africa could become a mediator between the ancestral world of collective wisdom and personal intuition on one side and the new world of quantified data and scientific analysis on the other. The sociology of knowledge is undergoing a change in Africa. And the computer is part of that process of change (Mazuri, 1977:303).

Structuring Interaction with Computers

Narratives sustain bureaucratic order. Cultural narratives are manifested in office work, professional education, and government policy. They demonstrate the similarities and differences between organizational cultures in Africa and the West. When these narratives are decoded by computer users, the link between narrative structure and everyday social practices is not problematic (cf. Haferkamp, 1987:187). These narratives contain basic assumptions motivating technological innovation. The daily social practices in automated African offices include using computers to compile inventories, ledgers, and statistics. The accumulation of these practices structures interaction with computers. Structuring practices, however, do not define the meaning of new technologies.

Pierre Bourdieu (1977:72) employs the concept of "habitus" to describe the social structures produced by everyday practices in particular environments. Both Bourdieu and Anthony Giddens (1984:28-36) are concerned with the ways in which these structures develop as products of human signification.[1] They assume that individuals monitor their actions based upon their immediate goals and means of access to knowledge. Bourdieu points out that strategies of interpretation devised in one habitus may be transferred by homology to other cultural contexts. At this point, cultural narratives resurface.

Cultural narratives are schema of interpretation that permit the transfer of knowledge from one domain to another. Thus, neophyte computer users in Kenya apply established social forms to their ritualistic interactions with the machines. Folktales about the magic of technology and the rituals surrounding sacred technological objects are part of a cultural narrative. I shall expand this approach by developing the narrative program for computer adoption.

Two Models of Narrative and Performance

Two models represent the symbolic progression of events. The first is the narrative program developed by A.J. Greimas and the second is Victor Turner's model of social drama (cf. Greimas, 1970:172-180; Turner, 1957:92-94). Although I shall draw insights from Turner's approach, Greimas' narrative program will be the primary source adopted to analyze African computer use. Greimas employs the narrative program to examine the roles of actors and agents, or senders and receivers of communication. He analyzes narrative unities and organization in terms of four stages. These stages extend from the confrontation of two subjects to a final contract that establishes a resolution. The four stages may be applied to human-machine interaction, decision making in firms, accounts of individual careers with computers, and public discourse about new technologies.

When computers enter the African scene, they are actors in an unfolding narrative. So too are organizations and entire countries in which computer policies are developed. Consequently, it is possible to classify nations in terms of their computer discourses and ideologies. Then we can examine the impact of these strategies on individuals, professional groups, and businesses. This approach to narrative encompasses the cultural understandings behind computer practices. The narrative program applies action categories to texts. In the section that follows, I shall employ these categories to analyze texts and work environments in Kenya and Ivory Coast.

There are parallels between Greimas' model and an earlier approach to social drama developed by Turner. In his anthropological studies of Zambian villages, Turner used social drama to account for the rise and resolution of conflict and change as a community performance. The first version of the model has four parts: (1) breach, or the initial appearance of conflict and change; (2) crisis, in which the conflict is played out to the point of a cleavage in social relationships;

(3) redressive action, through which attempts are made to resolve the crisis; and (4) resolution and integration (Turner, 1957:91-95). During the third phase of the drama, public performances, pronouncements, and ceremonies are used to resolve crisis. Public discourse emerges as a way of reinstating harmonious social relationships. Turner (1974:38) analyzes both the crisis of social confrontation and its resolution in symbolic terms. His breach phase and the semiotic category of confrontation are similar. Symbolic crisis emerges as part of the narrative.

Turner (1985:199) considers redressive action to be the source of "various cultural performances." At this point, participants treat conflict as a cultural narrative that is played out in public ceremonies and ritual actions. When the narrative closes, the source of conflict is redefined within a new interpretive scheme. This phase of Turner's model resembles the process of contractual resolution in Greimas' narrative program. In both models, confrontation, or breach, triggers a chain of social and symbolic events, a discursive litany about these events, and attempts to resolve conflict through public performances. Whereas Turner considers these events to emerge most clearly in the public performances and ceremonies that characterize the redressive phase of social drama, Greimas emphasizes the position of symbolic actors in a text and the processes of signification that occur at each stage. I suggest that at each phase of the narrative progression, a shift in subject-object relationships occurs for the actors in the scenario. Turner moves toward this position in his final version of the framework for social drama. He states (Turner, 1985:203):

> Now I see social drama in its full formal development, its full phase structure, as a process of converting particular values and ends, distributed over a range of actors, into a system (which may be seen as temporary or provisional) of shared or consensual meaning.

The semiotic narrative program may be used to interpret responses to the computer and its attributes in the workplace. This approach offers a method for analyzing the job titles, work tasks, and conflicting areas of responsibility surrounding the introduction of the computer into new business environments. Although this scheme applies to the progression of narrative texts, it also explains the social practices and visual displays associated with computers in Africa. The stages of the narrative program encompass the introduction, use, and institutionalization of the computer. They go beyond these processes to include discourse about the perceptions of new technologies.

Although the narrative program involves progressive stages, it is not a teleological method. No presuppositions are made about the outcomes and content of the narrative.

A Narrative Program for Computer Adoption in Africa

In the introduction, simulation was discussed as a process by which a new vision of society is created. The imprecision of social life is removed by the precision of machines. By endowing electronic machines with the intentionality of human users, machines may be made to take the blame for the breaches and failures in social relationships. On the other hand, new technologies may be viewed as the source for resolving problems of social inequality and bureaucratic inefficiency. Alvin Gouldner (1979:28-30) argues that the members of the new technocratic class, including high technology experts and professionals, share a common critical discourse. He asserts that critical discourse is "pattern-and-principle" oriented and that it acquires its rhetorical force, not from claims of veracity or references to the social status of speakers, but through the technical uses of language. Thus, language becomes the key to building a reputation in the technical field.

Three types of technical discourse and texts are pertinent to this study: (1) everyday speech of computer users in their local environments; (2) public discourse about computers found in political pronouncements; and (3) government and media reports on the role of computers. Both public policy and media reports comment on computers by describing instances of successful implementation. Everyday speech derives from and shapes the daily social practices surrounding computers. Political discourse creates representations of new technologies in the form of social policy, development goals, and images of postmodernity. The new forms of sociality surrounding computers are incrementally linked to the public discourse about them. In the African cases discussed here, the technological environment, or habitus, of small firms provides an incubation ground for government policies about computers. With this universe of discourse in mind, it is possible to apply the narrative program in finer detail to enterprises and individual computer users. To do so, I have modified and expanded the four phases of the semiotic framework to include the social and interpretive practices surrounding computer use.

Stage 1: Confrontation. Confrontation establishes a relationship between two subjects, the senders and receivers of messages in a narrative. These subjects, known as narrative actants, face complications and obstacles. The first narrative complication, or qualifying test, arises at the confrontation stage. The principal actants are the computer user and the machine, which is a second subject. The computer's simulation of events is reproduced as a representation, i.e., a computer program.[2] It is possible to modify this reproduced program by mapping out a series of alternative steps and procedures to reach a goal. Applying a computer-like process to the mapping of social activities permits the human actor to employ a cognitive short-cut to attain specific social and cultural ends. The manifestations of technological confrontation in discourse will be discussed in more detail in Chapter Two. The confrontation stage may be diagrammed as follows:

$$\textit{Stage}_1\text{: Confrontation } (S_1 \longleftrightarrow S_2)$$

Figure 1-1. Confrontation in the Narrative Program for Computer Adoption

The computer's self-referential capacity is distinct from what people make of the computer in the workplace. For example, the members of the Management Committee at a handicraft cooperative in Nairobi were reluctant to give the computer consultant accounting and inventory data. They insisted, however, that the computer remain in their front office as a sign of progress. Similarly, the computer operators at a Kenyan wholesale outlet placed "dead data" on file for future reference while hand-processing accounts and sales. This use of the computer reproduces old systems of bureaucratic organization. It points to the sign-function of the computer as a representation of unrealized change. The reproduction of old social forms by clinging to the computer as a symbol of potential, but unrealized, change typically occurs at the beginning of the narrative of computer adoption.

The computer is imagined to perform bureaucratic miracles, but the actual steps required to accomplish these goals may never be undertaken. Computer users remain in a state of confrontation with the machine and inhibit its use while invoking its efficiency. The symbolic presence of the computer represents a new social order.

To remove the computer would deprive the workplace of its newly found aura of postmodernity and the illusion of automated efficiency.

A senior programmer at a large data bureau in Nairobi described his experiences of initial contact with companies in which computers had recently been installed. He characterized the confrontation stage as "dusting the terminals."

1.1 A.I., senior programmer, Data Centre, Nairobi:
You know // like you know // with your // let's say desk-top micro // which can do so much // if you put it on-line then you realize that it can actually be like a terminal // you know? And now // okay fine // they decide maybe they'd like to hook it up to a bigger machine somewhere. Now the question of compatibility comes in // you know. They could even be // surprisingly // from the same product name // but still the compatibility may not be there // and besides that // it might be extra cost to put in the right bolt to get the correct connection and whatnot. It still comes down to the fact that computer users in this country maybe // are not // you know // don't know much about the machines as such. They again // are *dusting the terminals* // but relying really on whoever is selling the machine to them // and all they're telling them // they're taking it literally. You know // and I think you want a wider scope in computer use.

The procedure of dusting the terminals raises the question of exactly what a terminal is and how it is seen. This question is not to be taken for granted, as is evident in my exchange with a computer manager at one of Nairobi's largest data bureaus. She corrected my initial perception of how terminals are used in her enterprise.

1.2. F.C., computer manager, Data Centre, Nairobi:
We do stock controls. We do memberships // assets // registers. We've got about seventeen systems and they cover a wide range. But they're mainly stock. Accountancy packages // and things like that. We also provide on-line services to some of our clients. They've got terminals in their offices. They've got direct lines to our center and they can do what we call on-line interrogation.

B. J.-R.: So they can punch in certain things?

F.C.: They // In a sense it's punching as it were. They interrogate their master files.

In the computer experiment at the Nairobi Handicraft Cooperative, dusting the terminal became a strategy for justifying the computer's continued presence in the organization. The computer consultant at the cooperative lamented:

1.3 K.C., computer consultant, Nairobi Handicraft:
These guys saw a free asset coming to them, and it was the computer. Therefore they said // "Yes // bring the computer" // instant gratification. It gave them prestige. Our society now has a computer. We are computer-run . . . . No one there would bother learning how to use the computer . . . . The computer was definitely a prestige thing because they would not let anyone take the computer anywhere (cf. excerpt 5.23).

The computer simulated the improved office practices sought by the enterprise, but its use was forbidden because placing accounting information on the computer would uncover the craft cooperative's secret system of preferential bookkeeping. The computer existed in a state of confrontation with the cooperative managers, who used the image and presence of the computer to reproduce the collective's aspirations while hiding daily work practices from outsiders. Appearances versus the technological reality of the workplace are at issue. As long as the computer is in the front office of an African business, that business may be preceived as progressive. If the business is asked to use the computer, the technological reality of its situation is revealed, and the status of the business changes.

Similar experiences have arisen in Kenya's public sector. Introducing technologies that an organization or agency cannot afford may have tragic consequences. In 1978, a team of German consultants attempted to introduce microcomputers into the Ministry of Agriculture (Broadman, 1985:18-19). The Hewlett-Packard microcomputer that they provided had a limited memory capacity and required considerable programming experience. Spare parts were unavailable for the upkeep of the machine. Eventually, one official learned to use the microcomputer, but his activities did not affect the ministry's organization. The German advisors abandoned the project and left ministry officials skeptical about computer use until a second computer project was introduced to deal with a fiscal crisis three years later. In Kenya, moving from confrontation to the domination stage of computer use involves an elaborate scenario in which a crisis frequently precipitates turning to the computer as a last resort. Resistance from potential computer users typifies the first stage of computer adoption under the fragile computer contract.

Stage 2: Domination. At the domination stage, one subject negates and poses obstacles for another. A relationship of opposition develops in which the principal subject pursues a course of action while the other acts as an opponent. The main narrative test during which initial obstacles are surmounted takes place at the domination stage. The discourse reflects opposition and control. Computer users achieve and display their technological mastery. In turn, mastery of the computer permits users to engage in a social display of the efficiency of the workplace. This social display is achieved by creating a homology between the workplace as a habitus and the technical functions of the computer. During this process, which is characteristic of the new mechanical solidarity, computer users identify with the computer by interchanging referents ("we"/"it") and by describing their work in terms of the computer's functions. Image and act flow together in the computer's ongoing simulation of office activities. The symbolic identification with ritual objects, such as masks, talismans, musical instruments, and computers, is characteristic of African cultural and religious practices. Identification bestows the object's powers upon its users. Individuals do not merely use the object, they become part of it.

During the second movement in the narrative program, the subject (computer user) dominates the computer. This domination involves a mastery through which the user and the machine establish a symbiotic relationship. The machine also dominates its user. Applying the narrative program, domination may be diagrammed as follows:

$$\textit{Stage}_2\textit{:}\ \text{Domination}\ (S_1 \longrightarrow S_2)$$

Figure 1-2. Domination in the Narrative Program for Computer Adoption

Routine work is transformed through the use of the computer. The programmer dominates the machine by streamlining tedious work procedures, but the machine controls its user by structuring work tasks. A common rhetorical strategy at the domination stage is to reference the technological power of the machine.

My interviews at Kenyan automobile dealerships and data bureaus contain laudatory references to the unconditional efficacy of computers as recordkeeping and retrieval devices. A data processing manager at a Nairobi automobile firm explained:

1.4 M.M., DPM, automobile dealership, Nairobi:
Before we had the IBM 3400 system here // it was a matter of doing the job ourselves and sending it out to the data bureau. That was a bad system. What we call bad. You know // you send the information // you wait for some time // it comes back. But here it's a *real* // what we call a real-time system. When you ask the computer whether it has a particular part, it will answer you there and then that *it* has // or *we* have // or *we* do not. So there is a change which is instantaneous.

The DPM describes the change brought about by an in-house computer as "instantaneous." He refers to the data bureau as cumbersome and speaks of the computer as instituting a "real-time" search for spare parts. The computer transforms organizational work, and its efficiency remains unquestioned. The pronominal shift from "it" [the computer] to "we" indicates the workers' strong identification with the functions of the computer.

Computer mastery triggers an idealized description of the efficiency of work-site procedures. One young senior programmer at a data bureau in Nairobi analyzed the results of efficient computerization.

1.5 A.I., senior programmer, Data Centre, Nairobi:
We really begin to appreciate the value of the computer. Now the need arises from many things. Everybody wants to see the latest results // updated really // actually cutting down on processing time from the time the document is prepared to the time they've got actual results to affect // maybe // the decisions.

Working with the computer during the domination stage involves negotiating the status of the machine in relationship to the human user. This negotiation is manifested in references to the technological structure of accountability. In accounts provided of their activities, members of organizations describe the computer's functions as a feature of their work. Its technological features stand as "sufficient evidence" for organizational policies and decisions (cf. Garfinkel, 1967:33). Technological accounts serve as descriptors for the activities, tasks, and goals of the organization.

During the course of a joint interview, a data processing manager and a chief programmer at an automobile dealership in Nairobi explained their organization with reference to the technological accountability of their work practices.

1.6

B. J.-R.:	How many terminals do you have?
DPM:	We have fifteen terminals.
B. J.-R.:	How are they used? In invoicing? In sales?
DPM:	You want me to break them // down? We have some for accounts, some for sales, some for receipts, and some for workshop. Then we have full invoicing.
B. J.-R.:	You are the data processing manager and he is the programmer?
DPM:	Yes.
B. J.-R.:	How many titles do you have?
Senior Programmer:	We have a computer manager, followed by the data processing manager // who is here // and one programmer with two computer operators and twelve // uh // twelve video operators.
B. J.-R.:	I see.
Senior Programmer:	Actually // the computers // the way we divide the jobs is the computer manager is also the central parts manager. Because over 80% of the computer's work is for parts. But the day-to-day running of the installation is done by the DPM. He is here. If there are any accounting problems // the DPM talks with the chief accountant. If he needs assistance // he just talks with the accountants.

Workers give testimonies of their adjustment to computers through the mechanization of their daily tasks. Reference is made to increasing mastery as demonstrated by the enhanced flow of work and its automatic quality.[3] In her research on word processors in small California firms, Phyllis Baker notes similar human-machine interactions. She claims that mechanization leads to a condition of boredom, or computer trance, in which workers are able to filter out background conditions and focus on specific features of their work tasks in an automated manner. In this instance, the machine subtly dominates the worker. One computer operator describes this state (Baker, 1986:11):

1.7 Word Processing Operator, Southern California:
Another operator started laughing at the fact that I was typing so fast with my eyes closed, completely ignoring the screen in front of me. I have almost dozed off in that condition, and I think it's funny. It shows you how you can know a skill goes beyond intelligence and becomes just as plainly mechanical. It requires very little thought.

Stage 3: Attribution. At the attribution stage, a cultural value is attributed to the appropriated object. The computer remains a second subject but is transformed into a valued cultural object. When the main narrative test has been surmounted, technology acquires a new meaning. Technological successes and failures are recognized. In the successful scenario, the computer is associated with progress. It appears in the most active sectors of the economy and promises prosperity and upward mobility to those who master it. Attribution confers a new status upon computer experts. Distinct from intellectuals and priests who controlled knowledge during previous eras, the technocratic elite relies upon the positive social value of technology in order to establish its status (cf. Mannheim, 1936:259-260). The power of computer experts is fragile because it is dependent upon displays of mastery and talk about technology. Attribution projects technological successes into a utopian future. Returning to the narrative program, attribution may be diagrammed as follows:

$$\textit{Stage}_3\text{: Attribution } (S_1 \longleftarrow 0)$$

Figure 1-3. Attribution in the Narrative Program for Computer Adoption

The computer is now an object of value and confers a sense of value upon its users. They are surrounded by a halo of success and prosperity.

The director of marketing at the British computer firm, ICL, in Nairobi describes the impact of his work on the Kenyan economy in terms of the imagery of positive attribution.

1.8 R.T., director of marketing, ICL, Nairobi:
The business has transformed itself even in the time that I've been on board. And it's still changing and developing very // very rapidly . . . . ICL has done a lot of work in the U.K., and there are still problems. We just introduced a system in the United

Kingdom with a voice synthesizer in it // with a repertoire of about 170 words. It's an integrated telephone and PC and everything. // In Kenya // we concentrate on a number of industry segments. I mean // I'll sell our computer to anybody who comes in the door, but there are a number of areas that are key to the Kenyan economy. Banking and finance // agriculture is very important. Government and the parastatal organizations are key areas. These four account for about 90% of our annual sales.

The marketing director described to me how he conceived of the difference between "rhetoric" about computers and description of the attributes and functions.

1.9 R.T.: Well // it's all very good to have rhetoric about computers, but a lot of these organizations couldn't function properly without computers. I mean // the billing system of the Post Office // there's no way they could have an efficient billing system without computers. And there is a very good system. I don't know if you've received a telephone bill, but we actually get international calls itemized with the telephone number that you've called. Well // I know you get one in the States, but you don't get it in the U.K.

B. J.-R.: No // you don't.

R.T.: In terms of efficient management and organization and cash flow, these organizations couldn't exist properly now without computers. We've got the Nairobi City Commission. // They have a cash receipting system for bills which enables people to pay water bills and rate bills, and they should get that information into the computer very quickly // which otherwise they would // their rate collection would be much less efficient than it is now. So I think // as I said // the requirement has gone ahead of the rhetoric.

Stage 4: Contract. At the final stage of the narrative program, all previous conflicts are resolved by a contract. The contract brings to bear the understandings and obligations existing between the sending and receiving subjects in the narrative. Although contractual understandings are implicit in communications as early as the confrontation stage, the final contract relies on the attribution of meaning achieved as the narrative closes.[4] The meaning of contract in narrative differs from its sociological implications because it is based on communications between the subjects of the narrative. The contractual situation involves a mandate of acceptance or a violation of previous understandings. An analysis of contractual exchanges will be presented in Chapter Eight. In terms of the study of computers, the contract marks the moment at which the machine's role in an enterprise is clearly established. The situation may be either positive (conjunctive) or negative (disjunctive). The contract acquires its meaning through a reversal of the narrative program.

Stage$_4$: Contract (Attribution ——→ Domination ——→ Confrontation)

Figure 1-4. Contract in the Narrative Program for Computer Adoption

A positive contract links everyday practices to public discourse about computers by recounting computer-related success stories. A short article from Kenya's only computer magazine, *East African Computer News*, illustrates the successful uses of computers to meet the needs of peasants, workers, and the state. This statement reflects a positive attribution and a conjunctive computer contract.

> 1.10 *Introducing the Personal Computer*
> Before using a data-base program package, however, one must be very clear about what one wants to use the PC for. A farmer, for example, may want to use the PC to keep records of cows on his farm. He may want reports telling him at a glance how much milk each cow has produced each month. Over a period of several years, the farmer may want reports on all cows showing which one has produced the most milk and which one has produced least. Such reports will assist him to determine when to get rid of a particular cow (*East African Computer News* 1, 2 October 1985d:13).

Contractual understandings are evident in the government's efforts to regulate new technologies. Ambivalent orientations toward technological change indicate a fragile computer contract. Government restrictions influence the availability of computers and the conditions for their use in the business world. Ivory Coast exhibits a progressive computer contract in its efforts to promote research and development in the computer field and engineering. According to Ivory Coast's

Plan Informatique National, 1986-1990 (p. 27), the "battle for redistribution of productive forces in the world" will be won by developing computer technology. The plan quotes the head of the Commission Nationale pour l'Informatique (CNI) as stating:

1.11 President, CNI, Abidjan, December, 1983:
In this battle for the redistribution of productive forces in the world, I repeat, the opportunity to arrive at an equitable position emerges for any country, including Third World countries, without an industrial past. Ivory Coast, thus, has the possibility to enter fully into the informatics era and to escape from its economic crisis. It therefore appears that the countries that will remain in a state of "crisis" will be those that have failed to understand and to integrate the inevitable character of this evolution. These countries risk being relegated to an unenviable role in the future of the world economy for a long time, a little bit like the actual situation of developing countries now.[5]

This text asserts that certain developing countries lacking an "industrial past" can transcend their economic problems with computer technology. Embracing the informatics revolution is the basis of a progressive computer contract that promises to rid the country of economic crisis. The progressive computer contract may be contrasted with the fragile contract that implies skepticism about the technology as a threat to national development. An article in one of Kenya's most prominent news magazines, the *Weekly Review*, expresses reluctance to embrace technological change.[6]

1.12 Most of all we regret the policy of importing highly sophisticated machinery and technical (and technological) complexes, and incorporating these foreign influences into a body ill-adapted to them. Such complexes need correspondingly complex cities and a highly educated elite to support them, and so a dual society is created. Little of the capital goes to the countryside, and so the movement from the country to the city is perpetuated. But for the majority there is nothing (*Weekly Review*, June 7, 1976:25).

Ivory Coast's informatics plans and Kenya's *Weekly Review* commentary represent contrasting approaches to the integration of computers into society and different strategies for resolving the narrative of computer adoption. These contrasting strategies filter down to businesses where computers have been introduced.

Social and Economic Environments of Computer Use in Kenya and Ivory Coast

The historical and structural conditions of the business worlds in Kenya and Ivory Coast constitute the environments for computer adoption. Agriculture is the major source of living for approximately 75% of the work force in both countries. The colonial experience is the backdrop for contrasting types of government bureaucracy and economic planning. In both countries, the coffee boom of the 1970s heralded a period of unprecedented economic growth. The boom resulted in expansion of the manufacturing and industrial sectors, as well as the growth of government administration. Since 1978, both countries have experienced an economic slowdown because of falling prices of major commodities – coffee and tea for Kenya and coffee and cocoa for Ivory Coast. In both nations, this economic stagnation has a direct effect upon decisions about computer adoption in the government and private businesses.

The peak period for multinational enterprise growth in Kenya took place between 1967 and 1972 and was followed by administrative expansion between 1976 and 1978. From 1976 through 1978, during the boom years, both Kenya and Ivory Coast gradually began to introduce computers. During this period, mainframe and batch computing were introduced into government agencies, universities, and private businesses. Kenya's initial approach to computerization was unplanned, restrictive, and crisis-oriented. Computers were introduced into government ministries to resolve fiscal and organizational problems. In contrast, Ivory Coast took a leading role in organizing intra-African conferences on computerization and in devising government plans to implement new technologies.

By 1981, there were estimated to be 127 mainframe computers officially registered in Kenya and 275 in Ivory Coast.[7] By way of comparison, ten years earlier there were estimated to be over 200 mainframe computer installations in the University of California alone (Mazuri, 1978:335). The small number of computers in Kenya and Ivory Coast by 1981, however, represents the beginning rather than the absence of technological change. Mainframe computers required a hierarchy of workers ranging from sophisticated programmers to keypunch operators. The available labor pool in both countries was small.

Gaining momentum in 1983, a microcomputer explosion occurred in both Kenya and Ivory Coast. Free-standing personal computers began to flood the market. Computer schools, consulting agencies, and foreign advisors took advantage of these conditions and expanded their clientele. The microcomputer invaded the workplace. By 1981, Ivory Coast had initiated administrative planning concerning new technologies. Kenya took a gradualist and eclectic approach to these developments and remained without a computer plan. Ivory Coast encouraged technological regulation and planned growth of computerization as part of a calculated economic leap. At the narrative stage of confrontation, Ivory Coast openly accepted computers while Kenya only tolerated their presence. The differences may be traced to Ivory Coast's liberal attitudes toward foreign investment and Kenya's association of computers with the threat of foreign economic intervention and control.

The tension between foreign investment and insider control leads to a particular representation of the development process. Economic development depends upon productivity and the degree of local control of the economy. Ali Mazuri (1978:322) succinctly emphasizes this point when he states that development means "modernization minus dependency." But dependency is replaced by another configuration of economic control. The result in Kenya was an ambivalent attitude toward foreign investment and alien technologies. Official attitudes toward new technologies and the economic position of the insider elite determined the history of computer use in Kenya's business world. Economic policies provided both a social and a symbolic environment for computerization.

In Ivory Coast between 1976 and 1980, multinational corporations became increasingly important in the development of industry and exports. The country's open-door policy toward foreign investment is evident in the case of multinational corporations in the sugar industry (cf. Masini, *et al.*, 1979:8-10). During the 1970s, a situation emerged in which non-Africans occupied many of the high-level technical and managerial positions in multinational companies because of "massive investments of European capital" (Joshi, Lubell, and Mouly, 1976:1-2). Although investment strategies did not change substantially during the 1980s, the government launched a new push toward the nationalization of industries and multinational affiliate firms. This push was part of the redefinition of development in terms of state control. Although the role of a large insider elite was weaker in Ivory

Coast than in Kenya, the new concern became rescuing the country from economic slowdown by leaping over problems of industrial development in two ways: increasing agricultural productivity and entering the new information era.

Ivory Coast's development pattern resembles that of Kenya in many ways. Foreign, in particular French, sources of capital, however, were deeply entrenched in Ivory Coast until the mid-1980s. The Ivorian investment code, in effect since September 1959, makes no distinction between investments of national and foreign capital in the areas of real estate, industrial agriculture, and power and production industries, all of which contribute to national development. More recently, however, the government has been concerned with repurchasing foreign industrial assets and nationalization of managers and workers in private sector firms. This move toward establishing economic autonomy, beginning with Ivory Coast's 1976-1980 development plan, is clearly reflected as a goal in the first and second national informatics plans.

The fragile and progressive computer contracts that are part of the narrative of computer adoption in Kenya and Ivory Coast affect all sectors of the economy. One might argue that Ivory Coast's more liberal policy toward computer imports was initially intended to facilitate the transactions of foreign investors. This policy was accompanied by an educational focus on the training of local engineers and managers. This educated elite differs from the Kenyan owners of businesses and firms because the Ivorian group was targeted to be primarily managerial, rather than assuming the dual role of capitalists and managers. Nevertheless, new opportunities for local investment have emerged in Ivory Coast's computer industry.

Focus on Small and Medium – Sized Enterprises

The four phases of the narrative program for computer adoption may be employed to analyze how small and medium-sized enterprises situate themselves in the local economy. The confrontation and domination stages refer to the processes through which these enterprises redefine their technological and economic goals. The attribution and contract stages illustrate the results of these adaptations for workers and the government.

Several standard quantitative and qualitative criteria have been established for classifying enterprises. They include: (1) number of employees; (2) enterprise assets; (3) sales turnover; and (4) qualitative definitions of the financial, technical, and managerial capacity of the enterprise (Harper, 1984:2-7). Initially, I defined small enterprises as businesses with twenty-five employees or less. This definition proved to be inappropriate for Kenya and Ivory Coast because it excluded the firms' assets and qualitative criteria pertaining to their internal management. For example, a charge card company with eight employees in Nairobi has five times the assets of a handicraft cooperative with 750 employees. In the latter case, many of the employees are affiliates of the cooperative, and their productive work is sporadic.[8] In Ivory Coast, locally controlled private computer firms with fifteen employees have a quarter of the assets of smaller computer consulting companies that are multinational affiliates. Size, thus, is most effectively evaluated with respect to the organization, goals, and assets of the company.

If small enterprises are defined as those within the reach of a large cross-section of the local population, as Malcolm Harper (1984:5) suggests for Tanzania, the definition becomes even more complex. In Kenya, charge card companies and insurance firms are used by only a small portion of the population. Based upon criteria of size and employee numbers, they fit the definition of small businesses. An enterprise that opts for computerization as a method of dealing with payroll, ledgers, and other forms of paperwork in Africa is special. It must, first of all, have either the capital to invest in computers or the political visibility to receive a computer as a donation from a vendor or an international agency. These organizations have played an important rôle in introducing microcomputers to Kenya. Firms adopting computers need to have a volume of business transactions and managerial activity sufficient to warrant the investment. These criteria are important factors that shape the contractual exchanges involved in computer adoption.

Finally, the relative definition of "small" in Africa requires special attention. Enterprises, which by U.S. or Western European standards would be considered "small" subsidiaries, often are large businesses listed on the stock exchange in Kenya and Ivory Coast. This situation applies to certain car dealerships and to insurance, finance, and consulting firms that appear small in terms of their number of employees and managerial base.[9] The recent departure of some multinational subsidiaries from Kenya creates further definitional complications.

When the foreign staff and support have left, international distributorships may regroup into small, local enterprises, as IBM and Burroughs did in Kenya. These companies maintain direct contact with the original parent firms that supply their stock and fund the purchases of capital equipment. Consequently, the definition of small enterprises in Africa needs to be flexible and is directly influenced by the transitional nature of the economy and business structures.

In Kenya, the attitude toward private enterprises has been positive, although enterprise development has occurred within a rigid political economy. As with Ivory Coast, the encouragement of parastatal and "mixed" enterprises has increased since 1979 (Lecallo, 1982:10). The Kenyan government has devoted attention to stimulating the growth of small businesses and informal sector enterprises.[10] The government intends to encourage the growth of informal sector enterprises by providing pre-investment advice, concessionary loan facilities, building sites, extension services, marketing incentives, and preferential purchase programs for small enterprises. The Kenya Industrial Estates plans to subsidize the construction of 339 low-cost workshops with electricity and other amenities for small enterprises in the greater Nairobi area.

Small enterprises that formerly operated on a very rudimentary scale with inadequate facilities will become larger and more cost-effective firms. These businesses involve more complex inventory and accounting procedures that provide the conditions for computerization of small enterprises. Although the potential for the computerization of consumer-oriented small enterprises will become clear only over time, government incentives to foster small enterprise growth and liberalization of government attitudes toward microcomputers could reasonably lead to a rise in computer use, if regulations regarding computer importation are reviewed.

External Computing and Internal Change

Progressive stages of the narrative program for computer adoption may be represented quantitatively. Moving from confrontation to domination requires a critical mass of computers in an organization and a nation. Only at this point can changes in employment structure and rates be documented. In Kenya and Ivory Coast, the spread of computers in the private sector has taken place in contrasting ways. Banking and financial institutions were among the first to computerize. In both cases, the number of computers in the country was extremely low and grew rapidly with the advent of microcomputers during the 1980s.

Although the rise in microcomputers has transformed the appearance of computerization in Kenya and Ivory Coast, the numbers are still relatively small in comparison to Europe and the U.S. I have already cited the Kenyan and Ivorian figures for computer installations. These figures may be contrasted with data on computer installations in the United States, Western Europe, and Japan between 1970 and 1978. Figures for Africa are not listed separately in this comparison.

	1970	1973	1978	% of increase 1970-1978
United States	65,000	110,000	200,000	207
Western Europe	21,000	55,000	110,000	423
Japan	6,000	19,000	45,000	650
Others	18,000	46,000	95,000	428

Figure 1-5. Number of Computers Installed from 1970 to 1978
Source: *La Micro-électronique, la Robotique et l'Emploi*. Paris: OCED, 1982:25.

This table suggests that the confrontation stage of exposure to new technologies may be defined in terms of low technological density. The automobile industry was among the first to computerize in Kenya. During the 1970s, Cooper Motors (Volkswagen and Audi) and Marshalls East Africa (Peugeot) introduced ICL mainframe computers to handle inventory, stock, and ledgers. The computer branch of Cooper Motors expanded into an autonomous computer bureau known as the Data Centre in 1973. The Data Centre initially used an ICL ME-29 mainframe computer and introduced a Data General "supercomputer" in 1986. The Centre has on-line communications with terminals in clients' enterprises. In 1984, another branch of the computer division of Cooper Motors became Abacus Data Ltd., a small computer bureau housed within the Cooper Motors repair plant. It engages in data preparation for Cooper Motors and a small group of outside clients. The Computer Bureau founded in 1977 is an offshoot of another automobile dealership, Marshalls East Africa. It employs an ICL ME-29 computer in batch computing for several small enterprises.

Computing services handled by these bureaus include accounting packages, payroll, inventory, stocks and ledgers, incoming accounts, and membership assets. The Data Centre develops special programs. It employs sixteen standard computer systems for multiclient use. The Computer Bureau, a slightly smaller operation, relies primarily on standardized inventory and payroll programs and does little tailoring of these packaged programs. As of 1984, Computer Consultants, a small firm, and Price Waterhouse have operated in a similar capacity. Some of the enterprises, now using mini and microcomputers, initially started with a data bureau until they could afford their own computer installations. Even then, certain tasks such as preparing payroll information were left as the responsibility of a data bureau for both security and personnel reasons. Farming out such data limits the technological power of the new computer administrators.

Data bureaus and consulting firms represent an area of entrepreneurship. They service multinational subsidiaries and local businesses by providing a training ground for the process of computerization. In Kenya and Ivory Coast, these bureaus play an important role in the diffusion of new technologies. Through data bureaus, many businesses experience their first encounters with computers. By means of their feasibility analyses, the bureaus educate firms in the use of computers and designate the types of activities which can be most effectively computerized. Data bureaus, however, have a mixed influence on the evolution of computer use. They introduce firms to computers while limiting computer use to the types of services that they offer. If an enterprise can afford to computerize payroll but not stocks and ledgers, it may continue with a piecemeal orientation toward computerization when the machines are installed on the premises. Thus, dependence on a bureau leads to the computerization of an enterprise but may result in the *selective use* and *underutilization* of computers.

In Ivory Coast, computer consulting firms have played a greater role in the spread of new technologies than they have in Kenya. An outgrowth of French banks such as Crédit Lyonnais and Société Générale de Banque, the consulting firms began with a direct liaison to large private corporations in the business and finance sector. As with the Kenyan case, these firms originally specialized in batch computing and then moved into microcomputer consultantships.

Government support and a positive climate for the development of these computer firms in Ivory Coast resulted in their rapid growth and competition.

Combined with the government's positive policy toward the education of engineers, the rise of the computer firms has created a situation in which Ivory Coast is well prepared, perhaps overly so, to execute the computerization of government, industry, and private business. This situation is reinforced by the new images of the development process presented in government documents. For example, the *Plan Informatique National, 1986-1990* (p. 33) describes "progressive deindustrialization" as a strategy intended to overcome the problems of industrial development. Alluding to the Korean and Taiwanese cases, the plan states that industrialization is a colonizing strategy leading to the creation of forms of heavy industry that are no longer viable. According to the plan, this dilemma could be averted by the installation of new factories using robotics and high technology. The result would be a competitive advantage for African countries where these factories could be mounted at low cost. This reinterpretation of the development process places high technologies at the heart of progress and promises to free the country from economic dependence.

New Technologies as Representations of Development

Representations of development are collective images reflected not only by gross national product and other statistical indicators but also by the images of development found in administrative policies and reports. These representations situate entire countries as well as individuals and enterprises as narrative actants and agents of change. During the years immediately following independence in many African nations, nationalism and industrialization were key themes in the representation of development. Nationalistic slogans emphasized unity, hard work, and heavy borrowing from international sources in order to achieve a point of economic take-off.

Some economists have begun to question the effectiveness of U.S. and European multinational subsidiaries in the Third World (Kumar, 1980; Lall, 1983; Wells, 1983). Rather than viewing local entrepreneurs as part of a dependent client class deriving exclusive economic benefits from their close association with multinational corporations, these analysts have chosen to focus on the role of nationalism in shaping the fiscal and political goals of local businesses (cf. Warren, 1973: 109-133; Kumar, 1980:17-18). State control, they argue, inhibits the growth and influence of international subsidiaries and provides an opening wedge for the rise of local companies and new, regionally based multinational corporations. Although this approach may be overly optimistic, it sheds new light on the interrelated processes of economic growth and technological change.

The Kenyan case exhibits protectionism and extensive government scrutiny of external sources of investment that increase economic dependency. Under this model, foreign aid projects filtered directly through the government threaten less external intervention than other forms of investment that might circumvent government control. In contrast, reinterpretation of economic development in Ivory Coast is firmly rooted in belief in the power of new technologies to lead the country to prosperity. Problems of unemployment resulting from technological change are of less concern to the government than the possibilities for creating new investment opportunities through automized industry. According to this strategy, more businesses will generate more jobs in the long run, and these jobs will benefit the entire population. This development strategy actually upholds old investment patterns while attempting to diminish their influence through a policy of nationalization and deindustrialization. To the extent that the country's agricultural base has not changed, a paradox is inherent in the new representations of technology's role in economic development.

The two contrasting representations of development demonstrated by Kenya and Ivory Coast constitute different resolutions of the narrative program of computer adoption. Kenya's initial resolution is disjunctive. The country opts for high control and restricted official use of imported technologies. Ivory Coast's resolution is conjunctive. The government encourages the dynamic use of computers as a rhetorical and an economic tool integrated into the plans for national development.

In the course of devising a model for a postindustrial and state-controlled economy, Ivory Coast advocates assimilating and redefining new technologies. Kenya, on the other hand, has taken a gradual approach to computer adoption in its effort to curtail economic dependency. Both approaches demonstrate the ways in which computer policies influence the workplace and educational settings. A critical ethnography of work settings in which computers are used illuminates the relationship between public discourse and everyday practices. In Chapter Two, I shall begin this ethnographic description by examining the structure of discourse in the workplace during the first stage of computer adoption.

Chapter Two
Computer-Related Successes and Excuses: The Discourse of Confrontation

> The division of work . . . does not consist only of differentiated technical skills. It consists as well of differential moral value attached to the possession and exercise of technical skills (Harold Garfinkel, 1967:194).

Introduction

Two important discourses emerge in African settings at the confrontation stage of the narrative program for computer adoption: a modernist view that causally links computers to economic development and a gradualist position in which computers are associated with foreign intervention and the negative aspects of Westernization (cf. Lewis, 1987:12). I shall refer to these descriptions as the discourse of confrontation. These views propose contrasting subject-object relationships in defining human interactions with computers.

According to the first view, computers are positively appropriated by their users. Under the second perspective, computers dominate users and their culture by imposing new forms of thought and social relationships. Both positions represent polar extremes. In actual cases of computer adoption, a variety of interactions develop during the course of putting new technologies into operation. I shall analyze these interactions by examining computer-related successes and excuses, that is, language used to attribute to the machines positive and negative effects resulting from their use. This linguistic edifice generates a utopian vision of a harmonious working world produced with the aid of new technologies. It is the discursive manifestation of the cultural narrative of new technologies.

In order to place this linguistic process in context, it is important to consider the larger picture of computer use on the national scene. De Lagy and Calhoun (1987), for example, point out that in the case of Sudan, a small number of computers has been installed in government and international agencies. Many of these computers are donations, and various pieces of equipment are outmoded and incompatible. There is little local incentive to expand the market for personal computers. In this instance, a modernist discourse surrounding computers exists only insofar as computers are used as signs of power by local administrators. The actual functioning of the computers is secondary to their presence in the country.

In contrast, Suzanne Grant Lewis (1988:138-140) describes a more active situation in Tanzania. The presence of approximately 460 microcomputers in the country as of 1986 created a small but active market for hardware and software. A survey by Lewis and Sheya (1987:16), however, found only 51 of these computers to be individually, as opposed to institutionally, owned. Although Lewis describes the rise of a professional elite in Tanzania with the support of the national computer association and various marketing networks, knowledge about computers remains restricted to a small professional elite. I found a similar situation in Kenya where computer professionals and vendors continue to control access to information about new technologies. In the Kenyan context, however, training outlets ranging from government internships to commercial colleges have expanded the popular knowledge about and access to computers. In Ivory Coast, in spite of a positive government policy toward the development of computers and informatics, a 1985 survey by the Commission Nationale pour l'Informatique (CNI) claims that 84% of the country's population has had little or no contact with computers.[1]

The cultural settings described here are, thus, situations in which a new technology, literally a foreign cultural object, has been introduced. Novel ways of communicating about that technology are beginning to develop. Much of the vocabulary surrounding computers is imported with the equipment. Dictionaries and glossaries of computer terminology abound in English and French.[2] The first national informatics plan of Ivory Coast contains a glossary of computer terms. Everyday discourse is definitely changing as a result of the computer's presence in many aspects of contemporary African life. (See Glossary.) A lexicon, however, does not account adequately for the full range of discourse surrounding computers in specific social contexts. The narrative program for computer adoption delineated in Chapter One provides a method for understanding how talk about the initial impact of computers emerges. In the work settings that I examined, talk about computers at the confrontation stage is a discourse of power that may be interpreted with relation to the social organization of the workplace.

The Discourse of Human-Machine Confrontation

When the computer is introduced into the workplace, it is not merely a tool for the accomplishment of specific tasks, it becomes a second subject that stands in a position of confrontation with human subjects. When computer adoption begins, it is difficult to assess the machine's effectiveness. Its possible costs and benefits remain virtual rather than actual. They are expressed in the subjunctive mode of future probabilities. Experimentation with computers at this stage may be compared to what Richard Schechner (1985:38) has termed subjunctive performances, or privately rehearsed activities that do not become public. As mastery of the computer occurs, these subjunctive performances become indicative acts through which a set of attributions are made in order to interpret human-machine interactions.

Although the computer is an inanimate object, it simulates aspects of human thought and action (cf. Baudrillard, 1970:156; Lussato, 1981:119; Turkle, 1984:272-276; Roszak, 1986:68-69). Hence, its mechanical performance may be attributed with moral value.[3] For example, Sherry Turkle (1984:32-34) found that children learning to use computers become frightened when they discover the computer's uncanny ability to predict their thoughts. It is not a novel idea to argue that modern machines are simulacra for human comportment (cf. Baudrillard, 1968:153-158; Toffler, 1980:72-78). It is, however, of sociological interest to take this hypothesis further by exploring how such simulation is accomplished in bureaucratic settings and the ways in which new technologies provide workers with idealized descriptions of their organization's social structure.

The computer is an anthropomorphic object in a real sense. If we say, "the pencil writes," we are attributing human qualities to an inanimate tool. The computer's ability to respond on command and to preprocess information from its users creates a second subjectivity. Thus, the computer does, in a sense, "write." According to Michael Heim (1987:31), "the printed word no longer seems sacred or complete as soon as it is typed – improvements are always possible. . . ." Heim's description of the philosophy of word processing is reflected in talk about computers. The tendency to alternate between "we/it" references when speaking about the computer is an example. In the last chapter, this tendency was considered characteristic of the domination stage of computer adoption in which technological mastery has been achieved. The symbiotic relationship between the user and the machine, however, begins with the discourse of confrontation.

A complicity develops between humans and machines. I have referred to this complicity as the new mechanical solidarity. The machine serves as a filter and buffer for human decisions. When violating the reciprocal expectations of sociality, computer users resort to an underlying solidarity with the machine, which is ultimately held responsible for the outcomes of human action. Mechanical errors are, thus, treated as the source of social breaches that may be repaired by reference to their technological origins. If dossiers are lost, accounts left unbalanced, ledgers incomplete, or data irretrievable, the machine can, for a time, take on the onus of failed social and bureaucratic outcomes. The successes of an enterprise may likewise be described in terms of the miraculous powers of machines and the ability of humans to master and simulate mechanical operations.

From an organizational standpoint, technological successes and excuses cannot be left unscrutinized. The new mechanical solidarity must come under managerial control. If technological excuses are too numerous or overutilized by the clerical and data-entry staff of an organization, the productive capacity of the enterprise may be jeopardized. Therefore, it is of interest to the organization's management to see that technological power and the ability to make excuses are not randomly distributed and remain under the social control of technical and managerial experts. It is also advantageous to managers to attribute the success of enterprise activities to their organizational skills rather than to machines.

The use of figurative language, or tropes (cf. Burke, 1945:507-508), in devising strategies for referring to computers and their position within an organization is a means of balancing social form against spontaneity in the workplace (cf. Simmel, 1950:385-386). The machine stands in a figurative relationship to the human task for which it is used as an organization tool. The computer is used figuratively to represent decision-making processes in enterprises where an element of human error accompanies every choice. Alfred Schutz (1962:54) emphasizes that partners in interaction rely upon reciprocal expectations and mutually shared visions of the social world. Although reciprocity may be viewed as a prerequisite for sociality, its fragility leads to a lack of decisiveness in everyday conversation and an implicit balancing of social form against process in the renegotiation of the unstated terms of interaction (cf. Schegloff, 1980:104-152; Goffman,

1981:246-247). When participants in an interaction pursue conflicting goals and agendas within an organization, ambiguity leaves room for multiple interpretations of discourse and leads to the use of rhetorical strategies to repair these discrepancies (cf. Cicourel, 1958: 54-58).

Ambiguities in discourse may be interpreted as misunderstandings concerning the frame of reference (Goffman, 1981:247). Since a given action has multiple meanings, the frame serves as an anchor for maintaining shared expectations about the outcome. Interactants, however, engage in several courses of action simultaneously. Erving Goffman (1959:65) describes this process when he refers to performances in the workplace as occasions on which "activities oriented towards work-tasks" are "converted into activities oriented towards communication." These workplace performances provide ample opportunities for the rhetorical manipulation of technological power.

When the computer is introduced as a medium through which human interactions are accomplished, it becomes an excuse for failed outcomes and a justification for the positive courses of action that humans take. Just as human power may be underutilized or mismanaged in the workplace, the computer may be placed in an organizational position where it is used in circumscribed ways. I shall identify six rhetorical strategies through which this technological mediation occurs.

1. *References to the Technological Power of the Machine*: A common strategy for integrating computers into the organizational structure of enterprises is the allusion to the extraordinary technological power of the computer to synthesize information and to streamline work processes, or make them more efficient. My interviews at Kenyan and Ivorian data bureaus contain laudatory references to the unconditional efficacy of computers as organizational devices.

2. *Descriptions of the Mechanized Character of Work with Machines*: Workers give testimonies of their adjustment to computers through the mechanization of their daily tasks. Reference is made to the degree of mastery demonstrated through the automatic quality of work.[4] Jean Baudrillard (1970:156) has termed this process the absence of "analytic reflection" that accompanies the individual's performance of automated work.

3. *References to the New Mechanical Solidarity and the Prestige of the Machine*: Workers invoke their solidarity with the machine and their

dependence on its functions. Consequently, if the machine malfunctions, the organization's activities stop. Human shortcomings and mistakes are justified by allusions to mechanical errors. The computer, in this instance, stands as a metaphor for a wide range of humanly organized office activities and breaches in social interaction that are not dependent upon it.

4. *Technological Structure of Accountability*: In providing accounts of their activities, members of organizations reference the technological background and features of their work (cf. Suchman, 1984:1-2). These technological features stand as sufficient evidence for organizational policies and decisions. Technological references serve as descriptive devices for representing the activities, tasks, and goals of the organization, and they frame the accounts that office workers give of their tasks.

5. *Metaphor of Machine Error*: Although technological errors undeniably hinder the routine flow of work processes, the point of sociological interest is how these errors are interpreted within organizations. Following the principle of technological accountability, these errors may be transformed into excuses for a host of organizational activities and relationships. The social ruptures that occur when work cannot be completed or delivered on time may be blamed on technological errors without requiring an extensive description of the chain of social events causing the problem.

6. *Recourse to the Inviolability of Expertise*: All problems are not simply blamed on the machine. The prospects of technological and social repair offer ways of resolving difficulties. The computer programmer, systems analyst, or repair person may "debug" the machine and save the day. Waiting for mechanical repair is a way of invoking the expertise of these specialists and provides a means of restoring the rupture in everyday routines.

Descriptions of daily activities use the computer as a point of departure to represent the flow of action in an organization. Based on these strategies, it becomes evident that the computer is an anthropomorphized object that reflects the organizational structure of an enterprise. Consequently, organizational goals are framed with reference to the power of the machine and the level of technical expertise of users. In cases where such expertise is lacking, the imagined magical, or technological, power of the computer and a fantasy of mechanical resolutions to social problems persist. At this point, the tropes of technological control become effective rhetorical strategies for preserving the image of an ideal organization in the face of contradictory social evidence. Office workers constantly make adjustments by monitoring their own actions and those of coworkers (cf. Wynn, 1979:118). The distance that the computer establishes between humans and their actions accounts in part for the popularity of computer terminology as a way of describing problematic situations in the workplace and in everyday life. These descriptive strategies are evident in talk about technology in Kenyan and Ivorian enterprises where computers have been introduced.

Similar problems of office organization emerging from the technological control of computer experts who apply systems that do not fit enterprise or agency needs are evident in other African contexts. Based on studies of the initial stages of computerization in Nigeria, Tanzania, and Uganda, John Wallace (1977:13-41) observes that many firms begin by organizing their computer and clerical staff around inappropriately designed and cumbersome, packaged computer systems. These systems place managerial responsibility and technological power in the hands of computer experts, who have technological knowledge but are not trained to handle administrative problems. According to Wallace, this imbalance leads to ineffective efforts to resolve organizational problems through technological means at the confrontation stage of computer use. He argues that organizational strategies should be developed for replacing systems analysts by reskilled computer operators and clerical staff.[5] Wallace (1977: 39) states:

> Developing countries that rely heavily on vendor-designed courses for systems analysts will possess many systems analysts whose skills lie in solving problems by putting them on a computer. Managers faced with clerical and administrative problems have no other type of technician.

Computer-Related Successes and Excuses

Alvin Gouldner describes the culture of critical discourse (CCD) as a form of rational discourse based solely on the self-referential logic of its own justification rather than on the authority of the speakers. He states:

> As a result, CCD de-authorizes all speech grounded in traditional societal authority, while it authorizes itself, the elaborated speech variant of the culture

> of critical discourse, as the standard for *all* "serious" speech (Gouldner, 1979:29).

Critical discourse is a form of postmodern communication. Gouldner characterizes it as the prototypical speech of the new technocratic elite.

The discourse of confrontation is a form of critical discourse. It relies on the power of technology as a justification. The rational features of computer talk at the confrontation stage, however, are highly questionable. References to computer-related successes are semipropositional statements about belief in the computer's technological power and successes (cf. Sperber, 1982:74-80). Although these statements may be subjected to empirical proof, such proof is *not* the objective of the speakers. Rather, these statements operate as projective descriptions of an unrealized ideal and as justifications for a course of action.

The narrative program in which computer-related successes and excuses emerge is framed by a social context. The narrative's progression leads to terminal signs that attribute a particular value to the computer. This attribution involves references that define the human relationship to technology. The organizational context intervenes in this definition by reinforcing the social hierarchy in which computer use takes place. Although the significance of this hierarchy for computer users is recoded as they master the new technology, the basic features of this hierarchy are established as part of the discourse of confrontation.

An article in *East African Computer News* (1, 1, May 1985b:7-12) presents an idealized version of the computer hierarchy. The first three rhetorical strategies outlined above may be applied to this diagram of computer use at the confrontation stage. The user is depicted as having a general idea of the organizational and work goals to be accomplished. Rather than dominating or directly employing the computer, the user coexists in the same environment with the machine at the confrontation stage. Plans are implemented by the systems analyst who translates the user's abstract knowledge into technical notes that can be utilized by the programmer. Next, the programmer develops the language for communicating with the machine and the operator interacts with the computer. Repairs are handled by an engineer, and outcomes are obtained by a control clerk. The computer manager oversees the entire operation. Although this organizational pattern could be used with a microcomputer, it appears

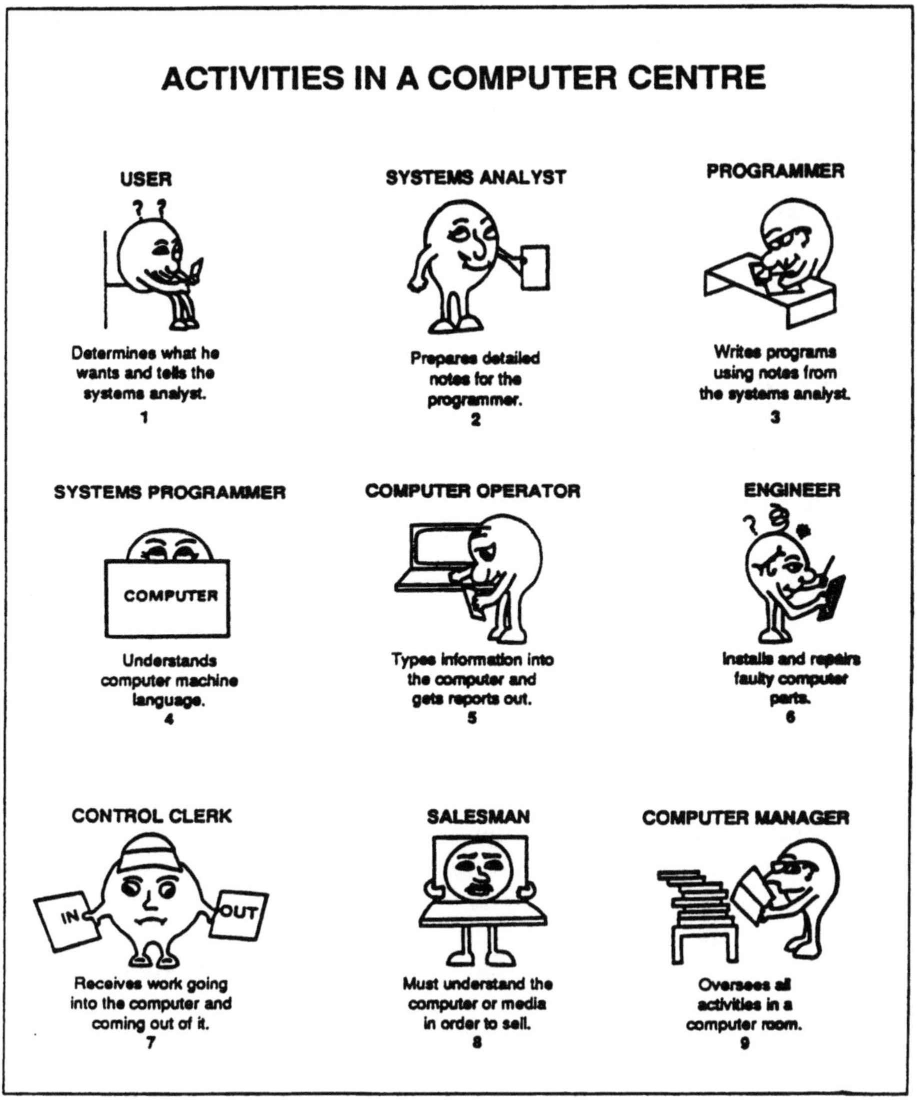

Figure 2-1. Hierarchy of Computer Specialists. (East African Computer News, 1,1,-May 1985b:10)

to be based on the more complex hierarchy of computer personnel surrounding a mainframe. The user's exact status is a mystery. It is unclear whether the user is a firm owner, a business manager, or a bureaucrat. The computer manager, who oversees the operation, stands in the terminal position in direct opposition to the user. The manager has technical control over the computer, while the user has conceptual and organizational control over the entire sequence of operations.

The user's head is crowned with question marks, but the computer manager, who looks back at the outcome, appears studious and self-assured. This media depiction of activities in an ideal computer center actually reflects the operations that I observed in several firms. In discourse set 1.6, these operations were described by a senior programmer at a major Kenyan car dealership.

2.1 G.K., senior programmer:
We have a computer manager // followed by the data processing manager // who is here // and one programmer with two computer operators and twelve // uh // twelve video operators.

B. J.-R.: I see.

In this dealership, the data processing manager occupied the position of the user – a rather unusual situation. He generated ideas from an abstract set of technical goals and allowed the programmer and the computer manager to implement the operations. The DPM was uncertain about how many computer terminals the enterprise had and called upon the senior programmer to explain the computer staff's operations to me. In my view, the DPM represented general management even though he held a technical title. He supported the old, overstaffed bureaucratic structure as its symbolic representative in the hierarchy. The terminal position, or real technological control of the operation, was wielded by the senior programmer while the DPM remained at the initial confrontation stage with respect to his knowledge about the computer's functioning.

According to some computer consultants, the confrontation stage is characterized by a lack of knowledge on the part of users about what can be expected from computers. A manager at one of Nairobi's largest data centers describes her clients as learning about computers by preparing data to be sent out for batch processing with her firm.

2.2 F.C., computer manager, Data Centre:
A lot of people had been with us as clients. And then the need for the computer grew and grew within their own company. They eventually transferred all their work to their own machines. But they've had quite a lot of knowledge on computers before they've actually got their own.

B. J.-R.: And being involved in the preparation?

F.C.: Preparation. Checking // printout. Actually telling us their requirements. Biggest difficulty is // people telling us exactly what they require from the computer // which means you have to sort of train them into what the computer can do. What it can't do. How it can do it. How it can't do it. It's a matter of feed in garbage // get garbage out. A lot of people don't understand that.

B. J.-R.: Yes. It can transform things totally.

F.C.: Yes. The biggest tendency in some cases today is "The computer can do anything," is the saying, but // it can only do it if you tell it how to do it. Right? And you've got to educate the user to what that actually means. What can the computer do. Which // I think with the publicity in magazines available // studies // just talking to them // that question // well // you know the computer has done it all wrong // is now dropping off. They realize that they're doing it wrong.

At the confrontation stage, the user may assume that computers are inadequate to accomplish specific organizational and data processing tasks. In the excerpt above, the manager blames the users rather than the computers when she states, "It's a matter of feed in garbage // get garbage out. A lot of people don't understand that." The computer manager's statement is a commentary on the use of rhetorical strategy 5: the metaphor of machine error. She argues that computer excuses in most cases merely mask the fact that computer users "are doing it wrong." In this case, as in Figure 2-1, the computer is treated as an idealized object with which users must interact in an appropriate way in order to obtain impressive results. The machine is seldom wrong according to the manager. The people are mistaken and inadequately trained.

In describing the relationship between managers and workers at the confrontation stage, Shoshana Zuboff (1984:88-90) asserts that new operators often experience a sense of mistrust toward machines as they question their own skills and competence. Managers attempt to eradicate the metaphor of machine error in order to reinstate trust in the new technology and increase the productivity of their enterprise. Remarks Zuboff (1988:89):

> Managers play an important role in this dilemma. They tend to perceive operator mistrust as an expression of resistance to technology change and so believe that their interests will be served by exhorting workers to put their faith in the new systems. They view the withholding of trust and the refusal to rely on the new technology as evidence of primitive thinking and look for ways to wean the worker away from a fixation with concrete reality.

A research representative from a large data processing and accounting firm in Nairobi explained the mistrust of computer technology with respect to "centralized systems." Although he alludes to the technological power of the machine (rhetorical strategy 1), he cites cases in which efficiency has not been the outcome of computer installation. The computer representative presented his view of the initial impact of microcomputers.

2.3 D.S., research representative:
I believe that the advent of micros will certainly have a marked effect, as they are obviously going into places where they never would have been thought in the last // recently. And it's both in business and also in development projects // and you're probably aware of this // that there are quite a few projects where they're being placed as part of the project. Umm // it will polarize somewhat // I mean this // minis will disappear and you'll basically have the micros and the centralized // the larger corporate centralized systems, which will still have a place. Certainly the micro area will move the fastest, and it is moving fast. The centralized one will continue to have the sort of constraints which it's had to date // which is a lack of knowledge and sophistication and understanding // that mainframe-type environment. It hasn't really been a very successful situation to date. Most of the large parastatals or government bodies in particular are not very effective, and there have been some pretty terrible disaster stories.

In this discussion, microcomputers and mainframe computers are juxtaposed. The "mainframe-type environment" that leads to "disaster stories" is viewed as cumbersome and potentially ineffective. The research representative does not place the blame solely on users, as does the computer manager. He shifts the blame to the social and organizational environment in which a particular type of computer is used. In this case, the computer technology is not considered at fault. Instead, the human environment in which computers are used has created "terrible disaster stories." At the confrontation stage, these disaster stories may be conceived of as micronarratives in which a negative and disjunctive resolution of human-machine interactions arises.

Disaster stories contrast with success stories, or positive accounts of the miraculous efficacy and efficiency of computer systems. In success stories, computer excuses are unnecessary because everything operates exactly as it should, and computers create miraculous changes. In media accounts, these stories are presented as expanded narratives of a series of positive computer interactions. The following success story is taken from an article by Gideon Nzioka on computerized banking published in *East African Computer News* (2, 1, 1986:16).

2.4 *Computerized Banking in East Africa*

> George Nderito wants to buy a house in Mombasa. He just read a newspaper ad for one that seems too good to be true. So George rings his bank manager and asks what his credit limit is. One problem – he left his check book at home, and he can only remember the last three digits of his account number. In fifteen minutes the manager calls George back to tell him how much money he can borrow and under what conditions.
>
> How did he do it? The bank manager summoned a junior officer to whom he passed George's name and partial account number. The junior officer queried the head office in Nairobi. In five minutes, the reply came back. The status of Nderito's account, his loan limit calculated from his monetary profile, the terms of the loan, deduced from factors such as the length of time he has been with the bank and the amounts of money that his transactions have involved, and finally, approval for the loan since his name does not appear on the "black list" of any bank branch. All these answers arrive courtesy of the bank's computer system.

In a total of fifteen minutes, Mr. Nderito's account can be checked and his loan approved based upon only partial information from the applicant. There are no mistakes and delays. The outcome is positive and the computer mediates between the applicant and the bank to make the loan possible. Transactions that might have taken days are reduced to a matter of minutes. Although a centralized computer system is used, no shortcomings in the system are evident.

D.S., the research representative cited above, voiced a different view of the implementation of computerized banking in Kenya at the early stages of developing this system. He saw the problem as one in which the new mechanical solidarity was fragile. According to him, the reduction of workers would eventually create serious labor problems.

2.5 D.S.: We are making the first real inroads of advanced banking // computerized banking // in Kenya. In fact // in this part of Africa // as far as I know.

B. J.-R.: How does that seem to have gone as an experiment? Or //

D.S.: Well // it's only now being implemented. The first bank has now got on-line tele-terminals, and potentially it could have // well // as most of the rest of the world has experienced // it could have quite dramatic effects on the nature of the clerical workers in the back office. I have no idea what the situation is. I know that they would be ultrasensitive // being a government bank.

The successes in finance and banking are documented by the media as the strongest areas of progress for computerization in Kenya. Another account of the efficiency of computers at the confrontation stage is presented in a description of the miracles that they have worked for Kenyan coffee and tea farmers. In this case, computers are depicted as making loan payments and debt recovery simpler and more efficient. Despite the fact that agricultural and craft cooperatives in Kenya have been reluctant to computerize for organizational reasons, this success story presents a case in which enormous benefits have resulted from technological change.

2.6 *Kenyan Coffee and Tea Farmers*

Coffee and tea form the backbone of Kenya's agricultural earnings. In the 1983/84 season, total payments to coffee farmers were of the order of 4 billion Kenyan shillings and tea farmers were paid 1.8 billion Kenyan shillings. Both these crops are grown by individual small-scale farmers or on large estates. The farmers sell their coffee indirectly through the Kenya Planter's Co-

> operative Union (KPCU) and tea through the Kenya Tea Development Authority (KTDA) . . . . In order to insure proper control of deliveries and payments, both KPCU and KTDA opted to use computers for this purpose. As a result, KPCU installed a Wang VS/80 in 1980 but KTDA had installed an IBM System/3 model 15 a year earlier. These systems are used to record quantities of coffee or tea brought in by farmers or their cooperatives, payments to be expected from these quantities, as well as actual payments made . . . . Both KTDA and KPCU give loans to farmers and recover them from payments [sic.] due to farmers. In both cases, the computer is being used to keep accurate records, thereby effecting prompt payments to farmers. (*East African Computer News* 1, 1, May 1985a:13 and 27).

The computer is described as the tool that makes keeping accurate records possible. It is an idealized object that has permitted a positive outcome to take place. Although this outcome is technically a feature of the attribution stage of the narrative program, this media account is presented in order to persuade potential computer users at the confrontation stage to employ computers in their daily activities. This account is pertinent because the Ministry of Agriculture experienced severe payment problems to cooperative farmers prior to its installation of a computerized recordkeeping system in 1982. When this article was written in 1985, the recordkeeping system of the combined Ministry of Agriculture and Livestock Development – although it had experienced demonstrable improvement – was still in flux. The article omits any discussion of problems of implementation of the system for recording farmers' loans and repayments and focuses solely on the computer's success in making the new system work. Reasons for installing a Wang VS/80 system a year after an IBM System/3 had been installed are omitted.

In order for accounts of computer successes to work, discussion must remain at the general level of subjunctive sources and performances. Descriptions of routine problems encountered in computer adoption are eliminated. Specific cases are presented to demonstrate how computer users move successfully from the subjunctive mode of virtual computer use into actual applications. We see "what happens if" the computer is installed by virtue of an interpretation of specific cases from which descriptions of the actual process of implementation have been excluded.

Discourse set 1.10, presented in Chapter One as an example of a positive attribution of the computer's effectiveness, actually describes a hypothetical case of implementation. This text demonstrates the benefits of computer use for local farmers and warns the user to select the appropriate data-base carefully. If such careful selection is performed successfully, the results on the practical level will be effective immediately and will allow the farmer "to determine when to get rid of a particular cow" (*East African Computer News*, 1, 2, October 1985d:13). As with the outcomes in the cases of Mr. Nderito's loan and the Kenyan coffee and tea farmers' financial arrangements, correct implementation at the confrontation stage unquestionably results in computer successes and a positive attribution later in the technological narrative.

R.T., the director of marketing at ICL in Nairobi, described problems that arise at the implementation stage from a commercial perspective. The marketing representative has to determine what the users' objectives and needs are. This process, however, involves ambiguities and the analysis of what factors, in the representative's view, may inhibit users in their plans for computer adoption.

2.7 R.T., director of marketing, ICL, Nairobi:
Well // we look at which organizations there are within various size categories by segments. What computer do they have // if any? If they haven't // why not? What their plans are // etc. It gives us a feel for what's inhibiting people from moving // what their development plans are.

One of the factors that inhibits change is the social organization of enterprises in which computers are adopted. The computer operators, who occupy image 5 of Figure 2-1, are portrayed in the idealized setting as executing the programmer's instructions without difficulty. In many instances, however, these operators are overworked and feel powerless as well as dominated by the organizations that hire them. The friction emerging between operators and other members of the computer personnel in an organization is an important feature of computer implementation as we move from the confrontation stage to the stage of technological mastery and domination. In some instances, as we shall see in Chapter Three, efforts are made to control operators' access to important organizational knowledge. In the computer dealerships that I studied and in an insurance company, payroll data were prepared for external processing in order to prevent operators from having access to classified salary information on other employees. The metaphor of machine error (rhetorical strategy 5) provides a convenient technique for blaming operators, who input data directly into the computer, for organizational mix-ups and errors.

Writing as an advocate for the operators, Maxwell Kareithi expresses his view of the operators' exploitation in an editorial published in *East African Computer News* (1, 4, December 1985/January 1986:5).

> 2.8 *Computer Operations Staff*
>
> It is the operator's task to monitor all the procedures in a running system, print reports, regulate the temperature of the computer room and, most of all, maintain a high level of data security. Due to their designation as operators, they are associated with shoveling dirt. And dirt they receive as well. Their wages are generally low and their work hours long and demanding. They get few breaks. They often work in very cold rooms sometimes waiting hours for programmers to give them the go-ahead to proceed . . . . Operators are therefore an undermined section of the working population whose contribution to the actual running of highly valuable data processing equipment is hardly taken into account. Their status within a company is established without consideration of the capital value of the system they run, and for sure, it excludes them from occupying a crucial role in a business that is changing the face of the world. Operators are perceived as the equivalent of semiskilled workers, shoved to the back of the queue when their employers plan improvements. They have more grievances than any others in the computer area. Many companies don't even bother to provide them with adequate rest room facilities . . . .
>
> The best way to encourage operators is to promote them to senior positions and let them face the challenge. They should receive intensive training, and, like any other employee, they need motivation.

This account differs from the micronarratives about computer successes. It describes the long hours that operators spend waiting for and processing data. The conditions of the operators' work and their economic circumstances are contrasted with the situations of their supervisors. The major subjects under consideration in this discussion are the operators and the companies hiring them. Computers are implicitly referred to in terms of the operators' literal tasks: "to monitor all procedures in a running system . . ." and their figurative tasks: "shoveling dirt." The author's ideal narrative outcome is the intensive retraining of operators leading toward their eventual promotion.

The computer's prestige does not bestow recognition on the operator in the same ways in which prestige may accrue to the company adopting computers (cf. rhetorical strategies 1, 3, and 6). Instead, the operators remain invisible. They are seldom mentioned in the hyperbolic discourse about computer successes. How does the enterprise actually conduct work with computers? Who deals with the data? The personal computer model of confrontation assumes that everyone at all levels of an organization will have the same access to and responsibilities for the computers installed. This model depicts the ideal office of the future. In contrast, the mainframe model of implementation is based on a hierarchy of general and technical managers in which the operator's proper role is to run the machine with programs suggested by systems analysts, computer managers, and senior programmers. In the case of a Nairobi car dealership in which the DPM had little technical knowledge, we saw that the chain of command did not function as projected. The case of operators – who aspire to be computer professionals but are treated as extensions of machines under the new mechanical solidarity – raises the possibility of further contradictions and omissions in the idealized discourse of computer confrontation.

Three bureaucratic strategies are employed to monitor and divert attention from the role of computer operators in enterprises: (1) dusting the terminals, whereby the computer is not used to full capacity; (2) focusing on the effectiveness of the computer system rather than on its operators; and (3) emphasizing the expertise of senior computer personnel in contrast to the competence and work of the operators. After the system has been installed, its successes are determined in terms of increases in productivity and efficiency. My discussions with the data processing manager at a Toyota dealership in Nairobi focused on the importance of "good" and "bad" systems in a comparison of the inventory and stock control systems before and after the installation of an IBM 34 system.

Before the computer was installed, it took several days to check on the availability of a specific part. Customers waited days to find out whether or not their cars could be repaired. After the system's installation, stock checking was immediate. The DPM complained about the old, "bad" system of relying on the data bureau and on manual checking for stock controls. The operation of the new system, the mastery of which is part of the domination stage of computer use, was described in discourse set 1.4. The "bad" system of sending the job out was contrasted with a "real-time" system providing immediate access to parts. The role of computer operators in implementing this system is not discussed in the DPM's description.

A similar case arose at another car dealership in Nairobi where accounts and payroll were eventually absorbed as part of the internal computer operations.

2.9 G.K., senior programmer:
The computer was principally bought for stock control // inventory control. The accounting was not done by the computer. Comp-Rite [the Apple dealership] was doing it. So they came to a point where they felt, "Why do we take some of the work to Comp-Rite? Why can't our machine do it?" So they designed the system // the system was designed locally // uh // internally // with our accounts people. To have an integrated system where even the accounting // our accounting books are inputted, so everything is integrated right now here at D.D. [pseudonym]. So we have now a complete on-line system that // even if a part is sold // our accounting book is updated immediately.

The effectiveness of this system is described as relying on its design rather than on computer operators. The accountants are presented as having a role in expanding and redesigning a good system for spare parts retrieval and bookkeeping. The system's strengths, rather than the work of its operators, become the focus of the computer's successes.

In emphasizing the expertise of senior computer personnel, the discourse of confrontation describes the personnel's initial efforts to master computers. Some of these efforts involve demonstrating competence in talking about, rather than using, the machines. Computer professionals contrast with the Kenyan carvers and craftspeople whom I studied. For the carvers, mastery means skilled performance rather than talk. Some computer professionals asserted that "talking confidently" about computers was an avowed goal of their training and a key to understanding their expertise. Talking confidently about the machines marks the progression from confrontation to technological mastery. One senior programmer at a data bureau in Nairobi described his training.

2.10 A.I., senior programmer, Data Centre:
We had mainly the four subjects. We had to do business studies // in particular a special course on management // English // business English that is // programming languages, and general computer studies. And the programming languages were mainly often BASIC. We had COBOL, and an introduction // maybe // to RPG II or FORTRAN. Nothing deep in it, but just an introduction more than anything else.

B. J.-R.: Did you find that this helped you a lot in being ready for work at the Data Centre and gave you a good background?

A.I.: Very much. // That's the idea. Because // okay // we never did anything really detailed // you know // but at least the introduction // you know // sort of sets you on the right path.

B. J.-R.: So it was like a general introduction, and then when you came here you went through another training?

A.I.: Yeah // that's right. // I remember the best thing I found about it is anytime anybody is talking about anything in computers anyway // you're not really lost. You probably might just have an idea what they're really talking about.

In another case, the individual in charge of training programs at a computer dealership explained the importance of talking confidently in order to promote his products.

2.11 R.O., computer trainer:
We use word processing to do all our documents, but you need // most of the applications that are on the Apple I've got to learn so that I can talk confidently about it . . . .

Similarly, a computer vendor described learning programming as "background knowledge" that gave him a way of talking about the product he had to sell.

2.12 S.M., vendor:
Well // I // After high school // I took a couple of courses in computer programming and so on. And I just did a bit of my own work.

B. J.-R.: And is programming one of the skills that you need on the job?

S.M.: For what I do? It's not an essential part // no. It's good background knowledge on the product we're marketing.

Talk about computers on the job is not a loose referencing process or a metaphorical "debugging" of the machines in the workplace. Computer users, programmers, and vendors feel constrained to *know* what they are talking about and are hesitant to enter into discussions with those who "know more." Talking about computer programs, operations, and products is a way of displaying mastery over them. Such talk is part of the technological structure of accountability (rhetorical strategy 4). Expert talk establishes the human subject's relationship to the machine as a user, an operator, or a programmer. It also positions the human subject with respect to other computer users within the social hierarchy of an organization. The ability to talk about computer programs or operations without having to input data distinguishes management from workers among the new computer professionals. Talking confidently establishes the user's competence and relationship to the computer ($S_1 \longleftrightarrow S_2$) at the confrontation stage.

New Technologies as Bureaucratic Justifications

James S. Coleman (1980:145) defines authority as a relation that arises "when there has been a transfer of control of a particular kind of resource: one's own actions."[6] In enterprises where computers are used, this transfer of resources is made with reference to the machine. The computer is used by management to support its hierarchy and authority structure through a rhetoric of technological accountability that justifies the relationship of managers to employees. Computers may be considered technological tools for accomplishing managerial goals. Discourse at the confrontation stage in Kenya, however, reveals that managers, or the users in Figure 2-1, are often uncertain about the capabilities of computers and about what they want computers to do. Computer firms and data bureaus intervene as middlemen who help managers to define these goals.

Researchers and programmers at these bureaus, however, repeatedly express their frustration with the unrealistic expectations of new users. Often, the indoctrination phase, during which a company sends data to a bureau, lasts several years before users feel confident enough to try an in-house system. Thus, the confrontation stage refers not only to a set of initial interactions upon first contact with computers but potentially to arrangements which may be in place for years while the trial-and-error use of computers remains restricted to specific organizational activities, such as payroll, accounting, or stock control. At this stage, computer users are in a subjunctive mode of anticipation in which they communicate about what the costs and benefits of a more sophisticated computer system might be. Media accounts (cf. discourse sets 2.4 and 2.6) are oriented toward computer users who are contemplating new technologies in this manner. These accounts illustrate the computer's unconditional efficacy as a recordkeeping and organizational device. They overlook problems of implementation and personnel.

Computers are treated as sacred objects over which management has ideological control. De Lagy (1987), for example, describes a case in Sudan where students in computer programming are not allowed access to the university's mainframe computer.[7] In Kenya, the initial phases of computer use are still characterized by the creation of a mainframe-type of computer environment in which priestly positions surrounding computers proliferate, even when microcomputers are in use. The proliferation of work functions goes hand-in-hand with the overbureaucratized office as a habitus. In terms of social control, this proliferation of functions involves clear status and socioeconomic boundaries. General managers remain in ideological control of the final uses made of the computer and of the funds available for computer operations, while senior programmers and computer managers oversee technical implementation. This elaborate bureaucratic structure creates a rigid status hierarchy. However, by virtue of their dependence on computers, all of these individuals are new technicians of the sacred, who stand in a privileged position with respect to the general public.

Manipulative space exists within computer hierarchies. We have already seen the ways in which computer vendors, senior programmers, and technicians emphasize, as a display of competence, the importance of talking about computer systems rather than using them. This type of self-reflexive talk is at the heart of postmodern discourse where the display of competence constitutes the exercise of social control.

The computer is an ethnographic descriptor used to create idealized visions of the working world. It is manipulated as a rationale for action when these images of the social world are disrupted during the course of everyday events. Manipulations of the computer as a sign of order and a disclaimer for rupture are strategies of technological control, invoked to sustain routine activities in organizations where workers describe the details and efficiency of their work tasks with reference to machines. The computer is the source for generating a new set of bureaucratic titles within an organizational hierarchy that utilizes the machine to justify the qualifications, work tasks, and responsibilities of employees.

Most of the firms that I studied already had computers in place when my interviews started. Computer managers and workers recalled the days before computers were installed and offered glowing accounts of the computer's success in streamlining work operations. At a large charge card company in Nairobi, I was able to witness the recoding of titles and tasks when a new multi-user system was installed. S.A., the computer consultant at the company, explained that work tasks changed before titles. Thus, recoding occurred as a feature of the firm's daily operations *before* it appeared on an organizational chart. Explained S.A.: "Presently, we have not established a computer department . . . . We have left the titles that way because we still have to start doing a lot of in-house development" (cf. discourse set 5.20).

Other cases, where computers were already in place, exhibited a modification of titles without a correlative change in work tasks. In one Kenyan automobile firm where the DPM was actually the office supervisor, rather than a technical manager, the DPM relied upon the unquestioned expertise of his chief programmer in order to explain and conduct computer operations (cf. discourse set 1.6). Nevertheless, his title and expertise were officially and economically defined with reference to the technological structure of accountability in the firm, that is, in relation to the computer, and not with regard to actual work tasks. The computer provided a bureaucratic rationale for the DPM's title, salary, prestige, and reputation outside of the company.

The new mechanical solidarity reinforces the prestige of enterprise titles associated with the computer. In a large Kenyan trading company, P.M., the computer controller, claimed that he deserved to be reclassified as a data processing manager. According to him, the new DPM title would reflect his current technical and managerial responsibilities and would more than double his salary. To support his bid for reclassification, the computer controller saw to it that his department appeared to run smoothly in spite of a tremendous backlog of accounting and recordkeeping work.

Accounts and inventories were months behind. To remedy this

problem, accounts were done by hand and, as time permitted, subsequently entered into the computer. The controller complained that the NCR 8200 minicomputer that he had inherited was obsolete and slow. He proposed the solution of hiring more data-entry workers. An increase in workers strengthened the controller's claim for promotion by broadening his supervisory responsibilities. P.M. explained: "I have one computer operator and four data preparation workers. I am also in the process of acquiring an additional computer operator and having internal promotions. In this case // I would like each section at least have a supervisor" (cf. discourse set 5.6).

In order to mask the inefficiency of the obsolete computer system while simultaneously making a bid for promotion, the controller maintained a silent complicity with the machine. He delegated computer tasks, such as accounts and inventories, to clerical workers who recorded the accounts manually. Although the computer system had word processing capabilities, a typist in the main office of the trading company prepared correspondence, accounts, and reports on an electric typewriter. The controller blamed the slow computer system for the inefficiency of this operation. As a strategy of organizational control, he protected his technological power by making certain that the daily computer work was completed by other means and eventually transferred to the computer. By fronting for the computer, the controller strengthened his managerial position in the organization while reserving the right to blame the technical features of the computer system for the shortcomings of the data processing operations in his department.

Eventually, the P.M. succeeded in acquiring a Wang VS/55 multi-user computer. He planned to implement this system by converting the keypunch operators into data-entry personnel. The acquisition of the system increased the computer controller's visibility and improved the chances that he would be promoted. The controller used his technological power to consolidate his position and expand control. By acquiring the new system, he improved his reputation outside of the trading company and became known as an active and innovative member of Nairobi's rising computer elite. Thus, an idealized view of the computer may be contradicted by its actual uses in an enterprise. Work titles are recoded in relation to the personal aspirations of computer experts within the organization.

A similar case of the manipulation of technological power occurred at the Nairobi Handicraft Cooperative in which I installed a Kaypro 4 computer. In spite of the reluctance of management to endorse the daily use of the computer, cooperative officials insisted on the efficacy of the machine and on its organizational power when in the hands of an expert selected from the cooperative's social network. With such an expert in mind, the management described a vision of the cooperative as smoothly operating with efficient inventories and books. They persisted in using the computer as a symbol of change that would bring about this ideal outcome. The computer's power was invoked to account for a myth that existed only in the minds of the managers. This cognitive utopia was sustained through the management's use of the discourse of technological confrontation to paint its portrait of the enterprise's organizational structure and relationship to the computer.[8] These observations on the discourse of confrontation illuminate the strategies through which workers orient toward the computer during a period of initial contact.

Conclusions

The rhetorical force of references to computers highlights ways in which they are integrated into firms as both essential tools and anthropomorphized objects. This focus is in no way intended to belittle the technological contributions of computers to the social and economic organization of firms. Comparisons with cross-cultural materials from non-African settings provide a means of further exploring whether these general observations apply to office environments in other cultures (cf. Rogers and Larsen, 1984:174-183). For example, in her analysis of computer subcultures in the United States, Turkle (1984:307) argues that computers offer a self-reflective way for individuals to interact while avoiding the entanglements of human commitment. Computer culture, according to this argument, is based upon new forms of sociality that broaden interactional networks among computer users while simultaneously reducing the depth of their personal relationships through psychological disengagement.

In contrast, the data from Kenya and Ivory Coast suggest that social relationships in enterprises remain complex and highly stratified but are filtered through the technological apparatus of the computer, which is treated as a rhetorical and mechanical interpreter of organizational structure. The computer's promise of efficiency allows it to be incorporated into the world of work as long as its technological powers are managed and assimilated into the indigenous structure of the organization. This management is achieved by making extraordinary claims about the computer's organizational capacity while under-utilizing it in the service of maintaining bureaucratic order and hierarchy. These extraordinary claims about the potential, or subjunctive performance, of computers characterize the discourse of confrontation. Under this discourse, new technology is, in Graeme Salaman's terms, "presented as an unavoidable and necessary means of obtaining the organization's goals" (Salaman, 1979:86). Salaman argues that the rationalizations for the necessity of technology "depoliticize" the role of computers and other new technologies in firms. I conclude instead that technology is, in fact, politicized through strategies of technological control in which computers are employed as justifications for organizational decisions.

Anthropomorphizing the computer in the Kenyan and Ivorian work contexts means subjecting it to the same sorts of bureaucratic limitations and constraints as those experienced by its human users. Thus, allusions to the computer's power mask the indigenous structure of the organization under the veneer of high technology and are used as justifications for change and as excuses for human conduct when reciprocal interactions fail. Technologies of control in the African workplace are not strategies of self-deception but rather are keys to understanding how computers are interpreted and integrated into particular cultural settings (cf. Maffesoli, 1979:137-138; Balandier, 1985:116-124). These strategies are ways of restoring reciprocity and a sense of sociality by using computers to sustain an idealized vision of a harmonious working world. This world is based upon a cognitive utopia in which disruption and conflict may be reduced to technological error, and the conventional social order may be transformed silently, with no blame placed upon the new technicians of the sacred who dare to change it.

Chapter Three
New Technicians of the Sacred: Technology, Belief Systems, and Social Control

> [T] hese peoples (and they are too diverse to be covered by a single name) are precisely "technicians" where it most concerns them specifically in their relation to the "sacred" as something they can actively create or capture (Rothenberg, 1968:xix).

Introduction

Exclusivity and doctrinal purity characterizing religious sects arise in the world of work. With regard to its tendencies toward utopian future-orientation, the computer industry exhibits sectarian closure. For example, Peter Halbherr (1987:20) analyzes IBM-France as a religious sect that proposes to transform the world through its technological and organizational innovations. Secularization and discourse are not the only sources of self-definition for the new technocratic class (cf. Gouldner, 1979:48-54; Eyerman *et al.*, 1987:1-15). Through secularization, a culture of critical discourse and rationality emerges. This form of rationality, however, requires doctrinal purity. In the world of work, we encounter enclaves based on hierarchies of technical competence and social inclusion.

Sectarian processes are evident when a valued cultural or intellectual object becomes the focus of work activities. Protection of exclusive access to valued objects is one of the roles of the professions.[1] Max Weber saw the process of professionalization as a consequence of strict control over education. He viewed such control as typical of the recruitment procedures of complex bureaucracies. According to Weber (1946:241):

> When we hear from all sides the demand for an introduction of regular curricula and special examinations, the reason behind it is, of course, not a suddenly awakened "thirst for education" but a desire for restricting the supply for these positions and their monopolization by the owners of educational certificates.

Status cultures insulate valued objects through education and indoctrination. Both exclusive professions and religious sects challenge the existing social and political order through their claims to privileged knowledge and the social hierarchies that support these claims. This challenge is represented by the rise of the new technocratic elite. Unsettling the existing social order, this rising elite may be compared to members of new religious movements. With reference to the narrative program for computer adoption, the rise of a new group of professionals occurs during the transformation from confrontation to domination.

New Technicians of the Sacred: Technology and Sectarian Processes

Validation of cultural objects as sacred is a central feature of all religious practice. By attaching a sense of charisma to leaders and their doctrines, religious sects reinforce their exclusivity. Religious practitioners create cognitive utopias in order to sustain their views of the sacred. Ideals such as bureaucratic efficiency in the world of work reinforce a cognitive utopia. Technology is used to implement these bureaucratic ideals. Technological implementation may be analyzed in semiotic terms as a correlation between valued cultural objects or signs, such as computers, and the social arrangements of office hierarchy. Social relationships are organized as an ecosystem surrounding the sacred object, and they are maintained by discourse about the object. As emphasized in Chapter Two, job titles, computer-related excuses, and high-tech jargon are manifestations of this universe of discourse (cf. Snow and Machalek, 1984:170). In my Kenyan computer experiment, we see how Kamba carvers used the computer as a symbolic justification for progressive office practices, even though the computer was never integrated into their work activities. In Kenyan banks, insurance enterprises, and wholesale companies, mainframe computers generated a new layer of bureaucracy during the 1970s. Data processing managers, as expressive and instrumental leaders empowered by computers, questioned the position and authority of general office managers. Dissemination of technology to more workers through the appearance of microcomputers and multi-user systems, in turn, challenged the authority of the DPM as both a technological innovator and a charismatic leader.

Computer experts are part of a closed world that relies on the possession of scarce and valued knowledge. This knowledge includes technical expertise with the computer as well as information stored in the machine. Knowledge about computers is framed by the fragile and progressive computer contracts. The credentialing crisis surrounding computer education in Kenya constitutes a gatekeeping process for selecting and controlling who has access to knowledge. Considered in this way, the opposition to the commercial colleges teaching microcomputer use in Kenya is not only a criticism of the schools' educational standards and economic motivations but also an effort to limit the number and types of computer users. A Nairobi executive's statement that he knew who was good and bad in the computer field illustrates this effort to restrict the domain of competent users through sectarian closure. Even the initiation rite of taking vendor-sponsored computer courses does not guarantee entry into the closed circle of experts whose competencies and influence are widely known in Nairobi's computer world.

Religious sects are upheld by a concern with purity of belief (cf. Bittner, 1963:937). In order to sustain itself, a cognitive utopia must be separated from other cultural realities by the conscious and concerted activities of believers. Daily practice, i.e., working with computers in the home and office, is one such boundary-maintenance activity. So too is talking about the extraordinary powers of computers. The audience for such talk, however, is limited in the African context. Although users' groups and computer magazines are widespread in the United States and Europe, *East African Computer News*, Kenya's only computer magazine, had difficulties maintaining its staff and circulation over a three-year period and nearly folded in 1986.

The task of the true believers in the new African informatics order is to persuade themselves and others that computers will improve conditions in a labor-intensive economy and promote prosperity without destroying the balance of power and information. Since the presence of computers in Kenya is due to the efforts of the multinational corporations that have introduced, imported, and distributed them, convincing the public that they are helpful, or at least harmless, is difficult because of their foreign origins. Jacques Ellul views state control of technology as a byproduct of the technology's evolution. In describing nuclear physics and aerospace research in Western Europe, he argues that current costs far exceed individual capabilities. Ellul (1964:236) states:

> Once a certain degree of technical progress has been achieved, continual improvements give rise to instrumentation so complex and large that the cost price is inaccessible to the individual. The present growth of cost price in all technical domains is unparalleled, even in recent history.

In the African context, microcomputers often exceed the cost of individual ownership. The computer profession emerges as a network that attempts to counterbalance state control of technology and personal career interests by giving voice to individuals with technical expertise regardless of their ownership or source of access to new technologies. The establishment of professional associations, such as the Association Professionnelle des Informaticiens de Côte d'Ivoire and the Computer Association of Tanzania (COMPAT), provides channels of access to new technologies. These associations define the emerging profession by creating standards of competence for their members and generating new sources of information. In both the Ivorian and Tanzanian cases, professional associations promote state support of computerization and establish the expertise of association members as brokers and shapers of computer technologies nationwide.

Computer associations have carved out a previously nonexistent domain of influence for their members. Professional associations encompass a broad cross-section of members, ranging from highly trained engineers to beginning programmers and computer operators. Consequently, the boundaries of the new profession are difficult to draw. Since these associations provide a means of self-definition and channels of upward mobility for credentialed computer professionals, they promote individual career interests while attempting to influence public policy and state priorities. Computer associations follow the pattern of postcolonial political groupings in several African nations. Kenneth Little (1974:30) describes this pattern as one in which "individual politicians jockeyed for position and . . . political groupings were mobilized for particular purposes but did not constitute permanent corporate groupings." In the Kenyan case where computer technicians and users have not formed a stable professional association, state importation restrictions limit access to computers, and a credentialing crisis threatens computer education.

The computer manager at one of Nairobi's data bureaus explained how she saw the process of computerization developing through the activities of a network of computer users and experts.

3.1 F.C., computer manager, Data Centre:
You get someone working in a company that's been using a computer, and he goes to another company which isn't using a computer. And they're spending days and days doing their accounts. And if he's got the sort of mettle of push // more than anything // he can probably get things computerized // which would mean that his firm's turnover would probably be reduced . . . . We're finding that more and more of our clients are people that have actually transferred or gone from one firm to another. And they've actually got the kind of experience with the computer to realize that // in actual fact // the computer can help quite a lot. And they are the ones who sort of make their new firm aware of computerization.

One computer vendor whom I interviewed analyzed the employment problems that result from increased computerization by affirming that "at this stage" in Kenya computers could never make a sufficient inroad to displace labor. Industrial automation is a long way off. Others within the network of computer users point out the advantages of reskilling and recycling employees into jobs using new technologies and the prospects of increasing the labor pool by creating computer experts. These arguments, however, are offset by the sectarian character of the computer world. Training that is too readily available is a threat to the priesthood of computer professionals. Hence, Kenyan computer professionals regard commercial colleges with suspicion because these institutions make technological training available to anyone who can pay for it. Dissemination of technological expertise challenges the subtle power balance sought by the new technicians of the sacred.[2]

Popularization of computer technologies depends on the expansion of Kenya's business sector. R.T., director of marketing at ICL in Nairobi, related the rise of computer sales to problems of foreign exchange. He explained:

3.2 R.T., director of marketing, ICL, Nairobi:
Of course, there are constraints // possible constraints // in the availability of foreign exchange which have hit people in the past // in 1980 and 81. If there's difficulty with coffee // tea // tourism // then the foreign exchange is just going to have an inhibiting effect. But there certainly hasn't been any difficulty in the last couple years. We've had no difficulty getting computer import licenses.

A senior programmer at an automobile dealership described the inevitability of business expansion in Nairobi and its consequences for computerization. He saw business expansion and computerization as synonymous.

3.3 G.K., senior programmer:
Well // I should say // computers have a very central place in a company that is bound to expand, and I think the central theme in every business // okay // is to expand // to have more and more transactions. So I should say in Nairobi we really need computers, and they are bound to increase.

Part of the problem of professionalization centers around whether business expansion requires a larger network of computer users with specialized qualifications. If so, do these professionals view themselves as a group with common interests? The answer to this question depends on whether computer experts are classified as part of the clerical or administrative labor force (cf. Crompton and Jones, 1984:30-36). Although computer vendors and distributors view themselves as separate from this labor force, they have a vested interest in the technological transformation of office work.

Public debates emerging in Kenya during the early 1980s focused on the stratification of jobs for computer workers. In contrast to the positive promises of the discourse of confrontation, these debates raise questions about the obsolescence of clerical workers who have not adapted to the latest electronic technologies. An article on keypunch operators in *East African Computer News* highlights these concerns.

3.4 *Punch Operators: How Secure Are Their Jobs?*
The result of these two trends (i.e., the swing towards personal computers and introduction of on-line systems) is to eliminate the work of a punch operator. What does the future hold in store for punch operators? For a start, the change-over from the existing systems to the new on-line systems is likely to continue for the next ten years or more. Therefore, people in established organizations who are currently doing punch operating work need not worry about the security of their jobs. However, as newer and newer systems are established and many new computers installed, existing punch operators must take the initiative to learn these new technologies. Their role, therefore, will change from copytyping to .either using small computers or operating

> computer terminals in an on-line environment. Failing to do so, they may easily find themselves out of work in a matter of five to ten years (*East African Computer News*, 1,1, May 1985c:6).

In this article, the burden is placed on workers who fail to adapt to technological change. If they do not conform, their jobs will be eliminated. These workers are not viewed as computer professionals or members of an interest group. Instead, they are treated as organizational subordinates who must change as the technology and economy shift. A similar view is expressed in Maxwell Kareithi's article on computer operators (discourse set 2.8) appearing in *East African Computer News* (1,4 December 1985:5).[3] Kareithi, however, shifts the burden of adaptation onto employers, who he thinks should involve workers in the introduction of new computer systems and, thereby, increase the scope of technical responsibility of computer operators. Both articles emphasize the rigid stratification of computer work in Kenya. Computer operators and keypunchers are excluded from decision-making positions and are on the margins of the computer profession. Seen from another perspective, these workers possess technical skills not available to the rest of the population, and they have the potential for vertical mobility within firms and across computer-related jobs.

Analyzing the emerging computer profession in Tanzania, Lewis (1987:10-12) emphasizes the critical role played by COMPAT in defining criteria of professional status. COMPAT members must hold educational credentials in the areas of computer and information sciences. Individuals without such credentials, including computer operators, keypunchers, and computer vendors, are excluded from association membership. Out of the forty trained professionals who founded COMPAT, only one was a woman. Lewis (1987:10) asserts: "Computer professionals are large system people who have moved into microcomputer use." In Ivory Coast, broader professional criteria are applied by the APICI, which admits computer operators. Although no computer association exists in Kenya, a tacit system of stratification based upon reputation operates to exclude computer operators, keypunchers, and lower level clerical workers from the ranks of computer professionals. Nevertheless, this exclusion cannot be enforced. Thus, vendors, commercial college graduates, and computer operators hold the hope of professional advancement by virtue of their association with new technologies and highly trained computer experts.

Regardless of their educational backgrounds, computer vendors in Kenya perceive their interests in terms of the emerging market for computers, even though they have been unable to coalesce into a united public force. One computer vendor in Nairobi lamented the fact that vendors and computer users had not been able to develop a lobby. He explained this situation.

3.5 S.M., computer vendor, Nairobi:
This is something that we have lacked. There are other pressure groups that have developed in other industries where they have got together and petitioned the government and done things. And they have been successful. Recently some paper manufacturers // people making stationery and what not // they lobbied about some of their raw materials, and their duties were reduced. They were able to get together and put up a case. We have not really done that // and I think one of the main reasons is that the rate of expansion has been so much. But you still do well with that. Although you grumble about it // you still do well.

The failure to develop an effective computer users' and vendors' lobby may be interpreted in at least two ways: (1) as a result of the loose organization of computer vendors who are in competition with each other in spite of sharing common professional interests and (2) as a consequence of the sectarian character of computer networks. These networks focus on doctrinal purity, internal organization, and professional standards at the expense of common interests. The first interpretation is an obvious one and resonates well with images of postmodern life. The second explanation, however, moves below the surface to uncover the relationship between the internal organization of the computer world and its external image. As with many religious sects, Nairobi's computer world is replete with internal distinctions and gradations but is less well equipped to defend itself against external criticism because of the absence of a collective identification (cf. Smelser, 1987:281-282).

Training is part of the sacrifice involved in entering the computer world. It reshapes the outlooks of computer users. Computer education creates a new world view based upon testable and alleged facts (cf. Polanyi, 1958:303). The user believes that mastery of work tasks will be made more efficient with the computer. Individuals entering the computer world are confident that their skills will lead to a vastly improved lifestyle. Fulfilling this dream requires years of commitment, work, and sacrifice.

The world of computer users, vendors, and experts restricts the entry of outsiders while expanding and popularizing its commercial base and influence. This delicate balance between closure and expansion resembles the process of recruitment into religious sects for which new members are essential, yet boundaries must be maintained with respect to the outside world. This enclosed network may be seen in perspective within the larger environment of computer use in Kenya. The government is fast becoming one of Kenya's most important computer users. The government establishes computer policies and technological practices. A comparison may be drawn between official efforts to regulate the diffusion of new technologies and those used to control new religions. Therefore, it is helpful to examine the Kenyan government's attitudes toward the enclosed circle of computer users in order to situate computer training and commercial activities in this comparison. Kenya's approach is characterized by ambivalence toward computers, desire for cultural capital, and fear of neocolonial domination through technology.

External Control of Technology and Religion

Negative responses reinforce the unique status and solidarity of the computer world. Throughout the 1980s, the prevailing government attitude toward computers in Kenya was one of suspicion. Editorials written by computer vendors and concerned consumers complained of political intolerance.[4] Government officials publicly argued that computers and electric typewriters were not appropriate technologies for Kenya because they increased dependence on foreign goods and threatened jobs (cf. Wescott, 1987:97). This rhetorical resistance to technology contradicts a complex process of computer adoption taking place within government agencies. The government advocates its own computerization while limiting technological diffusion in the country.

During the early 1980s, in Kenya, no public policy concerning computer use had been formulated. Nevertheless, presidential policy in other domains indirectly affected the computerization of the Ministry of Agriculture (MOA), which was one of the first government agencies to adopt microcomputers for fiscal planning. In a speech delivered on September 21, 1982, President Daniel Arap Moi stated that regional districts should become the administrative centers for all rural development projects: "I have instructed all ministries to insure that this new approach is put into full operation by July 1983" (cf. Brodman, 1985: 25). This directive required the MOA to develop a new fiscal reporting system to monitor all local-level expenditures as discrete budgetary entries. This process required the reorganization and compilation of so much data that computers became the only possible solution. Thus, computerization was mandated indirectly and in an unplanned manner.

After the coffee boom of the mid-1970s, the MOA experienced difficulties in adjusting to a new situation of budgetary austerity. These difficulties were accompanied by cumbersome procedures of financial reporting and problems of budgetary allocation to district-level agricultural projects. Pressure for reform came from both the President's Office and external donor agencies such as the World Bank and USAID.[5] During the 1970s, computing throughout the government was coordinated by the Government Computer Services Centre (GCSC). The operations of GCSC are similar to those of private sector enterprises reliant on mainframe computers. IBM and Wang computers are run by data processing specialists who receive much of their data from government agencies behind schedule and, consequently, experience a heavy backlog. A report cited by computer consultant Clay G. Wescott (1985:2) indicates that the GCSC had a three-year backlog of income tax data awaiting entry as of 1985, a four-year backlog of motor vehicle data, and a one-year backlog of migration statistics. Clearly, the cumbersome and underequipped central computing office was unable to meet all government agency needs.

In 1981, the MOA experiment began with a demonstration of microcomputers by USAID technological advisors. The ministry's officials were sufficiently impressed by the demonstration to decide to adopt the computers on a trial basis. A team of advisors from the Harvard Institute of International Development worked with the Kenyan financial managers. Their task was to devise a system for the use of microcomputers in the preparation of budget estimates and forward budgets to be submitted to the Treasury. The ministry's 1979-80 budget had greatly exceeded the Treasury's limits and, in the end, the ministry had received far less than it needed for operating expenses and ongoing projects in 1980 in spite of considerable development aid (cf. Pinckney, Cohen, and Leonard, 1987:73).

Initially, an Apple II+ microcomputer with two disk drives and a matrix printer were installed for expenditure monitoring. The ministry used packaged VISICALC accounting programs to compile the budget. Regional reports were entered into the computer immediately upon arrival. Totals were made, and budget entries subsequently rechecked. This process eliminated a cumbersome, hand-calculating procedure that required cross-checking and revision of each report. A 1984 report submitted by the Harvard team stated that because VISICALC was easy to learn and manipulate, financial officers and their clerical staff were able to use it after only a two-hour training session, provided that an experienced user was nearby for consultation. Consequently, the long wait for bureaucratic clearance of reports was eliminated, and clerical staff could be used more efficiently.

The final reports were far more streamlined and accessible than the 100 page outdated documents that previously circulated. The VISICALC program resulted in a standardization of the reporting format and made the ministry's budget more accessible to Treasury officials and other government functionaries. Financial reports no longer collected dust on desk tops. They were read and implemented. Corrections and modifications were simple to make. By 1982, the ministry had reformed its financial reporting practices and cut its expenditures. In his report on the use of microcomputers for budgetary analysis in the Kenyan government, Wescott (1985:2) asserts that "microcomputers seemed robust" in the ministry's environment and that "scarce Kenyan professionals and technical assistance staff were becoming more productive." This accomplishment is a remarkable computer success.

VISICALC, however, did not prove flexible enough to meet the ministry's needs and resulted in accounting and typographical errors that were difficult to locate and cross-check. Pinckney, Cohen, and Leonard (1987:75) explain: "[V]ISICALC'S reports were not flexible enough for the numerous ways in which the data were being used. In particular, VISICALC was unable to highlight problem areas." In 1982, an IBM program using D-Base II was introduced for the budget. It was subsequently modified to a D-Base III program. Preliminary agricultural expenditures for the districts were prepared and reviewed on computer. Problems with allocating funds from the ministry's central office to district agencies appeared to decline. Deadlines were met for the ministry's 1982-83 forward budget, which in previous years had been up to six months behind schedule. With the new program, the ministry could handle the allocation of funds to districts and review its budgetary expenditures.

The initial Harvard report does not, however, explore the bureaucratic reasons for which VISICALC may have been an inappropriate tool for the ministry's expenditure monitoring and allocation plans. Instead, it refers to the level of computer error that was perpetuated by the VISICALC program. The shift from VISICALC to D-Base II involved bringing in individuals with higher programming skills and upgrading the skills of the computer users already employed to monitor the accounts. Personnel problems arose as the agency was rebureaucratized. Technical errors open the door for a host of other bureaucratic and computer-related excuses. D-Base II reintroduced a more complex office structure.

In her 1985 report on the then combined Ministry of Agriculture and Livestock Development (MALD) experiment, Janice Brodman (1985:25-27) cites four reasons for the change in financial planning and budgets from VISICALC to D-Base II: (1) disaggregation of financial data required by President Moi's directive for a more intensive focus on district and regional control of agricultural projects and expenditures by mid-1983; (2) difficulties with coding records and reports using VISICALC; (3) problems with typographical and accounting errors; and (4) merger of the Ministry of Agriculture and the Ministry of Livestock Development in September, 1983. This merger resulted in increased responsibilities and a heavier workload for the combined ministries. Although Brodman's account presents the factors leading to this transition in a broader perspective than the first Harvard report, neither account devotes sufficient attention to the internal changes for clerks and computer workers as a consequence of adopting D-Base II. In publications of the results of their research interventions in Kenya, Pinckney, Cohen, and Leonard (1987: 88-89) and Wescott (1987:103-110) conclude that problems of communication between government administrators and computer staff along with difficulties in staff training and retention are stumbling blocks in the process of computer adoption. These problems, typical of the confrontation stage of computer adoption, are the source of personnel recommendations made by the foreign consultants.

According to Brodman, MALD financial officers refused several technical advisors' requests to learn D-Base II. A group of six clerks was trained in 1982. Because of difficulties in adapting the manuals to their budgets and an absence of financial and bureaucratic incentives, only one of these clerks had mastered the computer program after six months. A year later, a single clerk prepared most of the reports on D-Base II while the others returned to data entry using packaged computer programs, Brodman (1985:31) describes this situation.

> 3.6 *Training to Provide Expenditure Monitoring Reports*
> Thus, by the end of 1984, only one clerk had become highly skilled in D-Base II programming. She is able to prepare most of the reports, including fairly complicated reports on request. Nevertheless, she does not understand the implications of the financial information. For new kinds of reports, she requires someone with knowledge of the finance system and some understanding of the principles of microcomputer operations to explain the kind of programming needed. The expertise of a more skilled programmer is still required to prepare a few of the more complicated reports.

In a publication of her findings from the MALD project, Brodman (1987:122) describes mixed leadership messages as one factor inhibiting the use of microcomputers in government decision making. Since workers with primarily clerical duties were assigned to microcomputers, financial officers made use of the reports but resisted learning about the technology. The gap between the general management and technical staff occurring in private businesses affected the government as well. Clerical workers' titles were not recoded even though their tasks changed and their responsibilities increased with the microcomputers' installation. Financial officers did not acknowledge the technological power of their subordinates in spite of the support for computerization from top echelons of the ministry.

These problems are typical of an uneasy transition from the confrontation to domination under a fragile computer contract. An initial success establishes the effectiveness of the microcomputer in the workplace. The need for more refined and complex procedures leads to the introduction of new forms of training, the reliance on outside experts until mastery is achieved, and the demand for an experienced senior programmer to oversee more of the complicated programming and budgetary procedures. The technocratic experts stand in opposition to financial officers who refuse to learn about computers. The computer experts constitute a new technocratic elite, possessing the potential to challenge and redesign the basis of bureaucratic control. Without any hope of increased financial remuneration, clerks slip back to data-entry tasks. This process was described in Chapter Two as a computer user's strategy occurring when the simultaneous recoding of job tasks and job titles does not take place.

A computer success story from the MALD's Rural Planning Project is recounted by Wescott (1985:3 and 1987:98-110). In 1984, secretaries at the ministry were asked to prepare draft reports for revision by private firms. To everyone's surprise, they were able to complete most of the final reports in publishable form within a short period.

> 3.7 *Government Secretaries Outperform the Private Sector*
> The microcomputers introduced under the Rural Planning Project were used initially for the production of the 1984-85 District Development Plans. The previous District Plan production effort had required over two years to move from first drafts to publication. To avoid this problem in the current round, arrangements were made by the management advisor with six local word processing firms to produce the plans. At the same time, ministry secretaries were taught by two local consultants (a Ugandan and a Kenyan) to use the Wordstar word processing program, with the the thought that they might produce a few draft quality plans which would then be polished up by the private firms. As it happened, the ministry secretaries outperformed any one of the private firms, and produced 11 out of 40 publication quality plans and 25 out of 40 annexes. Because of their efforts, the plans were produced on schedule, in time for the 20th Anniversary of Independence festivities. The time needed between draft and publication had been reduced from over two years to six months (Wescott, 1985:3).

This account praises the efficiency of retrained clerical workers, not required to learn programming skills. In this case, the bureaucratic hierarchy of the agency was not challenged, and the efficiency of production procedures was upgraded. The computer consultants involved in the MALD project are justifiably proud of their success. Efficiency and effectiveness were improved by computerizing expenditure monitoring, budget preparation, and report drafting. This success story, however, has broader implications for official attitudes concerning computer use in Kenya. The ministry's initial success led to a similar microcomputer experiment in the Ministry of Finance and Planning in 1983. Kenya became one of the few countries in the world to prepare its entire national budget on microcomputers.[6] The Financial Secretary working with a team of advisors used Kaypro, IBM, and Osborne microcomputers to accomplish this feat.

Weekly monitoring reports, budget projects, and annual budget requests were placed on computer for cross-checking and revision. The MFP was able to coordinate its activities with the MALD's new Rural Planning Division, which was also being computerized. The result was the presentation of the 1985-86 estimated budgets ahead of time, with considerable interministry cooperation. The foreign consultants produced positive and detailed reports of all these endeavors.

These projects demonstrate the application of microcomputers to tasks for which they are not considered adequate tools in Western nations. Home computers were successfully brought into the workplace and the seat of national financial planning. The projection, reporting, and distribution of funds now came under full state control with prospects for further coordination and advanced planning based on more reliable and accessible data. Some of the computers used by the government were received as donations from USAID and from Kaypro and Osborne vendors. These computers entered Kenya duty-free to be used for the common good in the public domain. They were, however, the same technological tools sought by small entrepreneurs. The government had effectively appropriated the home computer as an instrument of management and control.

Kenya's use of the microcomputer in national planning and management has received high marks from international agencies, donors, and consultants. As far as the tasks at hand were concerned, an "appropriate" use of computer technology was made as a result of the USAID-Harvard experiments. These experiments led to a redefinition of the role and image of microcomputers in Kenya. At the same time that commercial concerns were domesticating the computer for work in the retail trade and the automobile, insurance, and hotel industries, government moved ahead with its technological agenda. Small entrepreneurs desiring access to desk-top computers paid top dollar while large government agencies received duty-free donations. During the same year that the MFP produced its first computerized budget, vendors agitated for reduced computer charges. Private and public educational institutions attempted to raise funds to meet import duties on machines for classroom use.

The foreign computer consultants, however, participated in a clever, although certainly beneficial, sleight of hand. Computers became instruments of government reform. How could the very same technology that reformed three government ministries be within easy reach of the individual entrepreneur? In symbolic terms, the answer is that this situation could not exist under a fragile computer contract without destroying the balance of technological power. In 1986, computer vendors felt more optimistic about their industry's plight than they had in 1984. They believed that government plans for more widespread computer use at the district level would eventually increase the demand for computer hardware and software. Nevertheless, an unvoiced technological competition had already developed between government and industry.

Wescott (1985:2) reports that the government lost senior-level programmers in large numbers to the private sector during the mid-1980s. He states: "As of February 1985, 13 out of 28 posts at the GCSC for job groups J to L [computer programmers] were vacant." This problem created chronic processing delays and the return to manually prepared information. Claims that a sufficient pool of computer experts is being trained locally have met with criticism, as evidenced by the complaints of both government and industry that qualified people are difficult to find. Since a national policy is not in place, it is too soon to speak of systematic government control of new technologies. The presidential committee for computer planning will undoubtedly need to examine the relationship between the credentialing crisis in computer education and the staffing of computer-related posts in government and industry. Mandates from the higher levels of the government have supported improvements made through the installation of microcomputers in selected ministries. The computer-related successes of foreign experts assisting Kenyan officials have been widely publicized outside of the country. These projects have received more attention in the United States than West African nationally planned programs in countries such as Cameroon, Gabon, and Ivory Coast.

The recommendations of computer consultants in Kenyan government projects focus on improving communication between senior management and computer personnel, retraining staff, and resolving logistical difficulties arising when computers are installed. These problems are typical of the confrontation stage of computer adoption. They occur in the private sector as well as in government. The consultants, however, do not analyze their role as actors in a technological narrative. The necessity for their intervention in government agencies is a byproduct of ambivalent political messages about new technologies. Fear of technological encroachment is expressed in government rhetoric while the need for technological innovation is recognized primarily at the moment of crisis. This crisis-orientation provides the government with computer-related excuses and rhetorical strategies.

Resolution of national crises through the use of new technologies justifies the introduction of computers in spite of the government's ambivalence toward their presence on the open market. Outside consultants remove the onus of technological change from the government by executing the transfer of technology. They can be blamed in the event of technological failure. The consultants, however, view themselves as facilitators of change and hope that their efforts will be institutionalized. In fact, the consultants are actors in the narrative program of computer adoption. Depending on the outcome of the narrative, they may be viewed as opponents or as facilitators of the government's technological agenda. Local computer experts, new technicians of the sacred, use the consultants' efforts to enhance their reputations as part of a rising professional elite.

Ivory Coast's approach to computerization contrasts with the Kenyan situation because of the government's aggressive efforts to foster computer training and use. The focus on training in Ivory Coast has resolved the credentialing crisis by creating a cadre of young, locally trained engineers. The private sector has been active in recruiting local engineers into research and middle-level management posts. The situation in Ivory Coast resembles the French case. Luc Boltanski (1982:63-69) has analyzed the development of a new social class, the entrepreneurial cadres in France, as a byproduct of recent engineering technologies and their influence on the division of labor in the workplace. These new technologies have transformed face-to-face interaction and the daily flow of activities in offices where the cadres are middle-level managers.

One young Ivorian engineer who headed the technical development wing of a private firm described his background and his company's recruitment policies.

3.8 J.R., research engineer:

I first worked in Belgium in electronics. I constructed smoke alarm systems. I returned to Ivory Coast in 1981 and came directly to this company. I became head of the research and development department as the individual responsible for technical development. From time to time // I continue to take courses on project analysis in France . . . . Most of the engineers in Ivory Coast before 1984 were trained in France. But recently the research engineers that we have recruited in this company have been trained locally at INSET // specifically at ISI // the Institut Supérieur d'Informatique // and also in Gabon at the Institut Africain d'Informatique. Now ISI is training *very* good engineers. In fact this company is part of their advisory board. . . . More and more // people here want to use companies known in Africa and in the Ivorian market. And now // we have some very good local engineers.

J.R. was emphatic about the shift from foreign-trained to locally educated engineers in Ivory Coast. He was, however, equally explicit about the distinction between highly trained computer engineers and computer operators. In the manner of the COMPAT officials of Tanzania, J.R. saw computer users with on-the-job training as a separate category from the professional elite. He was critical of the national computer association, APICI, for having lax entry requirements. J.R. asserted that professionals are those with engineering degrees and large-systems experience. Only they would be effective in guiding the technological evolution of government agencies and private corporations. As a trained engineer and a department head, J.R. underscored the need for rigid boundaries within the computer world. He advocated a hierarchy of priesthood among the new technicians.

3.9 J.R.:

With microcomputers // you know // you have to decide // a bit // whether you want to do professional informatics or to use computers just to say that you work with computers. A micro is fine if you have a small company that needs three, four, or five posts that one can interconnect. It's not necessary to have them spend 50 million francs to put in a large system. But if one has a thousand employees, a more complex system is needed.

Professionalism accompanies adoption of more complex systems. The case of the Kenyan government's computerization demonstrates use of an external advisory team to spur professional development. The Ivorian approach is more comprehensive. Engineers trained at INSET move into professional posts on a large scale, without external intervention. Even this approach has its limitations, as R.B., a young woman engineer with the Ivorian government, explains.

3.10 R.B., research engineer, Abidjan:

> It's a question of mentality in the workplace. The mentality of men and of women. In Ivory Coast today // there are very few women at a high level in informatics and engineering. All of the women are at a low level // keypunchers, data-entry people, and computer operators // that sort of thing. The higher one goes // the fewer women one sees. We are always outnumbered in decisions. And always // always // there is prejudice. They say, "Women don't work hard enough. Women don't produce. Women are always absent. Women go on maternity leave." . . . Everywhere, it's the same thing, but here it's even worse because of a resistant mentality toward women. In the United States it has changed a little, but here it has not. So what happens is that women find themselves in a ghetto. They are all assembled together without very important tasks. There is no chance to grow or to advance. Rarely is a position of responsibility offered to a woman. And we accept this situation by becoming less and less aggressive // because the battle is already lost in advance.

R.B., who had earned a master's degree in computer science in Canada before acquiring her post, complained about the limited access of women engineers to positions of responsibility. She saw two axes of closure in Ivory Coast's computer world: (1) the axis of professional training and (2) the barriers created through gender differences. R.B. concluded that women had limited vertical mobility by virtue of the quality of their on-the-job training in government bureaus and private agencies. She foresaw few remedies to this problem.

In her frustration, R.B. reiterated her problems through a micro-narrative in which she juxtaposed being a woman and becoming a professional engineer.

3.11 R.B.:

> We are always treated as women and not as professionals. For example // one day I said to a male colleague who is also a research engineer, "Come into my office, and we can discuss the problems with the program." He said, "I'm not moving. You are a woman. You come into my office." There's a story for you!

R.B.'s complaint highlights the sectarian characteristics of Ivory Coast's computer elite and the bureaucratic contexts in which computer professionals operate (cf. Jules-Rosette, 1988:16-18). In Kenya, the commercial colleges train women in large numbers to become word processors and data-entry personnel. Although these women are not prevented from advancing in the computer world, they tend to be retrained clerks and do not generally hold high-level technical positions. Lewis' study of the computer profession in Tanzania, presents similar findings. She states:

> The [computer] profession has been and continues to be predominantly male, a reflection of the gender differentiated participation in mathematics and engineering opportunities in higher education. Female employment in the computer industry has been restricted to clerical data entry levels. Very few high level females are employed in public computer installations or by vendors. There are no female computer consultants or educators. The profession has not indicated recognition of this disparity, and therefore no effort has yet been made to understand the problem or affect a solution (Lewis, 1987:21).

In spite of its closed character, the world of computer professionals is not stable. Kenya has not yet developed explicit professional standards while Ivory Coast has taken a step in this direction through attempts to plan large-scale initiatives in engineering and computer education. The precarious nature of the computer world reinforces attitudes among certain professionals, such as J.R., that clear lines must be drawn to separate engineers and highly trained professionals from the masses of computer workers.

Computers represent a major ideological challenge for Africa. New technologies resemble religious ideologies in their promise of an altered vision of the world. In their efforts to establish utopian communities, many of Africa's independent churches have confronted the authority of colonial regimes and new states. These communities were far more tightly knit than the emerging technocratic elite, and their beliefs have been more stridently voiced. The rise of a new proletariat, created by industrial and agricultural exploitation, resulted in socioeconomic and cultural marginality for individuals torn away from village life. In urban areas, independent church members have tried to build multiethnic villages that preserved customs in the face of modern social life and the new labor market. Postcolonial governments regarded these efforts with suspicion. In Zaire, Zambia, Malawi, and Kenya restrictions were placed on many new churches, and these groups were closely scrutinized for signs of political protest and dissension.[7]

My 1981 study of community change in Zambia demonstrated how two such groups achieved grass-roots modification of internationally funded development schemes.[8] Through their local efforts, these groups redirected development plans to meet their needs. They played upon interagency conflict in the implementation of development plans in order to secure plots for their homes in new relocation areas. A parallel may be drawn between these groups and the more loosely articulated Kenyan computer network. Although the computer vendors' lobby was unsuccessful, computer users have benefited indirectly from the government's experiments in computerization. These experiments provide a training ground for new computer users who subsequently leave government posts for the private sector. Liberalization of the government's attitudes toward computers may eventually lead to growth of the computer industry in Kenya and expanded opportunities for employment.

New religious groups and the computer professionals are part of the emerging class structures in contemporary Africa. The religious groups have historically been associated with the proletariat of the industrial era while the computer networks forecast the rise of a new informatics revolution and a technocratic elite. These technocrats, however, are not confined to government circles. Their ideologies of reform affect many aspects of life in the workplace. The same technological tools that render government budgets more accessible for interagency control increase the accountability of government to a larger portion of the population and expand the base of information control in society. Thus, the computer represents a challenge to government centralization. Belief systems that accompany technological change on the grass-roots level will, in all likelihood, be subjected to official scrutiny that resembles government control of new religious groups. Anxiety over transborder data flow in the Third World nations is a concern not only with the potential of computerized information leaving the country but also with the maintenance of internal order and official control over information (cf. Pichette, 1979; Secretariat of Informatics of Brazil, 1983). Both new religions and technologies contain ideologies of control that broaden the popular base of decision making.

The computer world monitors itself. Credentialed professionals work to raise entry qualifications and to distinguish themselves from self-taught computer users. They resemble the high priests of religious movements who govern their followers through canons of orthodoxy and threats of expulsion. Processes of professional entry and exclusion influence national standards for computer education and recruitment. This influence is evident in the activities of Ivory Coast's computer association and is emerging in Kenya through the debates about computer training publicized by the mass media.

Computers as Symbols of Change

A symbol of change, such as the computer, may be distinguished from a cultural sign with multiple meanings by specific reference to alterations in the social context of its users. Ali Mazuri (1978:321-323) discusses the computer as a source of economic dependency and Westernization. He proposes a gradualist approach to the indigenization of the computer in Africa in order to free nations from control by multinational corporations and external investors. He suggests diversifying sources for obtaining computers in order to prevent the development of overseas monopolies.[9] Mazuri's analysis raises the question of whether the business expansion that accompanies computerization is beneficial to the government and the private sector. These proposals closely resemble the Kenyan government's approach to new technologies. Mazuri (1978:323) identifies three processes involved in technological change: (1) secularization of education, (2) technicalization of work, (3) and future-orientation of government policy. He sees the computer as part of all three processes. Secularization and technicalization, however, are not equivalent processes, and they are not necessarily coterminous.

The professional closure of the computer world leads to the emergence of a new type of religiosity. In analyzing the new technicians of the sacred, I have argued that the computer's role in secularization may be superficial. The computer brings with it a technocracy and an enclosed circle of experts who become a new religious order. Granted, clerical workers may be reskilled as word processors and data-entry personnel in a matter of hours. As the case studies from the Kenyan ministries show, however, these workers are either replaced by more highly skilled programmers or forced to acquire more advanced skills as their jobs evolve. When they have mastered these skills, they become part of a network of computer specialists, a new sacred order whose services are in demand. In Ivory Coast and Tanzania, this new technological order is reinforced by professional associations and training outlets that serve as gatekeepers for high technology.

Technicalization is a symbol of change. It is associated with the belief that increased efficiency means progress and that machines can bring about organizational transformations. In the Kenyan cases discussed, we have seen a process of democratization of information at work with the introduction of microcomputers into government agencies. In contrast, a new technocratic hierarchy develops with the use of mainframes, minicomputers, and more complex programs. The DPM and the senior programmer are at the top of this hierarchy in a stand-off against bureaucrats and organizational managers. Technicalization enhances the mystique of computer experts as new technicians of the sacred.

In her research on the science of oceanography, Chandra Mukerji (1989:125-145) points out the contrasting roles played by scientists and technicians in the laboratory and the negative views held toward the contributions of technicians to the process of scientific discovery. When technicians wield conceptual power in a laboratory, they are often granted limited autonomy at work and receive little recognition for their efforts. With regard to their role in Africa's computer world, data-entry personnel and word processors may be compared with laboratory technicians. Although data-entry workers run the computer systems that make technicalization possible, their importance is overlooked by data bureaus, private consulting firms, and government agencies.

This attitude is not an accident. Honoring the role of computer operators as opposed to professionals, is not only a symbolic gesture but also an act that entails financial remuneration. As in the case of laboratory technicians, computer workers occupy a specific bureaucratic status that must be recoded if they are to be considered professionals. Although recoding is evident in vertical mobility within firms, acknowledging a new status for technicians on a national basis across firms is another matter. At this point, computer associations enter the picture as intermediary organizations and gatekeepers for their constituencies.

Technological innovations must be reviewed and officially recognized before they are acknowledged as accomplishments in the computer world. Although it is an essential activity for local organizations, adapting packaged programs and menus to individual enterprise needs is acknowledged as a technological innovation. Limited programming and the modification of packaged programs is not basic research and requires little formal training. The gatekeeping processes surrounding the computer profession resemble internal sectarian dynamics through which ordained leaders and high priests receive recognition for the support and innovations of rank-and-file members.

Mazuri's category of future-orientation stands at the core of beliefs about computers. The promise of a brighter future for those who learn to use computers is millenarian in its intensity. This ideal inspires young trainees who hope to become leaders in the computer industry. This promise also motivates government agencies in their attempts to technologize bureaucracy. When computers are introduced, administrators and consultants anticipate that overdrawn budgets will balance; files will become orderly; and work tasks will flow smoothly. These wishes, characteristic of the confrontation and early domination stages of computer use, are seldom realized with ease. Technological fantasies, however, induce real organizational changes. The computer as a symbol of change is a potent link between a technological artifact and a dream. The computer is a mirror of the process of change that it engenders.

Technology and Religion as Modeling Systems

Technology and religion share characteristics as semiotic modeling systems. Semiotic systems, or s-codes, contain all possible variations of elements within a particular domain (cf. Eco, 1976:38). A modeling system structures these elements in hierarchial order, such that the smallest element reflects the larger system. Zaliznjak, Ivanov, and Toporov (1977:47) consider language and religion to be systems possessing "a maximum modeling capacity." These systems coordinate the structure and content of expression on several levels. Secondary modeling systems restructure a code by adding to it new and spontaneous elements. Thomas A. Sebeok (1976:23) has defined secondary modeling systems as "complex macrostructures composed of a natural language as their infrastructure and having at least a text coded with prefabricated units and constructions as their secondary superstructure."[10]

In its application to religion, the example of modeling systems is further elaborated by the analysis of New Testament parables. Zaliznjak, Ivanov, and Toporov (1977:52) consider parables to be spontaneous expressions that initially do not form part of a religious system but may be transformed into symbols within the s-code. Parables are examples intended to explain the total religious system and its texts to members. Subsequently, they may be cited as messages in their own right and may alter the interpretation of the religious system. Thus, Zaliznjak, Ivanov, and Toporov (1977:52) state:

This view of the parable examines its original use. Later application of the parable as a ritualized sign and standardized citation transformed it or its fragment into its own opposite by including it in the alphabet of the religious code, and simultaneously reduced the information contained in it. The structure of some New Testament texts can be described in large measure by analyzing the alternation of standardized citations with parables in the original sense, whose abundance raises the possibility of interpreting the whole text as a parable. Other semiotic modeling systems such as verbal and figurative art display a similar alternation of standardized units and elements that do not enter into the alphabet of the code.

Computers recode cultural expression and social activity. This modeling process occurs in at least three ways with relationship to computers. (1) The computer operates with a fixed program or coded text into which information (a natural language) is entered. (2) The computer, including its hardware and software, constitutes a macro-structure about which discourse is developed. (3) The entire narrative program of computer adoption is a structure within which the processes of technological change are reflected. Although these processes of change vary from one setting to another, the progressive steps of the narrative program do not alter.

Regarding the computer in terms of coded texts leads to an analysis of the relationship between formal computer languages and data entered into a program. This topic is touched upon in my discussions of computer training and the business applications of the computer. Considering the computer as an object of discourse has been an important theme throughout this chapter. The computer operates as a descriptive device in the Kenyan automobile enterprises in which the work of compiling stocks, inventories, and sales is explained in terms of the computer's functions. Computer-related excuses emerge as a way in which workers displace human error onto the operations of the machine. The proliferation of bureaucratic titles surrounding computer use creates an increasingly complex social and linguistic environment in which computer successes and excuses develop.[11] Shifts in software and programs, such as the changes described in the MALD experiment, signal organizational problems that are only indirectly related to the computer's functions.

Organizational problems in computerizing firms emerge in perspective as part of the narrative program of computer adoption. Fantasies and fears about computer use appear at the confrontation stage. These expectations change when computer mastery occurs. Computers are then integrated into the everyday operations of private enterprises and government agencies. Mastery leads to a redefinition of the computer and to new technological successes.[12]

The computer transforms the social environment by requiring new relationships of dependency and control. This process emerges in individual enterprises and with respect to the image of the computer as a symbol of change in the development process. The individual's decision to pursue a computer career and the nation-state's decision to computerize government functions reflect the role of the computer as a sign of future-orientation, dream fulfillment, and millenarian fantasy, grounded in mundane activities surrounding the computer's use.

When we examine uses and images of the computer, we may ask whether it is employed to deal with the narrative organization of future scenarios or to handle the reporting of past events. The type of cognitive and narrative modeling that accompanies computer use influences the social organization of the settings in which computers appear (cf. Danow, 1986:245-268). In some Kenyan offices, mainframe and minicomputers are used to compile data and reports of past events. Developing spreadsheets and forecasting are not major uses of the computer in Kenya. Thus, the computer's association with a future-orientation may be superficial in certain businesses where it is used primarily as a documentary device that records and historicizes office activities (cf. Mannheim, 1952:45-47).

Religion encompasses sign relationships that resemble those connected with the introduction of new computer technologies. Ivanov (1969:218-221) describes the process of sign-symbol transformation through which existing social relationships are changed into symbolic and ritual representations. These rituals and representations, in turn, serve as justifications for perpetuating the social order (cf. Collins, 1987:201-205). Cultural representations are a symbolic distillation of social relationships. Thus, the myth of divine kingship is played out in the social order that sustains this cultural narrative and perpetuates it over generations. The computer's role as a system of justification is more subtle and mutable. Although the computer's use may reinforce existing social hierarchies, it inevitably modifies and challenges them. Office managers who promote the installation of computers may be replaced by a new group of technocratic experts. Paradoxically, the systems that the computer is introduced to streamline may become obsolete as a result of the presence of the new technology. Hence, computers are not merely secondary modeling systems that reflect their environment. They are technologies that are different enough from their original environment to induce radical change.

Conclusions

The closed character of computer networks has been examined in terms of their external dynamics and control. I have compared networks of computer professionals and sectarian religious groups with reference to their social organization and types of doctrinal expression. The status of such movements has been assessed in the context of national political policy and with reference to the intervention of foreign experts. Based upon this comparison, the computer's role has been described as an aspect of a cultural modeling system that reflects and modifies existing patterns of authority in organizations. Computer experts jockey for power in firms and national politics. In both Kenya and Ivory Coast, networks of computer experts filter out neophyte computer users.

Technology and religion are sources of new cultural images that are promoted within small enclaves. The resultant innovations are subject to state control by virtue of their ideological challenges to the existing social and political order. These challenges have been framed as symbols of change and analyzed in terms of how they are actualized in the sign-symbol transformation from ideal to act in social life. The computer is used as a tool of the new bureaucracy. It is also as a transformative symbol that shapes the technocratic elite's vision of progress and change. In the next chapter, the implications of these social and symbolic processes will be examined with respect to the case of computer adoption in Ivory Coast.

Part II

New Technologies, Work Organization, and the Administrative Revolution

Chapter Four
The Last Train of the Twentieth Century: The Computer Revolution in Ivory Coast

> Certain periods in history are periods of reflection. We have arrived today at one of these periods; the old countries are beginning to be aware and to do something about it. The others, in particular those who have recently achieved their independence, should be alerted this time not to miss the last train of the twentieth century (President of the CNI, Ivory Coast, December, 1983).

Abidjan On-Line

New technicians of the sacred – computer experts of the highest caliber – are hard at work throughout Ivory Coast. In a small office in the textile commercial center of Abidjan, a young woman – the assistant director of a computer consulting company – walks through the open door to the director's adjoining office. She carries with her a long computer printout with handwritten notes and diagrams at the bottom. The accounting program that he has suggested is not working. She wants to move sums around to different columns and to reconfigure them. She argues modestly but with authority about the efficiency of her approach. At first he disagrees and finally nods, "Go ahead, try it your way, but after lunch." She retreats to the adjoining office where their computers are linked, and the telephones are poised in readiness for daily calls to a Paris bank that finances their company's operations. She shuts off the computer and picks up the telephone to inform the children that they will both be home for lunch and will return to the office at four o'clock in the afternoon.

Neither the assistant director, born in Abidjan of missionary parents, nor her husband, misses the hectic life of Paris – the métro, the rapid business lunches, and the busy telephones. As they walk into the hot noonday sun of Abidjan, they nod to the Ivorian receptionist, who sits in a quiet front office. The receptionist arranges a stack of papers on her desk. They advertise the third annual *journées africaines d'informatique*, a series of symposia and sales sessions on computer technology at the Hotel Ivoire. Their firm has been well represented with a large ad reading, "Conseil, Assistance, Ingénerie, Formation – Terriblement Efficace!"

Across town, near the government center, a young member of parliament who received his training as an engineer in the United States sits behind his desk in another computer consulting firm. He has just given the company engineers their field instructions for the day. Although he too is anxious to leave for lunch, a long line of young men extending into the hallway awaits him. Diplomas and resumés in hand, these young engineers and computer programmers have come for job interviews, to drop off their applications, and to seek his advice. He retreats into the office and buzzes the Ivorian receptionist that he will stay in for lunch. He then calls his colleague at the nearby General Secretariat of Informatics to chat briefly about afternoon plans before receiving the next visitor. He is dismayed that the young engineers seem to believe that his thirty-person firm, C.I., has the capacity to absorb countless employees and interns. He too has a copy of the *journées africaines d'informatique* brochure on his desk. His company has a full-page ad on the last page: "Compagnie Ivoirienne d'Études et Réalisations en Informatique et Automatisme – des références internationales." His name appears on top of the ad as "President Directeur General" (PDG) of a company with $129,032 of capital assets, 23% Ivorian public sector, 49% French capital, and 28% Ivorian private sector investment. He is proud that the controlling interest in his company is Ivorian.

Back in the commercial center, a French consultant sits before his computer terminal in the IBM building. He has just sent his first BITNET message from Abidjan to the United States. It has reached New York via Montpellier and Paris in eight minutes and reaches Pittsburgh in four more minutes. Another message has just arrived from Tokyo. The consultant is satisfied that the system is working. He sends a message to France that the status of the system is good and then logs off the computer. On his nearly bare desk beside the computer console are a stack of printouts, a few letters, and a copy of the *journées africaines d'informatique* brochure. IBM has taken a full page ad with photographs. The ad depicts a young Frenchman in a business suit reclining in a comfortable, modern chair with a pen and closed appointment book on his lap. He is relaxed and confident. At his feet is an IBM 36 supercompact computer. The ad reads simply: "Ordinateur IBM 36: Avec lui, jouez la carte de la tranquilité." The young executive indeed appears tranquil and confident in the efficiency of his IBM system. At the bottom of the ad in small red letters, the post box and telephone number of IBM Abidjan has been printed, as if it is an afterthought. The consultant closes the brochure, shuts off the lights, and leaves for lunch in a nearby *maquis* café. He is anxious to return in the late afternoon to read his messages from America.

Computers are an important part of the fabric of daily business life in Abidjan. They are found in banks, hotels, travel agencies, the university, and, of course, in the computer consulting and engineering firms that are growing at a rapid rate throughout the city. Government figures indicate that approximately 3,000 Ivorians worked directly with computers as of 1984 (*Plan Informatique National, 1986-1990*: 55). Customs taxes have been reduced to 11.11% of the cost for the importation of computers, and no taxes are charged on software imports. More and more, computers are found in the homes of the affluent, alongside television sets and video recorders. Computer consulting firms promise to transform businesses, large and small, with their specially modified *progiciel de gestion* – business and accounting packages designed for the Ivorian market.

As the glimpse of the C.I. firm suggests, optimistic young engineers are in abundant supply. The country is primed for technological progress, but many of the larger establishments – banks, hotels, and multinational firms – already have their own computers and in-house computer staffs. Although they operate in an atmosphere of affluence, vendors worry that the local market for computers will soon be saturated. They compete fiercely to convince the government and private enterprises to change allegiences and adopt the latest equipment and software from their companies. Indeed, some vendors have been successful in obtaining large, long-term government contracts. Another hope for expansion lies in providing service agreements, spare parts, and training sessions, but none of these activities is as lucrative as sales. Consequently, the vendors seek new markets – in famine parched Burkina Faso, Mali, Niger, and as far west as Sénégal. They are determined that Abidjan will become the regional center for computer sales and distribution in French-speaking West Africa.

Much of this optimism and expansionism results from an aggressive government policy toward computer adoption. The positive politics of informatics has as its goal the stimulation of industry and private enterprise. Entrepreneurs even speak of launching computer assembly plants in Ivory Coast, and many engineers and planners speculate about new developments in the production of software. Local experts speak of Ivory Coast as having reached the second stage of computer development that includes broad public awareness, good distribution, and quality training. The next stage is regional leadership in software and hardware production.

Interpreting Computer Adoption in Ivory Coast

Indigenous implantation of computers is a real possibility in Ivory Coast. This case offers a startling example of the ideological acceptance and appropriation of new technologies. American theorists of different stripes have been eager to dismiss the presence of new technologies in Africa as alien, neocolonial impositions (cf. Perrolle, 1987:208). Rather than viewing computers as a threat to national development (cf. Carim, 1983:9), the Ivorian government has expressed confidence in the transformative potential of new information technologies as a means for improving the position of African nations in the world economy.

In January 1978, Simon Nora and Alain Minc submitted to the French government a report entitled *L'Informatisation de la Société,* translated into English as *The Computerization of Society*. This report describes an electrifying vision of the future in which on-line computer communications, satellite connections, and rapid data transfer change the face of a sluggish French economy and society. Although the report warns readers about the drastic changes in training and employment patterns that result from computerization, it extols the virtues of rapid data processing and advanced networks of telecommunication. The report covers the computerization of government, the impact of computerization on employment, the political consequences of technological change, and the advantages of improving telecommunications networks. In his introduction to the English translation of the report, Daniel Bell (1981:xv) states: "The importance of the Nora/Minc report is that it calls for and prescribes a unified national policy to utilize the new technology of *télématique*."

Immediately upon its publication, the Nora/Minc report, which became a best seller in France, was the subject of much public debate.

These heated arguments revealed the concerns of labor unions, the anxieties of workers, and the optimism of the champions of new technologies. Echoes immediately crossed the Atlantic into French-speaking West Africa, where new developments in Paris are a source of constant curiosity and occasional criticism among government officials and intellectuals. Therefore, it is not surprising that less than a year after the report's publication in France, a conference on computerization was convened in Abidjan.

Through its Office Central de la Mécanographie (OCM) and the Bureau Intergouvernemental pour l'Informatique (IBI), Ivory Coast sponsored the first Conference for Integration of African Informatics on November 22, 1979. Held in Abidjan, this international conference outlined five major objectives for the development and application of computer technologies throughout Africa. The themes covered by the conference were: (1) strategies of computerization in Africa; (2) telecommunications and satellite connections for African countries; (3) uses of computers in government administration and taxation; (4) applications of computers for national archival and documentary purposes; and (5) the future of microcomputers in Africa.[1] These concerns were eventually integrated into the first and second national informatics plans for Ivory Coast.

Dating from the mid-1970s, Ivory Coast's positive national outlook on computers has led to many local changes. Applying the narrative program for computer use to Ivory Coast, one finds that efforts to regulate and promote computerization are evident at each stage. The confrontation stage of computer adoption is characterized by the acceptance of computers into the operations of government enterprises and large business. Centralized control over this process through the OCM and other government bureaus charged with overseeing computerization may be construed as a formal effort to monitor the environment for computer use. This growth-oriented strategy of regulation has three important components at the domination stage of computer adoption: (1) economic access to computers increases; (2) local standards for training and credentialing computer professionals are established; and (3) a controlled although competitive environment emerges for the distribution and sale of computers. These three developments have a direct impact on small enterprises by making computers affordable for them and increasing the pool of locally trained professionals to staff computer operations. Therefore, small enterprises are able to select the technologies appropriate for them during the early stages of computer use and mastery.

These developments lead to a positive attribution, or interpretation, of the value of computers. Not only are computers seen as affecting enormous changes in business and government, they also become the source of new strategies for economic development. Ultimately, Ivory Coast plans to launch its own computer industry as well as to increase the automation of existing industrial and agricultural enterprises. If these plans for automation are executed, problems with labor displacement may arise. Although the government plans emphasize the training of Ivorian engineers and the reskilling of employees, farmers and unskilled industrial workers will not be in this labor pool. These economic obstacles, however, have not prevented the government from championing its plans for new technologies.

Positive attribution leads to a progressive computer contract at the final stage of the narrative program. Contract, as used here, is the semiotic term for the resolution of a relationship between two acting subjects. This progressive contract is not necessarily realistic. Just as mixed attributions toward computers result in a fragile and problematic computer contract in Kenya, as we shall see in Chapter Five, the progressive contract masks apprehensions about possible shortcomings of computer adoption. The progressive contract is a byproduct of three conditions of computerization: (1) consistent administrative and economic support of computerization by the government; (2) competitive control of foreign computer vendors, investors, and expert consultants; and (3) national promotion of computer training programs for administrators, government personnel, and engineers. One consequence of the progressive contract is the selection of appropriate computer equipment at the initial stage of adoption. With lower import duties and fewer restrictions on equipment purchased in Ivory Coast than in Kenya, it is more likely that small companies can begin with the equipment they need. Anxiety over the selection of equipment, however, surfaces in other ways. The market for computer sales is small, and the competition is high. Computer vendors and consulting companies abound. In fact, computer consulting companies are among the primary computer users in Ivory Coast, and their clientele is limited.

Another consequence of the progressive computer contract is internal control of government computerization without depending upon the assistance of donor agencies and outside experts. Some government computer experts criticized the excessive regulations surrounding computerization in their agencies and the government's practice of accepting bids for computers and services from the same "insider" firms regardless of whether their materials were the most appropriate or effective. From another perspective, vendors complained of a closed government bidding system that they feared might drive their companies out of Ivory Coast if government contracts were no longer available to certain firms. Government control of computerization has a trickle down effect on small businesses by lowering equipment prices and increasing the range of available goods. In Africa, areas of ambiguity surrounding the computer contract have the potential to generate corruption. Regulation, however, leads to stability in most cases. The effects of government policies regulating computerization in Ivory Coast may be evaluated by analyzing the first and second national informatics plans.

The Ivorian Informatics Plans

After the Conference for Integration of African Informatics in 1979, President Félix Houphouët-Boigny circulated to Ivorian government offices an official letter in which he stated, "information is power" and that the countries neglecting development in this domain risk finding themselves at a further disadvantage in their economic competition with the rest of the world. The letter counseled against proceeding with dispersed computerization (*l'informatisation dispersée*) and suggested that a commission be formed to allow state agencies to collaborate in developing a computer plan. At first, computerization was to be handled by the Office Central de la Méchanographie. In January 1980, the National Commission for Information Processing was appointed, and in 1983, the Commission Nationale pour l'Informatique (CNI) was established along with the corresponding Sécrétariat Général à l'Informatique (SGI).

The first computer plan is a 141 page document that seeks to define the role of computers in Ivory Coast and the appropriate government mechanisms to implement computerization (cf. Figure 4-1).[2] The plan begins by acknowledging that informatics can become a tool for administrative reform (*Plan Informatique, 1981-85:28*). It emphasizes the importance of telecommunications with the aid of computers (*la télématique*). Computers and telecommunications are depicted as a means of enhancing regional development throughout the country. Three major domains of computer use in Ivory Coast are outlined: (1) administration and finance; (2) industrial development through computers and robotics; and (3) scientific and technical applications.

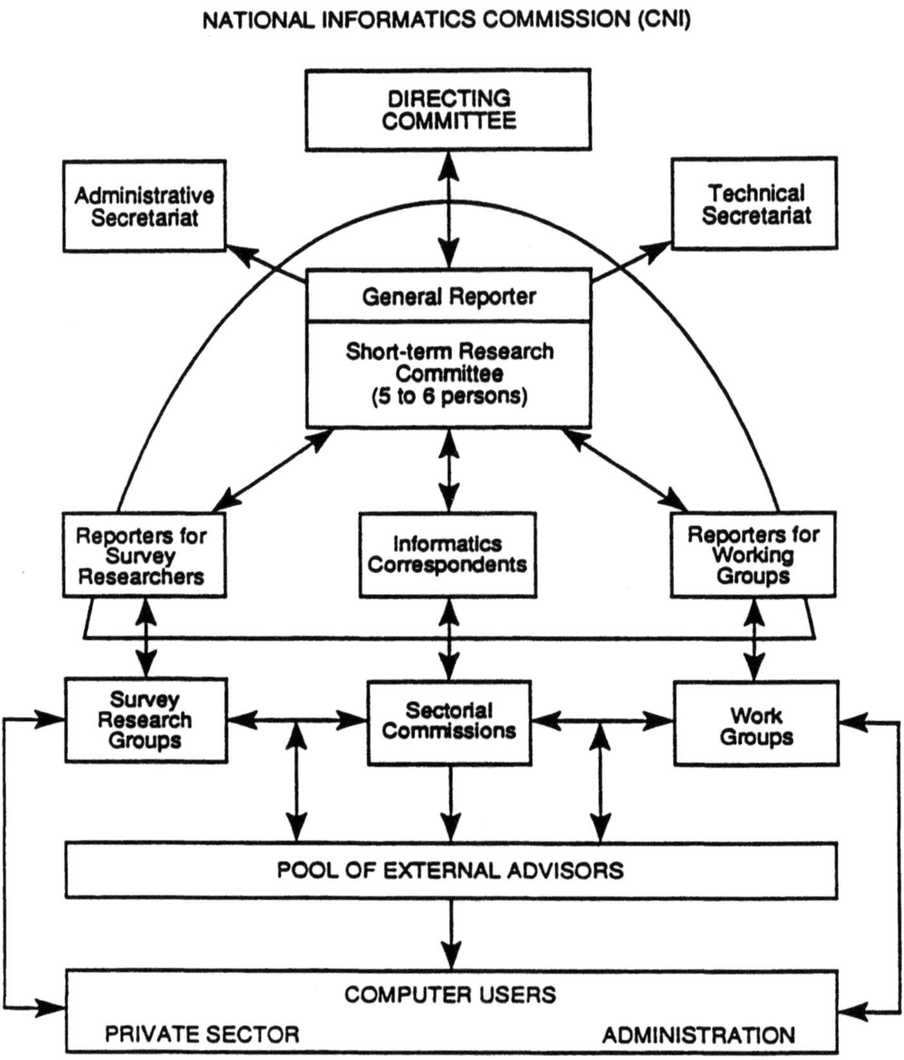

Figure 4-1. Government Structure for Implementing Computerization in Ivory Coast.
This figure is reproduced from *Plan Informatique, 1981-1985*, Abidjan, Ivory Coast: Commission Nationale Pour le Traitement de l'Information, 1981:20 bis.

Originally, according to the plan, computers were to be used by the administration to assist with the large tasks of tax collection, state accounting, and import and export statistics. The plan provides for the eventual computerization of customs services. This process was actually completed in 1986. The plan also suggests extending the use of computers to all government ministries. Although full government computerization was not achieved by the first plan's 1985 closing period, it became a driving force behind the second computer plan.

The first plan centers around five foci: (1) financial organization of the administration and private enterprises; (2) improvement of administrative services; (3) upgrading of productivity in industry and the private sector; (4) advancement in information processing and telecommunications; and (5) international economic cooperation. These objectives were to be achieved by increasing the awareness (*la sensibilisation*) of administrators and the general public concerning the importance and uses of computers, by developing high quality national computer education programs, and by putting in place an administrative structure to oversee computerization (the Commission Nationale pour l'Informatique, or CNI). The plan outlines projects for implementing changes in these five areas in general terms. It projects that the number of computers in Ivory Coast would more than quadruple between 1980 and 1985, and it proposes that the government be at the helm of this inordinate growth.

A major area of concern of the plan is the lack of qualified local personnel in the computer industry and private consulting firms. The plan states that only 26% of the management personnel in computer service and engineering companies was Ivorian as of 1980 (*Plan Informatique, 1981-85*:87). Therefore, considerable attention is devoted to establishing national criteria for engineering and computer-related education. Although an official educational plan does not remedy the computer credentialing problem, it is the first step in sorting out computer experts from general users as well as standardizing job qualifications in the computer world.

The first informatics plan projects use of INSET, the Institut National Supérieur de l'Enseignement Technique, and the national university as key resources for the training process. Computer and engineering titles are attached to each level of training offered by these institutions. Passing the baccalaureate exam (BAC) for secondary school and two years of postgraduate training are required for programmer-analysts.

The baccalaureate and four or five years of training are essential for engineers and project heads. INSET is charged with devising several diploma programs to meet these requirements. Thus, an INSET engineer could be relied upon to assume a technical position in government, industry, or private enterprise. The plan's approved goal is to create enough Ivorians to fill the major posts in computer management and engineering by an unspecified date. Since the first INSET computer engineering class did not graduate until 1984, this objective obviously could *not* be fulfilled before the termination of the first planning period in 1985.

Government statistics estimate total computer expenditures in Ivory Coast as of 1980 at the equivalent of U.S. $40 million. The government's plan sets aside the equivalent of U.S. $20.5 million for the implementation of the first national computer plan, which includes computerization for business, industry, and education. This sum represents a substantial commitment to bring Ivory Coast to the forefront of computer development in Africa. The plan self-consciously compares Ivory Coast to Brazil, South Korea, and India in its efforts to make rapid postindustrial strides and literally to create a computer revolution from scratch.[3] To conclude the plan, a glossary is presented defining such terms as bit, unbundling, optical fibres, modems, software, and intelligent terminals for somewhat bewildered laypeople and administrators. Managers and government administrators are urged to attend three-to-five-day seminars that initiate them into the procedures for computer adoption in their offices and indoctrinate them concerning the appropriate uses of computers and informatics for national development.

The rhetoric of information as synonymous with international power becomes even more pronounced in the second computer plan. This plan is 232 pages long with two annexes of over 200 pages each. If the sheer volume of planning material is any indicator, the computer revolution has indeed been launched already in Ivory Coast. The line of argument in the second plan reveals a collaboration of engineers, economists, and politicians. The plan begins by outlining the substantial technological evolution in the country since 1981 and discusses the glowing possibilities for increasing industrial productivity through robotics and initiating national research projects in artificial intelligence and telecommunications. Young engineers are clearly making their voices and visions of the future known through the vehicle of political planning.

The second informatics plan proposes four areas of national development to be aided by computers: (1) state finances; (2) agricultural production; (3) education; and (4) industry. These areas will be advanced by focusing specifically on the administrative regulation of customs activities and tax collecting by computer, on the development of robotics for the industrial sector, and on the expansion of jobs in the computer world with projections for the eventual mounting of a microcomputer assembly plant in Ivory Coast.[4] The second plan continues the emphasis on computer education and creating a national cadre of engineers and computer operators in order to meet administrative and industrial needs. A host of other administrative usages for the computer are listed. They include organization of public health statistics for the country, development of data banks for socioeconomic and agricultural surveys, computerization of the Ministry of Justice, crime monitoring through the use of the computer, and documentation of climate and rainfall in the country. In effect, computers are used to highlight every area of national need and are proposed as the keys to solving major problems of national development.

The plan begins by redefining development as an internal problem rather than an externally motivated condition. "Development," states the second plan (p. 32) "is thus an irreversible phenomenon. Seen in this way, every developing nation does not have the same point of departure. At the same time, there are a certain number of objective considerations that apply to them all." Not only is the computer the key to development, so too are the human resources that make computerization possible. In an altruistic tone, the plan (p. 33) states: "The essential factor for the future of developing nations is the conviction that brain-power constitutes the most important resource of any country."

Through advances in technoscience, the plan proposes improvements in the administration's fiscal practices, in the private sector's accounting and communications activities, and in industry's productivity. The plan outlines ten specific goals to be reached by 1990 (cf. *Plan Informatique National, 1986-1990*:77-87). These objectives follow the political and economic strategies already described in the first informatics plan. They include the politically harmonious use of computers in step with Ivorian economic planning, optimal exploitation of human resources for advancing and operating computer technologies, technical and scientific mastery of computer technologies, and intensification and expansion of the use of computers for national economic development. The tenth objective, which summarizes the previous nine goals, includes an exhortation "to mobilize the collection of actors necessary for the development of a national informatics" in order to encourage "synergy and progress" by making Ivory Coast capable of autonomy in its choice of new information technologies. Although these goals remain general and are presented as provisional, specific strategies are outlined for accomplishing them by 1990.

These strategies of implementation distinguish the second computer plan from the first (cf. *Plan Informatique National, 1986-1990*:183). Foremost among these strategies is the nationalization of computer engineers and operators by 1990 through the use of local and international educational resources. The second informatics plan (p. 186) outlines a goal of 55% Ivorian nationals among high-level engineers by 1990; 92% of programmers and analysts; 61% of systems analysts; and 99% of computer operators and information and word processors. By nationalizing the cadre of computer personnel, the plan proposes to accomplish four key objectives simultaneously: (1) providing national direction for upgrading computer technologies in the country; (2) expanding the market for high-level engineers; (3) increasing public and private sector computer use; and (4) moving toward local microcomputer assembly. As a technical institute of higher learning, INSET plays an important role in pursuing these goals.

INSET: The University of the Future

INSET is an impressive and well-equipped institution. Located in the President's birthplace, the rural town of Yamoussoukro, and spreading over the several acres of tropical land, INSET consists of nine *grandes écoles*. Mr. Seri-Bi N'Guessan, the head of the Central Informatics Service at INSET, provided a detailed interview on the institution's complex organizational structure. He explained that the nine schools constituting INSET specialize in the education of industrial engineers and commercial technicians in short programs, or cycles, lasting two to three years, and in more comprehensive five-year programs. The institution trains technicians for the business sector in two-year and five-year programs. N'Guessan outlined the curriculum for me.

Figure 4-2. INSET: University of the Future.
The Institut National Supérieur de l'Enseignement Technique (INSET) is an impressive structure located in Yamoussoukro, Ivory Coast.

4.1 Seri-Bi N'Guessan, INSET:
INSET is a grouping of nine *grandes écoles* that are subdivided into schools preparing students for the industrial and the tertiary sectors. And within each school of engineers, technical specialists [techniciens superiéurs], and computer experts, the same division applies between the industrial and the commercial sector. It is exactly the same thing. That is to say // here we don't speak of engineering schools, we speak of long training programs and short training programs. The engineers are trained in cycles of five years after the BAC [secondary school completion]. The long cycles last five years for both the industrial and the commercial programs. Voilà! That is how INSET is organized. It is a *grande école.* But actually it consists of nine schools in the industrial and commercial sectors with long and short training programs. The short programs are for technicians and the long programs are for engineers.

INSET's schools include (1) an industrial engineering institute that trains electrical engineers and industrial experts; (2) a school for electrical engineers, electrical and mechanical engineers, and industrial technicians; (3) an institute training engineers in scientific and research-oriented informatics; (4) a training institute for electronics technicians, industrial technologists, and computer maintenance personnel; (5) an institute preparing senior programmers and computer specialists for data management positions; (6) a commercial management school with a focus on the business uses of computer technologies; (7) a financial management and accounting school; (8) a school offering a diploma in insurance, management logistics, accounting, and business; and (9) an institute devoted to technological and management training for government managers, assistant directors, and other officials.

Although the industrial engineering training is highly technical, the commercial training cycles emphasize a combination of management and technical skills. This combined training follows the guidelines of the second national informatics plan, which highlights the importance of advanced technological training for government administrators and business managers. This planning philosophy assumes that computer technologies will be successful only when adopted by both the administrators and the computer operations staff in an enterprise.

At INSET, the two most important institutes for computer and informatics training are: ESII (École Supérieure d'Ingénieurs Informaticiens) and ISI (Institut Supérieur d'Informatique). Both of these institutes train engineers and computer experts in three-year cycles. ESII focuses on scientific and research training for computer experts and software engineers while ISI emphasizes computer training for business and finance. Normally, an engineering diploma takes five years of university level instruction to complete in Ivory Coast. ESII lasts only three years because students are recruited after completing two years of university study with a scientific specialization. When they have finished the three-year ESII program, the engineering graduates are ready for immediate employment in business, industry, and government. N'Guessan described the practical basis of the training program to me.

4.2 Seri-Bi N'Guessan:

Our school has the following philosophy. We want to train people with practical skills in industrial and commercial informatics. They begin practical work while in school // so that they are able to work in an enterprise immediately. Whether they are in the five-year, the three-year, or the two-year cycles, they complete internships with enterprises // either in the middle of their studies or at the end of their studies. We have introduced informatics into all of these programs. So // for example // a general engineer in the last year of the cycle might take an internship at a computer bureau and // in the end // this engineer will become a computer expert, even without specialized computer training.

As previously mentioned, ESII is oriented toward training software researchers and specialists. These individuals are referred to in French as *génies logiciels* (software whizzes). They are schooled in the development of new programs and applications specifically tailored to the business and administrative requirements of Ivory Coast. The ESII graduates are trained to be at the vanguard of the computer revolution and at the helms of the new software industries and computer consulting firms in Ivory Coast. Entry into the program is highly competitive. Candidates are required to exhibit excellent skills in mathematics, one field of science, logic, teamwork, and mastery of both French and English. The competition is open to candidates of the age of 24 and over, if they have successfully completed their baccalaureate examination and have become certified programmer-analysts. However, students who are 27 years old and more, with two years of university training after the BAC, are the preferred candidates for the engineering cycle at ESII. N'Guessan emphasized the special characteristics of ESII graduates.

4.3 Seri-Bi N'Guessan:
Those who are a little snobbish say that we are training the students to become software whizzes (*génies logiciels*). Well // that's a bit of an exaggeration // but in reality their training has a highly scientific basis in order to give them the possibility afterwards to enter industry // to do industrial informatics // or to enter business as directors of computer bureaus, directors of development centers, or directors of software research, etc. So it's for that reason that we say that ESII is an institute for scientific informatics. In their first year of instruction, for example, students take courses in mathematics, logic, and theoretical physics before we give them specialized training in informatics during their second year. Voilà.

ISI trains students in a three-year cycle to be computer experts in the areas of accounting, commercial management, and computer systems management for business. If these students wish to continue beyond the basic courses, they may receive a master's degree in business applications of informatics (MIAG). Equipped with this degree, graduates are considered to be ready to work in engineering design for commercial computer applications and eventually to become heads of projects. Mr. N'Guessan qualified that these students should be capable of working directly with a project head (*chef de projet*) and that their eventual ambition should be to become project heads.

This training program emerges in perspective when we re-examine the stated objectives of the second national informatics plan in which the government forecasts that 55% of the project heads in Ivorian firms and enterprises will be drawn from locally trained personnel by 1990. Many of these computer experts will receive their degrees at INSET. The first class of twenty students graduates from ESII in 1989. ISI now has an enrollment of thirty students per year and over seventy in residence. The engineers completing these programs will become part of the new computer elite as software whizzes and new technicians of the sacred. Whether this elite group will be sufficient to meet the technological demands of government, industry, and private business in Ivory Coast is open to question. There is, however, the possibility that the training institutions will produce *too many* experts and an insufficient number of computer operators and data-entry workers. According to the second informatics plan, local training of experts will be supplemented by training of Ivorian nationals sent to France with special scholarships and by instructional programs at the IAI (Institut Africain d'Informatique) in Libreville, Gabon.

Employment placement expectations projected by INSET officials were, however, contradicted by a computer firm director who explained in detail how young engineers are hired and progress through the ranks of his company.

4.4 Computer firm director, Abidjan:

We employ some young engineers just as they graduate from school. They generally have a master's degree or a diploma in computer science. We call them student engineers. After two or three years of experience, they can become research engineers (*ingénieurs-concepteurs*). Then, if they reach a fifth year of experience, they are able to become principal engineers. This is really after five to seven years of experience. A chief engineer and project head // that requires twelve years of experience. But

> you must recognize that the computer field is a sector of high mobility. Sometimes we recruit young Ivorian engineers who have a master's or a Ph.D. in computer science from the United States as beginners in a firm, but they do not stay for a long time because of the high mobility rate.

One French computer consultant praised the quality of INSET interns hired by his firm. He explained: "INSET is a very, very good school, especially ISI. We have had ISI interns who are in general very well trained." The quality of INSET training appears to be beneficial to the private sector. Directors and managers are already able to find highly qualified personnel among the INSET apprentice students and graduates.

INSET's objectives are the training of scientifically qualified, competent, and marketable Ivorian engineers in a fairly short time. The institute's directors hope to produce a continuous flow of engineers to meet the country's information processing and industrial demands. The result will be the emergence of a new professional class in Ivory Coast. This professional class will have been trained locally for the commercial and scientific market. Training too many engineers could result in over-professionalization of the computer industry and lead to a proliferation of computer consultancies. These new engineers will be prepared to assume full leadership in business and management regardless of the available positions.

The Role of the Commercial Schools

The objectives of the commercial colleges differ from those of INSET. Although they are not as numerous as the commercial schools in Nairobi, several colleges, secretarial schools, and commercial institutes in Abidjan provide computer training. Among these institutions are CETI (Centre d'Enseignement Technique en Informatique), which is a computer training school for government administrators and office workers, and SIEPAT-Ordinateur, a training institute also specializing in computer repair. SIEPAT-Ordinateur was originally established in 1984 by two young Ivorians, Blagué Valet and Raymond Kragby. It began as an agricultural equipment firm in which computers were imported for recordkeeping on large plantations. Although the import business was soon overshadowed by computer training, Valet and Kragby were at an advantage because they were able to import their own personal computers. Valet worked as a computer instructor and engineer and Kragby as a programmer and instructor. Within a year, their business was transformed into a small computer college. In the Ivorian Chamber of Commerce listing, SIEPAT is recorded as having capital assets of 20,000,000 CFA francs, or approximately U.S. $62,516.[5] The school's staff is entirely Ivorian. A large blue and white poster adorned with photographs of computers hangs on the wall near the entryway at SIEPAT-Ordinateur. In bold letters, the poster announces the school's computer training program.

4.5 SIEPAT's Advertisement:
Professional training and continuing education for adults. Entrust the management of your routine tasks to the machine! Engage in intensive practical training that demystifies computers! Training for young people starting at eleven years old, secretarial programs, and microcomputer repair. Engineers with years of experience at your service!

Seated behind his desk with another copy of the poster before him, Valet elaborated upon the objectives of his school.

4.6 Blagué Valet, SIEPAT-Ordinateur:
We have three principal foci. They are computer training, maintenance of microcomputers, and preparing computer programs for small and medium-sized enterprises // that is // the PME. We write stock, payroll, and accounting programs. Our students are periodic. Most of them come from companies that want us to provide training programs for them. Our target audience is the workers in the enterprises. We train them in small, ten-person seminars that last anywhere from two weeks to three months. We teach them basic programs for business // such as for example // Lotus 1, 2, and 3, D-base III, Symphony, Framework, etc. And then there are seminars that are primarily concerned with the DOS System // MS/DOS, PC/DOS, and all of that. We have eleven IBM microcomputers and nine Apples. The Apples include an Apple IIE, IIC, and MacIntosh. In the IBM line, we have a normal PC, an XT, and an AT. Well // overall // that's what we do.

Groomed as an electrical engineer in the United States, Valet repairs SIEPAT's microcomputers and takes in repair work from other schools and small enterprises. Part of the support for the school's activities comes from his computer maintenance contracts. In addition, Valet assists with instruction in the small seminars for which SIEPAT charges $200 per person. Kragby develops the lesson plans and instruction programs and shares the teaching duties with Valet. SIEPAT has seven other employees, including instructors in programming and word processing, two repair assistants, an office administrator, and a full-time accountant. Valet is proud of his staff's professional qualifications. As an example, he outlined his own background and training for me.

4.7 Blagué Valet:
My own training is as follows. First of all I did undergraduate work in electronics and received a B.T.S. diploma in general electronics in France. I got a master's degree in electrical engineering at Georgia Tech, in the United States. Then I worked there at an enterprise called Northern Telecom for four years. During those four years, I took night courses in informatics and computer science. So // uh // I am an electrical engineer who has become a computer expert. I am a specialist in computer hardware while Monsieur Kragby is an expert in software and programming. He studied only informatics. Maintenance permits me to explain the operations of the machine. I often go to enterprises and international agencies to give courses. For example, I spent three months at the World Bank to teach them the Lotus program. Sometimes // during a course // the machine breaks down. You have to repair it first before you can go on. And the boss is going to lose time. My training permits me to repair the machine right away. And I think that's an advantage.

Valet's background makes it possible for him to offer an effective, flexible curriculum at SIEPAT and to keep the school's computers in running order. His repair room is filled with spare parts and partially assembled machines. Valet studies each new model and often takes it apart in order to remain abreast of the latest developments. Eventually, he hopes that this repair work will pay off with the possibility of opening a microcomputer assembly plant.

On the pedagogical side, SIEPAT's activities include three-month classes on two afternoons per week for high school students and a computer camp during the vacation period. According to Valet, "Fifty percent of these students have a micro at home, so they are able to learn very quickly. They come from affluent families." In spite of the emerging interest in computers among these students, only three secondary schools in Abidjan – the Lycée Marmose, the Lycée Professionnel Commercial de Youpougon, and the American School – offer computer courses as part of the curriculum. Some of their students come to SIEPAT for tutorials and practice sessions. Other students come to learn about computers on their own as an extracurricular activity. The large population of secondary school leavers in evidence at Nairobi's commercial computer colleges is absent among SIEPAT's clientele. Most adult students enroll through special arrangements with their employers, while the secondary school students are funded by their parents. Several of SIEPAT's adult students are professionals – lawyers, doctors, and businessmen – who have recently purchased computers for their offices and want brief, initiatory training. SIEPAT's professional clientele is both Ivorian and European. Valet described the training sessions provided for adults.

4.8 Blagué Valet:

Apart from individuals sent by companies, we have people who come in on their own // for example university professors, lawyers, doctors, and instructors // who come to be initiated into computer use. They investigate what software they should use in their enterprises // for example to run a pharmacy. We advertise our night courses in the newspapers. When we have ten people // which is our maximum capacity for one per computer // we start a group. The course lasts two-and-a-half months. . . . For adults, our lessons focus on practical applications for work // accounting, spare parts inventories, and word processing. We also have a calculation program that is called TABLEUR in French. That permits the preparation of budgets and the verification of finances. Lotus 1, 2, and 3 are also included in the course.

On the occasions when I observed and videotaped SIEPAT classes, the seminars for secondary school students were in progress. Each student had an individual computer terminal. The students worked on forty-five minute exercises in applications and were then asked to solve programming problems on the blackboard. While the instructors moved about the room to assist them with problem-solving, the students worked at their own pace. On a daily basis, Kragby and

Valet updated the lessons to remain in step with the students' progress. Added Valet:

4.9 Blagué Valet:
We begin by familiarizing the students with the machine // what's in the machine. We open up the computers and printers and let them look inside. Then we teach them some general principles of information processing. After two familiarization sessions, the students begin to learn about programming. We offer two sessions on the software, then we attack programming itself // programming in BASIC. We don't give them any written texts or papers. That is // everything is done *orally*, and they take notes. For example // if the course session lasts one hour, we have forty-five minutes of exercises. It's more practical application than theory.

Using the revenue from the classes and the computer maintenance contracts, Valet and Kragby hope to accumulate enough capital to

Figure 4-3. SIEPAT-Ordinateur.
A computer training session at SIEPAT-Ordinateur, a commercial school in Abidjan, Ivory Coast.

open a small microcomputer assembly plant. Some of Valet's repair jobs already involve reassembling entire units. With the appropriate equipment and support, he believes that SIEPAT could move into full assembly and that the regional market for locally built computers would be sufficient to make this venture profitable. The master plan for an assembly plant discussed in government circles, however, describes a computer assembly operation that is larger and more ambitious than Valet and Kragby can hope for at the moment. Their dream indicates that when computer assembly becomes a reality in Ivory Coast, some small entrepreneurs will be eager to enter the market.

The case of SIEPAT demonstrates the positive attitude toward entrepreneurship held by small firms in Ivory Coast's computer world. INSET's strong leadership in providing quality education for engineers has set the tone for computer training in the country. The close integration of ESII and ISI with Ivory Coast's business environment results in a healthy respect for highly trained computer experts capable of dealing with the practicalities of work in government, industry, and small enterprises. The restrained growth of commercial training institutions, such as SIEPAT, may be a reaction to INSET's domination of the educational field. Moreover, this slow growth may indicate that the market for new programmers and computer experts, especially in the private sector, is more limited than the national informatics plans suggest.

Commercial training institutions in Abidjan play a much smaller role in the education of new young computer operators than similar institutions do in Nairobi. In comparison with the carefully planned encouragement of computer development in Ivory Coast, Kenya's computer world appears subterranean and loosely organized. Vendors in Nairobi struggle against high taxes and duties while the commercial colleges fight to establish the legitimacy of their credentials and curricula. Ivory Coast's official computer policy removes the necessity for these struggles. Eventually, the result of this aggressive policy, however, may be overproduction of qualified computer experts and workers in an environment that cannot absorb them.

The Computer Consulting Firms (SSII)

The rise of computer use in Ivory Coast has been accompanied by the rapid growth of computer consulting firms.[6] Known in France and Ivory Coast as *sociètès de services et d'ingènerie en informatique* (SSII), these companies conduct data processing and provide software development, engineering consultation, and systems analysis for both government and the private sector.[7] Some of these companies sell both imported and locally produced software, while others do not act as vendors. In 1985, a lobby of computer consulting companies and computer-related firms was established in Ivory Coast. The lobby, GIMI (Groupement Interprofessionnel du Management et de l'Informatique), represents the interests of computer companies vis-à-vis the government and sets professional standards for private computer-related firms. As indicated in the introduction, this group contrasts with the APICI (Association Professionnelle des Informaticiens de Côte d'Ivoire), which is a professional association consisting of practicing computer experts. In addition, the government is particularly concerned about the regulation and growth of private computer firms. *Annexe II* of the 1986-1990 informatics plan is devoted to the use of computers in the private sector and begins with a section on the SSII. The opening section of the plan outlines these concerns.

> 4.10 *Focus on SSII*
> This phenomenon [the rise of microcomputers in Ivory Coast] has given birth to a software industry, which has led to the creation of the service societies in informatics engineering (SSII). The activities of these societies are now the object of particular attention from the authorities responsible for developing a national informatics policy (*Plan Informatique National, 1986-1990*:53).

Although part of this attention is regulatory, it is intended primarily to encourage the growth of these companies as a source for employing and adequately utilizing locally trained engineers, analysts, programmers, and computer operators. In 1984, the Secrètariat Gènèral à l'Informatique surveyed forty of these companies and analyzed their functions and employment patterns (cf. Appendix A). I interviewed enterprise heads and workers at fifteen of these forty companies. Based on these interviews, I shall provide an overview of Ivory Coast's computer consulting firms and access their role in the local computer world.

Through faithful translation of all the interviews from French, I have attempted to preserve their original flair.

I conducted research in Ivory Coast in 1987. The studies of computer training institutions and the SSII in Ivory Coast were intended to provide a comparison with companies that I had previously surveyed in Kenya. The unique character of the Ivorian setting dictated the types of companies that were studied. The SSII in Ivory Coast are the equivalent of the data bureaus in Kenya. I obtained a listing of the SSII by consulting the national informatics plans and Chamber of Commerce publications. In most cases, I interviewed enterprise heads, technical managers, or systems analysts. Enterprise heads honored prearranged appointments and facilitated my access to their companies. Follow-up observations were conducted in each case. I focused on the attitudes of managers and workers toward computer use in their settings and on their opinions about the impact of national computer policies on local enterprises.

The computer consulting companies may be divided into three types: (1) larger companies with combined Ivorian and foreign capital equivalent to between U.S. $120,00 to $200,000 assets in which up to twenty-five workers are employed; (2) smaller engineering and consulting firms employing between six and fifteen workers and listing capital assets between U.S. $30,000 and $100,000; and (3) smaller companies producing computer-related equipment, such as surge suppressors, with U.S. $30,000 to $40,000 capital in which between fifteen and twenty-five workers are employed. Companies in the first two categories specialize in data processing, systems analysis, and the adaptation of software for accounts, inventories, stock, and payroll. Apart from size, the major difference between Category I and Category II companies is the source of their clientele. The larger companies count among their major clients banks, administrative ministries, and large hotel chains. The small consulting firms focus on small and medium-sized enterprises (PME) with a few banking or insurance companies in both Ivory Coast and neighboring countries. Companies in the third category, of which there are only three in Ivory Coast, provide services to large and small clients, but these services are highly specialized and periodic.

1. C.I.: A Locally Controlled Computer Firm

In the first category of SSII, three companies among those that I surveyed emerged as important: C.I., S.G., and S.L. I have already introduced C.I., which is one of the most dynamic companies controlled by Ivorian interests. Located in a modern brick building near Abidjan's administrative center, the offices of C.I. are generally crowded with clients and young computer technicians seeking employment. Fifty-one percent of C.I.'s controlling interests come from the Ivorian public and private sectors. The director, Z.F., in addition to being a member of parliament, is the Secretary General of the APICI. Founded in April of 1980, C.I. is listed in the Chamber of Commerce records as having capital assets of 40,000,000 CFA francs, or approximately U.S. $129,032. It has between thirty and thirty-seven permanent employees of whom twelve are Ivorian, including the director. C.I. develops software for business and accounting services, produces feasibility reports for computer installations, and engages in computer consultancy and data processing for its clients. C.I. is equipped with an IBM 4341 minicomputer. Z.F. outlined his company's projects and recent successes for me.

4.11 Z.F., director, C.I.:

C.I. is a computer consulting firm. // So // principally we set up computer systems for our clients, mostly business systems // although we have also set up a few technical and scientific systems. At present // about 80% of our work is in computer applications for businesses. And we work a great deal with the Ivorian government, with the governments of neighboring countries, and with the private sector. But now // most of our work is with our government and parastatal organizations. For example, we organized the fiscal accounting system for the Ivorian Treasury. That was a very large five-year contract that has permitted us to computerize all of the fiscal operations of the treasury. And the results have been impressive! Previously, the treasury budgets came out three or four years late. For example, the 1976 budget came out in 1980. This year, the Ministry of Finance closed the fiscal year on December 31 and the final budget was available by the end of January. And Ivory Coast publishes reliable figures for international agencies such as the World Bank. Voilà!

Z.F. continued to explain other government projects completed by his company.

4.12 Z.F.:

We have also computerized the customs with a system called SYDAM // Système de Dédouanement Automatique des Marchandises. The SYDAM project began three years ago and cost five million CFA francs. When you recognize that customs revenues represent 60% of the national budget // that means that our system raised national revenues by 30%. We have also developed a computer system for the Ministry of Transportation. So we can automatically produce and verify driver's licenses, gray cards, and transport permits. All of this is controlled by a master file. C.I. has also computerized the Ministry of Justice. All legal files are now on computer. Voilà! These are a few of our projects to give you a little idea of what we do.

C.I.'s director is a highly trained computer scientist. Z.F. explained that he acquired his expertise through long years of training and practical experience.

4.13 Z.F.:

I received a diploma in statistical engineering at the École Nationale Supérieure de Statistique et Économie Appliquée here in Abidjan, and I did additional engineering studies in France at INSTAT (l'Institut National Pour la Sous-Traitance et l'Assistance Technique). Then I went to the United States for my computer specialization at the University of Illinois, Champagne-Urbana. I received my master's degree in computer science in 1970 and returned to Ivory Coast. I worked in the Ministry of Planning and then I was transferred to OCM as an engineer, and I conducted feasibility studies and installed their computerized accounting system. Then I went back to the Ministry of Planning to work on their master plan as head of an accounts and auditing department. I have done other computer work here and there. Then I became director of C.I.

Z.F.'s impressive qualifications reinforce his technical leadership at C.I. He requires high-level qualifications of his engineers as well. Most of them begin as student engineers with at least a master's degree in computer science and engineering. Then they are able to work up to the title of research engineer, principal engineer, and project head. Y.K., one of the research engineers at the firm, explained this process to me.

4.14 Y.K., research engineer, C.I.:

I came here with a master's degree from the University of Abidjan and three years of study at the Polytechnic Institute of New York in Brooklyn. I started working at C.I. in 1983. Here we have several titles // project head, principal engineer, research engineer, and student engineer. I am a research engineer. There are several grades within these titles // for example grade one and grade two. Moving up in grade depends on your competence. For example // if you are an engineer when you arrive, you start as a student engineer. Then you can move up according to your ability // usually pretty fast. So // if you perform // you get promoted.

Both the director and Y.K. emphasized that upward mobility is rapid at C.I. and that the turnover of engineers is constant. They explained that a computer consulting firm requires engineers to have a working knowledge of several operating systems and to change equipment frequently in accordance with their consulting contracts. Z.F. conceded that these rapid changes might be disconcerting for engineers and could lead them to seek employment with the government or with computer vendors who distribute a single line of products.

Nevertheless, C.I. remains the leading company in its field. Government investment may give C.I. a competitive edge in bidding for large contracts from the Ivorian administration. The director is active in formulating government computer policy and in encouraging collaboration between the government and the private sector. The high qualifications of C.I.'s staff play a role in the company's success. In presenting an overview of consulting firms in Abidjan to me, the director classified C.I. among the top three with its two principal competitors, S.G. and S.L. – firms that place a greater emphasis on consulting for the private sector.

2. *S.G.: A Computer Firm Focusing on the Private Sector*

Founded in 1976 as an offshoot of the Société Générale de Banque in France, S.G. is one of the first computer consultancies to enter Ivory Coast with French backing. Chamber of Commerce listings record S.G. as having 55,000,000 CFA francs in capital assets, or approximately U.S. $177,419. The company lists 50% of its assets as French and 50% as Ivorian. S.G.'s director has developed a regional clientele in Mauritania, Niger, and Sénégal. Although consulting is conducted on BULL and Goupil computers from France, most of the firm's in-house work is done on its IBM 34 system. J.R., the head of research and technical development at S.G., explained that most of the company's income is derived from contracts with banks, hotels, and large enterprises. Among the clients are, for example, two banks in Sénégal, the Central Bank of Mauritania, and a large hotel chain in Abidjan. J.R. outlined the consulting firm's activities for me.

4.15 J.R., research engineer, S.G.:
Well // S.G. is an affiliate of the Société Générale de Banque in France. Our major goal is to provide computer and business services. Because of our origins, our first focus was on banks. Okay // in Ivory Coast // we distribute business and accounting software for banks and industry. So // our programs cover payrolls, accounting, stocks, billing, investments, and we also do auditing and financial planning. That's about it. So // to sum it up // you could say that we have two // pardon // three main foci // standard business programs, development of specialized programs, and auditing and financial planning. We have been in Ivory Coast and other countries in West Africa about four years. And we are continuing to pursue contracts in Togo, Sénégal, Niger, and even Kenya.

At S.G., the banking, accounting, and payroll activities are divided among several staff members consisting of a project head for computer systems, a research engineer, an engineer for microcomputer systems, a financial officer, and two clerical workers. J.R. described these titles in terms of "productive" and "nonproductive" positions. All engineers, data analysts, and computer operators were classified as productive workers while the support staff was described as nonproductive.

4.16 J.R.:
All of the productive workers begin as programmer-analysts. That is // for the most part they are engineers // but they are recruited as analysts. And you have a person in Dakar who occupies a hybrid post. He is our representative in Dakar, but he does not have the title of project head. He is a research engineer. He is also our commercial representative in Dakar. // Okay // After this level of employees // you have the accountant, the secretary, and the receptionist. That's about it.

S.G. is typical of the Ivorian consulting firms engaged primarily in banking and financial affairs. J.R. explained: "Well, our work is '*zero*' per cent with the Ivorian government, 40% with banks, and 60% with private industries." He contrasted the operations of his firm with those of C.I., which he saw as working primarily with the Ivorian government but utilizing strong financial backing from the Banque de Dépôts et de Consignations in France. S.G.'s clients include the largest hotel and tourist resort operations in Abidjan. In addition, a Kenyan firm uses S.G. to do programming and data processing for its local operations in Ivory Coast. Added J.R.: "Well, this is not very interesting for us because after the programs are written, we have to hire a person to process their data. It would be better if we had on-line communications with them." At the time of the interview, however, satellite computer connections between Ivory Coast and Kenya did not exist.

J.R.'s position as research engineer resembles similar functions in other firms with partial French control. The director, who is French, travels between Paris and Abidjan while J.R., who is Ivorian with training in France and Belgium, represents his firm locally. J.R. explained this arrangement.

4.17 J.R.:

> It's a delicate situation. Officially I am in charge of technical research. In fact // I also act as the assistant director for a very simple reason. If you look at the organigram // the director is in France supervising external contracts. But when he's not here // it's really necessary for someone to replace him. Well // what I mean is that I replace him often // not only in a technical capacity but also as an administrator.

Although J.R.'s description of his administrative and technical functions is echoed in many of my interviews with data processing managers and systems analysts in both Ivory Coast and Kenya, the pattern of local representation for foreign affiliate companies is more common in Ivory Coast. This pattern is reinforced by a tendency for high-level technical personnel in small firms to handle management duties in addition to their specialized functions. Thus, at high levels of technical expertise, the assumption that computerization leads to overspecialization does not prove to be the case for many small firms.

3. S.L.: Affiliate of a French Computer Consulting Firm

Established in December of 1978 as an affiliate of Crédit Lyonnais in France, S.L. has 50,000,000 CFA francs, or U.S. $161,290, in capital assets, of which 70% are French and 30% are Ivorian. The firm has six permanent employees and hires several consultants from France to work on special projects each year. Its French director, R.M., collaborates with his wife on developing accounting, inventory, and payroll programs for clients. As the director, R.M. not only handles the administrative work and international contracts for the company, he participates actively in devising special software applications for his clients and in marketing GTX business software (progiciel de gestion). In France, S.L. specializes in accounting packages for the French VISA card (*Carte Bleue*). These activities are not carried over to Abidjan, however. In Abidjan, S.L. focuses on the development of special business programs for private enterprises. Most of the work at S.L. is done on an in-house IBM 34 system.

The brochure for S.L.'s international operations opens with a photograph of its director in France. The director is dressed in an impeccable, steel-blue business suit and stands in front of a blue postmodern, abstract painting. He wears glasses tinted grayish blue. His pose is one of authority and total composure. Blue, which we already observed on SIEPAT's computer poster, is the color of dreams and of a future that appears in the form of a photographic negative that is not yet fully developed. A caption reading: "The industry of computer consulting and engineering (SSII) has become one of the most dynamic modern industries in France, and it is here to stay," accompanies the director's photograph.

R.M. began our discussion by contrasting the functions of S.L. in France and Ivory Coast.

4.18 R.M., director, S.L.:

S.L. is an affiliate of Crédit Lyonnais in France and was established in Ivory Coast in December of 1978. It has several areas of activity. The first area // representing about a third of the company's profits // is means of payment // in particular credit card payments because we are the leading representative for credit cards in France. Our second area of activity is assistance with on-line computing with centers equipped with computers and connected to terminals or clients' computers. Okay // and the third area of activity is research consulting

// which includes a lot of the work done here in Ivory Coast. Okay // we cover consultancies with enterprise managers, the installation of financial systems, and writing business programs. So we cover all of the activities involved in commercial programming. We have about two hundred engineers and technicians at S.L. in France. Voilà! That's overall what we do.

In contrast, R.M. described the activities of his company as considerably reduced in Ivory Coast. With periodic assistance from consultants in France, his six-person staff focuses primarily on developing commercial programs for private enterprises and organizing computerized accounting and financial systems. R.M. outlined his firm's operations in Ivory Coast to me.

4.19 R.M.:

The African affiliate of S.L. // which is called S.L.-Côte d'Ivoire // has very, very reduced activities in comparison with S.L.-France. Theoretically, we represent S.L. here. But that is in theory because the Carte Bleue // VISA card in Ivory Coast // is useless. It does not function here. Okay // euh // so we primarily conduct research and consulting, adapting programs, developing specialized programs, and distributing business software. Until last year // we focused on large computer systems. At the beginning of this year, we started to distribute business software for microcomputers. So // if you will // for the past year, we have reviewed and adapted software from France for distribution in Ivory Coast. That is // we follow Ivorian regulations for small and medium-sized enterprises [PME] and for small and medium-sized industries [PMI]. We focus on packaged programs for stocks, billing, accounting, payroll, and sales statistics. Here is an example of a payroll document for you. Voilà. There are our current activities in Ivory Coast.

On occasion, R.M.'s permanent staff works with as many as twenty consultants flown in from France for special jobs. R.M. explained that his role as the local director for S.L. gave him considerable administrative autonomy within the commercial guidelines of the parent company. As with many other managers in small computer firms, R.M. stated that he was "responsible for everything," that is, a combination of administrative and technical duties.

4.20 R.M.:

It's difficult to define the responsibilities of a director. In my opinion, a director is responsible for everything. If you wish to speak of autonomy from France // euh // our autonomy is rather great. We are obliged to follow commercial and financial instructions from France and to submit our accounts and profit-loss statements to the French office. But in terms of day-to-day decision making // everything is relatively flexible. I also serve as the head engineer here at S.L. We have a systems engineer and a programmer-analyst. These engineers are naturally supplemented by the arrival of consulting engineers from France.

I asked R.M. to compare the climate for computer experts in France and in Ivory Coast and to reflect on what his position might have been if he had remained in Paris as a computer expert.

4.21 R.M.:

Et voilà! You mean the problem of re-entry (*réinsertion*). I could always go back to Paris. // euh // And in Paris // I would be a head engineer like so many others at S.L. or in the Crédit Lyonnais group. I would live in the Défense suburb of Paris with everything that entails // that is // taking the métro every morning and all of that. That would be a Parisian lifestyle. // Good // the Parisian lifestyle is all very well if you are a student. But if you are a salaried worker in Paris, it is totally different. You realize right away that you should go somewhere else!

Indeed, R.M. runs his firm in Abidjan with an Ivorian flair. He and his wife share administrative and project design decisions. Their pace of work fits in with that of the local businesses with the obligatory break for an afternoon siesta. Above all, R.M. knows most of his clients and competitors personally. He is part of Abidjan's enclosed world of computer experts. He sees his major task as adapting French computer and business systems to Ivory Coast rather than imposing the French system on Ivorian businesses. The bottom line, however, is technological and financial accountability. The monthly profit-loss statement must always go back to France.

Both S.G. and S.L. acknowledge C.I.'s success with government contracts, and they instead concentrate on the banking institutions and large private enterprises. There is a considerable economic gap between these three firms and the other computer consulting firms in Abidjan. Although some of the vendors such as BULL, IBM, and

UNISYS offer computer services, sales and maintenance are their focus. Thus, the big three computer companies dominate a fairly restricted consulting and data processing market, but they are certainly not alone.

4. Three Smaller Computer Consulting Firms

In the second category of SSII, I have included smaller computer and engineering firms that specialize in consulting, systems analysis, programming, and software distribution. They include V.I., O.E., and I.I. Of these three companies, V.I. is a general engineering and computer consulting firm that resembles the big three companies while I.I. and O.E. began as software distribution companies that earn a large portion of their income through sales in addition to consulting.

Trained as an engineer in several courses with IBM-France between 1955 and 1959, J.D. came to Cameroon in late 1959. After working in assorted West African countries, J.D. and several partners founded V.I. in 1979. V.I. is a small computer firm with J.D. still at its helm. He has fifteen employees, including his wife, who works as the administrative secretary and his son, who is head of microcomputer software development. Of the remaining employees, eleven are Ivorian. Data processing is done on a Burroughs B100 minicomputer with software packages developed for accounting, inventory, and payroll. These software packages are marketed to clients who are given assistance with in-house data processing in their firms. Seated in a small room next to his wife's administrative office, J.D. provided me with a brief description of his company.

4.22 J.D., director, V.I.:

We are a computer consulting firm with a Burroughs B100 system. We do data processing for our clients, including accounts, payroll, and stock. // Voilà! // we have approximately forty clients, but some of our clients are accounting bureaus who ask us to handle their clients. So // in reality // I deal with more dossiers than I have clients. In addition to our primary data processing activities, we provide assistance // euh // to companies with their own computers // consultation, programming, and assistance with their systems. I am the managing director // which doesn't mean a lot in terms of salary. ((Laughter.)) What I do is manage the company and I handle the technical

> responsibilities // systems analysis and // if necessary // programming. I was trained with IBM-France in short courses. I spent about one year with IBM-France at different times between 1955 and 1959. But I received most of my training through practical experience.

V.I. has 10,000,000 CFA francs in capital assets, or U.S. $32,258, according to Chamber of Commerce listings. Thus, there is a gap between this company and the big three firms in terms of its financial base. J.D. explained that most of his clients are small and medium-sized enterprises with whom he has short-term software and maintenance contracts. In the future, the firm plans to expand into the large-scale marketing of software as the representative of a French company.

A head engineer at one of the big three companies explained that the larger computer firms, in his view, were primarily affiliates of foreign companies attracted to Ivory Coast "when the economy was flourishing before 1980." He saw the smaller SSII as newcomers taking advantage of the economic niche developed by the big three firms. J.D. seemed to share this opinion, but from another perspective. In assessing the position in his company in Ivory Coast's computer world, J.D. frankly stated:

4.23 J.D.:

> Do you want me to make you a list of our competitors? We cannot be compared with S.L. or S.G. as firms. No // we are not very well placed on the list. Our work is primarily with programming, consultation, and software distribution. We are like I.I. // In fact, they are our colleagues.

Established in October of 1974, I.I. is primarily a software and microcomputer distribution company that provides engineering advice and training for some of its clients. The company's permanent staff of forty-two employees is large for a firm that lists capital assets of 30,000,000 CFA francs, or U.S. $96,774. Twenty-five of these employees are Ivorian, and several of them are INSET graduates. According to the company's assistant director, 100% of I.I.'s capital is French. In addition to distributing computer software, I.I. sells and rents IBM personal computers and electric typewriters. Most of its large staff engage in program adaptation for business, installation, and repair of equipment. Hence, while I.I. is a French supported firm, it is far from a family-run enterprise resembling the V.I. company.

Many members of its large staff are part of a maintenance crew. The company's major focus is on sales and rental of hardware and software rather than on data processing and consulting. The assistant director, Y.M., a young French engineer, summarized his company's activities for me.

4.24 Y.M., assistant director, I.I.:

I am the assistant director of I.I., which means that when the director is absent, I do a little bit of everything // technical programming, commercial affairs, administration, and importation. Things like that. // Good. // We have two departments here. The oldest is data processing and batch computing. Now we use the IBM 4341. We do payroll, accounts, and stock // things like that // for enterprises that don't have their own computers. Okay // that's the oldest department. And for the last three-and-a-half years, we have become an official distributor of IBM microcomputers and typewriters. We also distribute other brands of printers. We have developed standard programs for microcomputers // payroll and accounts // adapted to Ivory Coast.

Y.M. lamented the fact that the competition for batch computing and distribution was on the rise. The market appeared to him to be filled with newcomers grasping for profits in a domain where his company was once the uncontested leader. Complained Y.M.:

4.25 Y.M.:

There's much more competition now. Before there were only a few other companies doing what we do. Now // there are over *one hundred* companies distributing computer hardware and software in Ivory Coast! They distribute hardware and programs, and they modify programs. So our turnover has gone down. Also // our batch computing clients have a tendency to // well // they are declining. Voilà. We are losing them because of the sale of microcomputers. Now we concentrate more on sales.

The changing market created by the increase in availability and sales of microcomputers has shifted the commercial priorities from batch computing to software sales and development in smaller companies such as I.I. and V.I. The smaller companies are also forced to follow commercial trends more closely than the large SSII that look for long-term contracts with the government or large private enterprises. Both the consulting and batch computing activities of the Category I companies are on such a large scale and require so much consultation that the frequent changes in the microcomputer industry have not lowered their profits. Yet, the possibility of over-saturating the local market may begin to affect the larger SSII that now consider themselves immune to such changes. One factor influencing these changes is the liberalization and reduction of taxes on computer hardware and software. When I interviewed Y.M., he expressed relief at the impending reduction of taxes on computer imports from 25% to 11.11% and the abolition of all taxes on software imports. "It will be good for business!" he exclaimed. These future-oriented changes in taxation regulations will probably have the side-effect of increasing competition among all of the SSII.

O.E. was founded in June of 1980. Its managing director from Sénégal is an electrical engineer and former head of the Sénégalese railways. He has attempted to establish a West African regional marketing firm linked to NCR. One of the head engineers at O.E., however, qualified that the exclusive distribution link to NCR is no longer in effect and that the company is an affiliate of the French firm, Eurosoft. O.E. has fifteen permanent employees and a capital base of 33,000,000 CFA francs, or U.S. $106,451. In the Chamber of Commerce guide, O.E. lists 55% of its capital as "African," but not exclusively Ivorian, and 45% of its capital as European. C.B., a research engineer at O.E., outlined his firm's principal areas of concentration.

4.26 C.B., research engineer, O.E.:

We specialize in computer engineering and consulting. Our principal activity is consultation. So // we are one of the SSII. We send out computer experts. We also conduct training sessions and do software development at our clients' offices. But that's at the individual client's request. Previously // we trained programmer-analysts. In the future // we hope to hold short training seminars again. We used to distribute computer hardware for NCR // microcomputers and computer systems. But we don't do that any longer. Now we conduct feasibility studies and suggest the best possible equipment for our clients // IBM, IBM compatibles, BULL, etc., etc. We don't insist on the choice of a single vendor. Now we negotiate with the vendors for our clients and help them to select the best software with all the guarantees that presupposes // essentially // maintenance contracts and all of the adequate software. Although we are

> not an affiliate of an overseas group // if we face special problems that we cannot resolve // we call on Eurosoft in Europe. They have over 350 engineers specialized in different domains of computer use.

C.B. explained that the research, as opposed to the training and distribution wing of O.E., is now expanding its offices from a regional base in Dakar to the Republic of Congo, Gabon, and Equatorial Guinea. As a smaller company in competition for the Ivorian market, O.E.'s management views regional expansion throughout West and Central Africa as an effective economic strategy. Thus, O.E. is a firm whose organizational goals and target market are in transition. As the firm develops its specialization in consulting, it is moving away from software and hardware distribution. Instead, O.E. operates as a broker negotiating contracts between clients and computer vendors. Five engineers out of the permanent staff of fifteen employees work on project development, program adaptation, computer installation, and customer training. In addition to his work as a research engineer, C.B. substitutes for the director during his absence at the regional offices.

Of the smaller SSII contacted, O.E. is the most heavily involved with government contracts. Due to its controlling interests, the company projects itself as "African" and emphasizes its commitment to government planning and development projects. The company's management compares O.E. favorably with C.I., but both its changing focus and the level of its capital assets place O.E. squarely within Category II of my framework. Nevertheless, O.E. contrasts with both V.I. and I.I. because of its involvement in both government and private sector consulting projects.

A third category of companies provides basic support services necessary for computer installation. One of these companies, G.R., formed in May of 1982, sells and repairs electronic materials used in computer installations. These items include surge suppressors (*onduleurs*), regulators, and transformers employed by large offices and enterprises in both the public and private sector. Although some of the major computer vendors, such as UNISYS, also provide these services, G.R. is the only company specializing uniquely in electronic support materials in Ivory Coast. J.F., the company's dynamic young French founder and director, began his business at the age of 25 and was able to capture nearly the entire market for electronic computer-support materials. He was able to secure the contracts for surge suppressors from the Office Central de la Mécanographie and from the Secrétariat Général à l'Informatique. His clients in the private sector include most large enterprises that have installed mainframe and minicomputers. J.F. provided an account of his activities in Ivory Coast for me.

4.27 J.F., director, G.R.:

The hardware that we sell here is called surge suppressors (*onduleurs*), or UPS. // Good. // We began our activities selling surge suppressors, but people had problems because they didn't know how to install them. So they asked us to do the installations. And // since then // we have tried to sell // euh // complete computer rooms. That is // absolutely everything *except* the computer itself // regulation of electric current, installation, floorboards, fire protection, access control, air conditioning, lighting, and partitions. We also provide computer paper and bindings // everything in the computer environment.

J.F.'s background differed from that of the research engineers interviewed at the SSII. He was trained in mechanical engineering, construction, and design.

4.28 J.F.:

My training is in metallic construction and design. I am more a technician than an engineer. I am not a specialist in electrical engineering. I have taken seminars on all of the equipment that I sell, and I have also worked in computer and electronic enterprises in France.

In carrying out his activities in Ivory Coast, J.F.'s major problems have centered around importation and shipments rather than technical difficulties. With a staff of three engineers and technicians to support him, J.F. asserted that he could deal with complex technical problems. He referred to his engineering staff as "hypercompetent," but he complained about the red tape involved in acquiring his equipment.

4.29 J.F.:

Now the TVA tax on computers has been reduced to about 11%. But what goes down is the cost of the computers themselves, not the surge suppressors. We have tried to contact the General Secretariat of Informatics about the surge suppressors // *because* why not reduce taxes on them as well since they are indispensable for running large computer systems. And then there are the

> transportation and importation problems! Sometimes our equipment is held up for months in customs. On the other side, all that the expediter has to do is to forget one of our containers in Europe, and that holds up our orders from two to three weeks. Shipment and customs delays cause us a lot of misery!

In spite of these problems, J.F. reported favorably on the positive economic situation of his company.

4.30 J.F.:

> We are not alone on the market. There are three or four other enterprises that do what we do // and they were here long before me. Since I have arrived, we have taken over 80% of the market for surge suppressors in Ivory Coast. Our profits have gone down a little recently since I had to lower my prices. But // you could say // our competitors gave us a lot of help in getting off to a good start!

G.R. has capital assets of 10,000,000 CFA francs, or U.S. $32,258 with a total of twenty-three employees, seven of whom are Ivorian. Three of the seven Ivorian employees are engineers and highly trained technicians. In addition to directing the company, handling customs clearance matters, and searching for new contracts, J.F. works directly with his engineers and crew on installation projects. He supervises each installation operation in Ivory Coast and has been involved in consultations on installations with the regional offices of G.R. in Cameroon and Togo. Although J.F. stated that his sources of capital investment were French, figures concerning investment sources and division of capital for his company are not included in the Chamber of Commerce listings.

Overview of the Computer Consulting Firms

Although differing in terms of their commercial focus and economic standing, the companies in the three categories of SSII have similar organizational structures. They have direct commercial links to foreign sources of capital, generally from France. The big three companies are supported by or affiliates of French banks. Yet, they proclaim local autonomy at the level of day-to-day decision making. In some cases, the company heads or presidents (PDG) are absentee landlords flying in periodically from Paris or Dakar to survey their holdings while chief engineers fill in for them. In other cases, the management is handled locally by enterprise heads. Both the large and small companies have a minimum of three engineers, or highly trained technical employees. In fact, staff size is not the major criterion separating the three categories of companies. Rather, they are distinguished by differences in capital assets, turnover, and clientele. Most of the enterprise heads had a realistic view of their company's standing in the market. Although companies in the big three were hesitant to claim first place in the market, they acknowledged their relative position of dominance with respect to the newer and smaller firms. Likewise, the small companies "knew their place" and, in some cases, complained about the economic success and competitive edge of the large SSII.

Managers repeatedly referred to their firm's organizational charts, or organigrams. Based on the interview data, a composite chart of organizational positions in the SSII may be devised. This organizational chart, which reflects workers' and managers' descriptions, may be compared with Figure 2-1 on the hierarchy of computer specialists presented in Chapter Two. Women are frequently found among the programmers and administrative personnel of the SSII, but there are few women engineers.

The political and economic environment for the SSII is shaped by a progressive computer education policy, on the one hand, and, on the other hand, by the competitive presence of computer vendors from France and the United States. These vendors engage in computer consulting and, in some cases, limited training. Therefore, they infringe upon the territory of the SSII. More importantly, however, most of the vendors were *there first*. They introduced and popularized computers for business and personal uses. The SSII have capitalized on the ideal conditions generated by the government and the vendors. Therefore, the Ivorian picture would not be complete without a discussion of the role played by the major computer vendors.

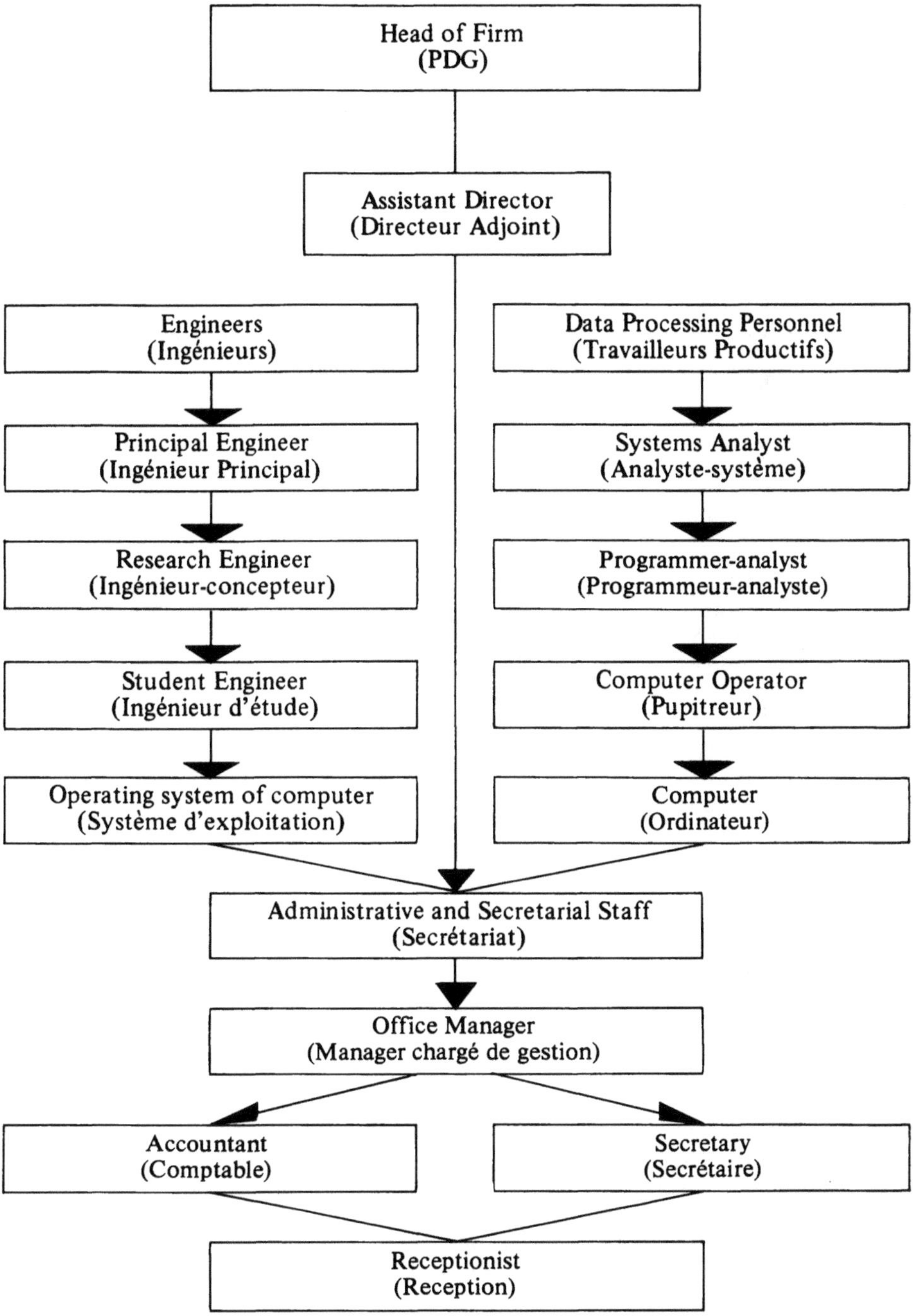

Figure 4-4: Organizational Structure of the SSII in Ivory Coast

Computer Vendors

In an article appearing in the June 1987 edition of *Jeune Afrique Économie*, the competition of large computer vendors for a piece of the African market is described as "a battle of giants."[8] Among the major contenders are IBM, Digital Equipment Corporation (DEC), and UNISYS (formerly Burroughs), all of which are represented in Ivory Coast along with BULL/Honeywell, Hewlett-Packard, Goupil, and Wang. The article presents a ranking of the top twenty international computer vendors, referred to in French as *constructeurs*. IBM and DEC occupy the top two positions in this international ranking based on sales turnover and the annual percentage of turnover for 1986 (cf. Figure 4-5).

Rank	Computer Vendor	Turnover for 1986 (in millions of dollars)	Total Turnover (in %)
1	IBM	51,250	95
2	Digital	7,590	100
3	UNISYS	7,400	100
4	Fujitsu	7,800	65
5	NCR	4,880	95
6	Hewlett-Packard	7,100	55
7	NEC	10,760	35
8	Control Data	3,400	100
9	Hitachi	24,200	12
10	Wang	2,640	100
11	BULL	2,580	100
12	Olivetti	4,700	53
13	Siemens	22,000	11
14	Honeywell	5,380	40
15	Apple	1,902	100
16	Nixdorf	1,840	100
17	Philips	22,560	8
18	ICL	1,315	100
19	Data General	1,268	100
20	McDonnell Douglas	12,000	10

Figure 4-5. Ranking of International Computer Vendors by Sales Turnover. This comparison is based on international sales and turnover of the top twenty computer companies as of 1986. These figures are reprinted from *Jeune Afrique Économie*, 98, June 1987:79.

By distributing computers and software as well as providing periodic training seminars, computer vendors shape public attitudes toward computers in Ivory Coast. Vendors influence the SSII that serve as brokers in the distribution and consultation process. They also affect government policy through direct consultation and sale of equipment to the Ivorian administration. The government is one of the largest customers for computer vendors. A long-term government contract including the purchase of computer hardware and software, maintenance contracts, and post-sales consultation is a prize eagerly sought by most computer vendors.

In terms of capital assets and business transactions, the largest computer vendors in Ivory Coast are BULL, IBM, and UNISYS. The major West African regional offices for these three companies are located in Abidjan. BULL and UNISYS have fifty-five employees each, approximately half of whom are Ivorian. IBM has forty-seven employees and lists forty-two as Ivorian. Although these companies are overseas affiliates of their French offices, only BULL/Honeywell is a company in which the controlling capital is French. Consequently, in the competition for government bids in French-based development projects, according to several of the vendors, BULL is often favored. BULL has the contract for the new Ivorian telecommunications system (SYTRANSPAC) and its equipment was used in the administration's computerization of the customs and taxation system (SYDAM). Although IBM in Ivory Coast administratively reports to IBM-France, IBM representatives claim that because the company is American, it is not in first position locally and is occasionally at a disadvantage in the competition for government contracts. The same complaint was voiced at UNISYS and NCR, both American companies.

American computer vendors in Ivory Coast use their French offices as an administrative base. Software packages are developed in the French offices for marketing in francophone Africa. Nevertheless, these companies depend on American capital and are not considered to be French even though their directors, engineers, and sales personnel are dispatched from France. A managing director at one of the American affiliates explained the complexities of this situation to me.

4.31 Director, computer vending company:

Most of our sales are of minicomputers. There are very few government contracts for larger systems. And when calls for bids (*appels d'offres*) go out, // such as the announcement that you see here // they are often for projects financed by the CCCE (Caisse Centrale de Coopération Économique) in France. The projects demand French computers. That is // they require the vendors that respond to be part of firms with French capital. So // for example // IBM and UNISYS can't respond because they are firms backed by American capital.

B. J.-R.: The situation is not completely clear to me. Could you explain it some more?

Director: It's up to *you* as a researcher to make sense of that! I can't. But what happens is this // France finances a development project for the Ivorian government. And France requires that French equipment be used. So // well // French equipment // we have it. Some of our computers are manufactured in France. But the catch is that the call for bids requires firms with French capital. That makes our situation impossible!

A commercial engineer at another American company expressed similar concerns when he described competition for the Ivorian computer sales market.

4.32 Commercial Engineer, computer vending company:
Well // we have lots of competition in Ivory Coast. Many different computer companies are registered here. // Good. // We are favorable to competition. In fact // we *welcome* it. There are companies such as Hewlett-Packard, UNISYS, Wang, NCR, and DEC. But the principal source of competition is the government and the public services. If I may express myself this way // the Ivorian administration is one of our big problems. We have a sense of injustice now. Between 20% and 30% of our business has gone // as a general rule // to the government. But in the last four to five years not even .1% of our contracts are with the government. That poses a major problem for us!

One might argue that the vendors are merely describing the effects of normal government regulation and monitoring of their activities. By controlling competition, the government has modified the international rankings for computer sales within the local market. It has forced vendors to focus more actively on private sector sales and on research and development activities. This clever strategy has created order and cooperation. Research teams from the national university and INSET have begun to collaborate with vendors on telecommunication projects, development of scientific data banks, and installation of on-line computer connections. If a vendor participates in a government or university-sponsored research project, its services and equipment will be utilized. Although regulation may appear unjust and threatening to the vendors, it is one of the key government strategies for achieving the development ideals expressed in the national informatics plans.

Regulation also permits newer companies with a small local base to compete more effectively in the Ivorian market. One example is DEC, represented in Abidjan by a new company known as Inforafric. This company has ten employees and an Ivorian general manager. DEC, which previously had only two African outlets – Nigeria and Zimbabwe – has been successful in selling its VAX equipment to the Ivorian government, educational institutions, and large private corporations. An article in *Jeune Afrique Économie* (1987:84) predicts DEC's continued success in the battle of computer giants for the West African high-tech market.[9]

As already emphasized, vendors occupy the overlapping terrain of consultation and software development and distribution with the SSII. This overlapping terrain is reflected by an increase in the number of microcomputers in Ivory Coast during the 1980s. Between 1981 and 1987, the number of microcomputers in the Ivorian private sector was projected to rise from less than 1,000 to over 10,000 (*Plan Informatique National, 1986-1990*:118). In the public sector, including the government and parastatal agencies, the number of microcomputers available increased from less than 100 to over 1,600. The second national informatics plan (p. 112) warns against a slowdown in government computerization in comparison to the private sector and encourages that more public funds be earmarked for the purchase of hardware and software. The government's focus on the SSII contributes to setting professional standards and is a means of controlling vendor competition for new contracts. The second informatics plan describes the computer and engineering firms rather than the vendors as the touchstone for the direction that computer technologies will take in the private sector.

The SSII too are subject to both direct government scrutiny and indirect competitive control. Even when they move toward regional expansion, the consulting and engineering companies appear to be anticipating a "computer revolution" that is still at an early phase of development. Therefore, the companies attempt to promote extraordinary growth in private sector computerization by taking advantage of each new technological development. The prospect that expansion of computer use in the private sector might be arrested or peak out when large enterprises have computerized does not abate the growth-orientated strategies of the SSII.

Conclusion: The Ivorian Picture in Comparative Perspective

Ivory Coast has indeed caught "the last train of the twentieth century." The country's aggressive computer policy sets it apart from many other African nations. Using the national informatics plans as vehicles for goal-setting, administrators have defined computers as the tools of development rather than exploitation. In a subtle way, through direct and indirect state control of a free enterprise system, Ivory Coast has lured major computer vendors to the country and has fostered competition for the benefit of development. Contracts for the ongoing computerization of government ministries are prizes sought after by the vendors and the SSII. Low duties and taxes on computer importation and the promise of boundless growth in the field of engineering stimulate both the vendors and the consulting firms to plan for commercial expansion.

In describing the 1976-1980 Ivorian development plan, Masini *et al.* (1979:9) state:

> The creation of the Office de Promotion de l'Enterprise Ivorienne – Office for the Promotion of Ivory Coast Companies – is specifically designed to group together small and medium-sized undertakings to facilitate their access to the market and to enable them to respond to invitations to tender from state corporations, which often present their invitations to tender to foreign firms and importers.

The state, in other words, has long been the intermediary facilitating transactions with foreign investment sources. Through this form of state regulation, private sector firms, including computer consulting companies and vendors, grow under the state's purview with the related goal of nationalization of the majority of the managers of these firms. Thereby, the country benefits from external capital and technological contributions while ultimately controlling them in the emerging postindustrial economy. In turn, government holds out the promise of more local jobs in the computer industry and high technology.

From the government's perspective, education is closely linked to the rise in computer use and the nationalization of private sector firms. INSET, in particular the ISI and ESII subdivisions, focus on the training of computer experts and engineers for the government and the private sector. Through education, the government plans to indigenize the computer world and create a new professional class for the postindustrial state. Yet, when I asked one young engineer at a computer consulting company what he thought the future of computers would be in Ivory Coast, he responded: "It all depends on the world price of coffee and cocoa." His somewhat ironic answer was, at the same time, an admission that computer vendors and consulting firms feed off of the rest of the economy. Although the government mentions plans for the development of a computer-assembly industry in Ivory Coast, such an industry is still a long way off. Instead, what is currently taking place is the education of young engineers along with a controlled effort to guide multinational computer corporations toward serving the country's development needs. This new electronic society embodies the referential illusion of progress.

In Ivory Coast, data bureaus have proliferated while in Kenya commercial training schools appear to be on the rise. An odd mutation has occurred in both cases. At first glance, there seem to be far too many Ivorian computer firms to meet the basic computerization demands of the country. Everyone wants to play a leading role in technological change. In the case of Kenya, the commercial schools have been publicly accused of overproducing underqualified computer workers. If both of these cases are examined from a symbolic rather than a purely economic perspective, a pattern emerges. Growth takes place in arenas where the promise of postmodernity appears to be the most appealing and where government regulations are in flux. To be sure, the government's role in promoting technological change and encouraging certain images of postmodernity has a direct impact on both enterprises and individuals. In this light, the attention devoted to consulting firms in the second informatics plan may be interpreted as either encouragement or restraint. In Ivory Coast's case, the informatics plans also contain an economic promise for engineers. On the other hand, in Kenya, where such a promise is not embedded in government policy, the absence of goal-setting creates an open frontier for educational institutions. The assumption in Kenya is that if properly trained, each individual should be able to enter and take advantage of the computer revolution.

Ivory Coast is of particular interest to students of technological change worldwide because, through its computer policy, the country has made a controlled attempt to leap from agricultural and industrial development into a postmodern age. The government is able to exercise direct control over the multinational corporations that market computers and over the educational strategies used for training an indigenous cadre of computer experts. In contrast, by viewing computers and automation as a threat to development and employment, Kenya has discouraged the expansion of international computer marketing companies – a situation that ultimately led to the withdrawal of IBM and Burroughs from the country during a period when the popularity of microcomputers was increasing. This policy resulted in restricting government supported education in computing and engineering. Nevertheless, the promise of the new world of information and computers has remained intact for those individuals bold enough to pursue it.

Reactions within the computer world in both countries mirror national development strategies and individual responses to them. The process of economic nationalization in Kenya has been left, in part, to an insider elite willing to challenge and work around external investors. In Ivory Coast, multinational corporations and external investors have been encouraged to compete within a national framework. The government hopes that controlling national interests in Ivory Coast will emerge in some foreign firms while the others will be so embattled with competition that none will achieve dominance.

Before the computer revolution, the Ivorian liberal economic strategy was criticized by some political commentators and scholars as a conciliatory and ineffective approach that would lead to neocolonial domination.[10] If, however, the goals of the 1986-1990 informatics plan are met, a challenge favoring further autonomy will have been made through the nationalization of the computer field and the launching of a postindustrial economy. The Ivorian train will then be well on its way to the twenty-first century.

Appendix A: Computer Consulting Companies Surveyed by the General Secretariat of Informatics (SGI), 1984

Name of Company	Engineering Informatics	Management	Distribution and Sales
ARTHUR Andersen	X	X	
CAI	X		X
CIERIA	X		
BULL CI	X		X
IBM CI			X
IVOIRE Système	X		X
LIFTEL Electro-hall			X
ORGATEC - Eurosoft	X		
ORGARIS - Syris	X	X	
PANNEL - KER		X	
SECT Ivoire		X	
SINORG	X	X	
SITEL			X
ECR	X	X	
SG2 C1	X		
CECOF		X	
TIME	X		X
IMIS			X
EXPI	X	X	
DATA TECHNIQUE (WANG)		X	X
C2A	X		
COOPERS & LYBRAND		X	
Ivoire Informatique	X		X
Organization et Informatique	X	X	
SVPI	X		X
SOPRA C1	X		
SLIGOS	X	X	
VRIDI Informatique	X		
DELOITTE HASKING	X	X	
HELIOS	X	X	
PRICE WATERHOUSE	X	X	
INFORMATIQUE de C1	X		
INFORMATIQUE et GESTION	X		
Ivoire Digital Système	X		
SADMIL	X		X
MICRO C1			X
NCR			X
SIGO (BAIO)	X	X	
SITI	X		X

Data from *Plan Informatique National, 1986-1990*, Annexe II, Abidjan: Commission Nationale pour l'Informatique, 1986:68.

Chapter Five

Display, Domination, and Mastery: Computers in the Kenyan Setting

The machine itself is designed to make life a bit easier and more exciting!
Computer Programmer, Nairobi, Kenya, 1985

Decentralization vs. Centralization: The Background for Computer Adoption

Administrative decentralization has been a recurrent issue for the national government since Kenya's independence on December 12, 1963. Kenya's fourth development plan for 1979 through 1983 emphasizes upgrading agricultural growth on a regional basis and shoring up the country's industrial and business sectors by continuing to promote tourism.[1] Increasing autonomy for the nation's forty districts and their provincial administrative offices is a key feature of the plan. President Daniel Arap Moi's July 1983 directive encouraging administrative decentralization, and indirectly influencing computerization in three government ministries, shortly preceded the fall national elections.

This political climate set the tone for changes in official attitudes toward computer adoption and established the need for a new technology capable of facilitating rapid reporting from district offices to the central government. As recordkeeping devices, computers have become tools for political change in Kenya. The government has expressed interest in bolstering private sector businesses and has announced the opening of sales of public companies to the private sector (cf. Wilson, 1988:26). Private sector growth offers motivations for computer adoption, but in Kenya, these incentives have been mixed. Government plans do not include a program for the adoption of new technologies. Reliance on outside computer consultants and an attitudinal association of new technologies with foreign business interests has created suspicion toward those individuals and firms too anxious to opt for technological change. The result has been the association of new technologies with areas of growth and change, such as coffee and tea farming (cf. excerpt 2.6) or fiscal planning, along with an ambivalence toward increasing the standardization of these technologies and increasing public access to them. Computers are associated with both the desire for cultural capital and the fear of colonizing control.

Modernity means more paperwork. The display of modernity involves the control of documents, records, and accounts. Heather Menzies (1981:15-16) quotes the results of a study on the paperburden of modern office work in Canada: "One consulting firm (WPI, 1977) . . . estimated that the average office worker is supported by between six and ten cubic feet of filing." In Kenya, computers have been introduced at crisis points where the flow of documents has exceeded management's ability to keep track of them by manual methods. Part of the rhetoric of computer adoption has been the hope that computers will "clean up the mess" and restore order to the cluttered modern world. How and why do technologies and strategies for creating domination and control potentially dominate the very individuals who introduce them? In the Kenyan firms discussed here, the technological order has reshaped the world of work by rendering outcomes more efficient and by harnessing workers to a new bureaucratic rationale. My interviews explore the attitudes of computer managers and workers toward these routines as they begin to master new technologies.

Application of the Narrative Program to the Kenyan Setting

The four phases of the narrative program may be used to examine responses to computers in Kenya and to frame the case study materials. At the initial confrontation stage, the Kenyan government has exhibited ambivalence toward the importation of computers. High taxes and import duties suggest the Kenyan government's desire to limit rather than regulate the process of computerization. The illusion of free market investment dilutes this goal by encouraging foreign computer vendors to enter Kenya on a competitive basis. The major vendors competing for their share of the Kenyan market include ICL, Wang, Olivetti, Apple, and, until 1986, IBM. Up to 1986, Apple, IBM, and Wang dominated the Kenyan microcomputer market (cf.

Price Waterhouse Associates, 1986:11). At the domination stage of computer use, local successes characterize the adoption of computers in Kenyan government agencies and the private sector. Computer-related successes in government, industry, and private sector businesses are based largely on individual initiative in learning to master new technologies. These local successes are characterized by a pattern of minimal investment in order to achieve maximal resuts. Microcomputer systems are utilized for the large tasks of computerizing national budgets and complex private sector operations. These computer systems are appropriate technologies at the time of application but do not always have the potential for adaptation to growth.

Consequently, mixed results emerge at the attribution stage of the narrative program for computer use. Computers are portrayed as capable of creating enormous changes in the efficiency of government agencies and private enterprises. Nevertheless, the reluctance of administrative personnel to learn about computer technologies and the ghettoization of the technologies among clerks and data processors relegate computers to an ambiguous status. Computerization is feared but not regulated.

Mixed computer attribution results in a fragile contract at the final stage of the narrative program. This fragile computer contract is a consequence of three conditions of computerization: (1) the tension between management and an emerging group of technocrats; (2) the uneasy relationship between outside computer experts and the internal staff; and (3) the choice of specific technologies, i.e., computer hardware and software. One outcome of the fragile contract is the elimination of computers as useless luxury items. Another outcome is the case of donor-sponsored computerization, in which a firm or agency uses equipment donated by an external sponsor. The case of government computerization in Sudan is an example in which incompatible equipment has been donated and cannot be used effectively. In the Kenyan MALD project and in one of the local charge card companies, the equipment available was adequate for the initial stages of computerization but lacked potential for adaptation. Although foreign consultants serve as catalysts in the process of computer adoption, they foresee short-term benefits that can be reported within a few months and often have less interest in the long-term potential of their experiments. When short-term technological solutions are accepted without full support of the local administrative staff, the success of the technology may last only as long as external advisors are present to monitor the project.

Problems abound if a technology is selected solely for its accessibility and low cost. The processing capabilities of the computers may be inadequate to the tasks, such as maintaining complex records and processing large numbers of accounts. The agency may be unable to retain or hire sufficient data-entry personnel. These problems are characteristic of a fragile computer contract and an ambivalence toward new technologies. Lack of regulation of computerization by a national policy creates a manipulative space for misapplications and potential crises. These possibilities surface in the areas of computer distribution and educational credentialing in Kenya. Computer adoption in Kenya contrasts with the more centralized planning of technological change in Ivory Coast.[2] An incipient national computer policy in Kenya provides the framework for the six case studies and the computer experiment that I conducted at the Nairobi Handicraft Cooperative.

Six Computer Case Studies

A case study approach is critical for understanding the impact of new technologies. By case study, I mean an approach that uncovers in ethnographic detail the organization of a particular enterprise and the managers' and workers' responses to new technologies. Such an approach contrasts with the surveys of work output, productivity, and effectiveness presented by management analysts and economists because it provides a description of the social and cultural environments in which computer applications are made.

My access to the management and computer divisions of Kenyan firms was not automatic. I proceeded by compiling a list based on Chamber of Commerce information and personal leads given to me by individuals working at firms where computers had recently been installed (cf. Appendix B). The fact that each interview had been approved and scheduled ahead of time did not assure that I would be able to see the right people. During the early phases of my study before my research was known within the network of Nairobi's computer experts, I often had to thread my way through a maze of managers before reaching the company's computer department. Waiting for hours to meet computer personnel was a typical occurrence. When I was known to the firms and the computer users during the second year of my study, on-site visits became routine and my access to the firms' computer operations increased. Along with two Kenyan research assistants, I spent hours on the shopfloors of car dealerships, in repair rooms, and in accounting offices.

Although this approach to ethnographic research differs from a conventional anthropological study of a village or a small urban community, I used similar procedures of long-term observation, in-depth questioning, and immersion into the daily operations and routines of each enterprise for the collection of data. My objective was not to produce a business productivity report or to conduct a technical analysis of computer systems but rather to examine the human element involved in adopting, interacting with, and talking about computers in Nairobi's business world.

I made three field trips to Nairobi from 1984 through 1986. I interviewed forty-two managers and data processors at twenty-five private firms and conducted in-depth ethnographic observations at six firms and six computer schools. On occasion, I was able to videotape the work procedures at the firms' computer centers. The materials presented in this chapter summarize these findings.

The six firms had differing degrees of foreign support for computerization. Some of the firms installed computers as a locally supported effort and others, such as the car dealerships, received programs and financial assistance for computerization from foreign offices. Companies studied but not discussed here included transnational corporations with Kenyan subsidiaries, such as Bata Shoes, in which Kenyan computer personnel implemented technological changes mandated abroad. In these cases, the processes of technological change resembled those taking place within Kenyan companies with the crucial difference that foreign companies were often able to supply more sophisticated, up-to-date computer equipment and consultantship during the initial stages of technology transfer. As a result, the management avoided problems of upgrading equipment at an early stage and recruited qualified computer operators rather than reskilling available staff.

1. A Toyota Dealership in Nairobi (T.D.)

Turning off of the crowded Uhuru Highway, one reaches a labyrinth of dirt roads in Nairobi's industrial area. At the end of the maze are the one-story, whitewashed offices of one of Nairobi's largest Toyota dealerships. The driveway is lined with cars awaiting repairs. Inside the air-cooled central office is an IBM 34 computer installation.

Figure 5-1. Nairobi's commercial district is the location of several data bureaus and computer vending companies.

The Toyota dealership, T.D., started approximately twenty-five years ago as a gasoline service station in the Westlands suburb of Nairobi. From car repairs and used car sales, the operation gradually grew into a major Toyota outlet with two branches in Nairobi and one in Mombasa. In 1974, the company began computerization by using the Data Centre for payroll processing. An IBM 34 system with six operating stations was installed in the area next to the spare parts warehouse in 1982. At the time of my first contact with T.D., the new system had been in operation for a little over two years. The software for the spare parts program was sent from Japan. A data processing manager, trained by IBM and the University of Nairobi, was hired to supervise the firm's data processing operation. The rest of the computer staff was drawn from the firm's existing sales personnel and received on-the-job training. Recycling of personnel was particularly effective in the case of T.D. The company used existing personnel and retrained workers to employ the computer in areas where they already had practical expertise.

A variety of factors make computer operations at T.D. successful.

Although the company has moved gradually from batch computing to its own system and retains payroll records with the Data Centre for security reasons, it has chosen a single system for all internal computer operations. An operations superintendent is in charge of the six computer stations manned by seven employees, consisting of a computer programmer, a sales operator, three salespeople, a service counter operator, and a serial print operator. Of these employees, only two – the sales operator and the service counter operator – are women. The T.D. computer system is interactive, controlled by a central minicomputer. Thus, a problem or error at one of the operating stations can be traced through the central computer. Although the computer operations are essential to stock control, inventories, and invoicing procedures at the warehouse, the computer division of T.D. functions as a semiautonomous unit that is responsible for its own management. All sales and spare parts requests are entered into the computer.

IBM provided one-week training courses for the computer users at T.D. All of them with the exception of the data processing manager

Figure 5-2. A computer professional attired in a white technician's robe to display his status at a Nairobi automobile dealership.

were recycled from sales and secretarial positions. Their prior knowledge of work operations was important because the computer reproduced all of the dealership's sales functions. J.W., the operations superintendent, explained the basic functions of the system to me.

5.1 J.W., operations superintendent, T.D.:
I am the controller for all of the operations in the system as well as the purchases that come from sister companies. Previously // we used the Data Centre // then we got the IBM 34 system. We have six terminals // including the system console. Costing is done automatically by the machine. Once we receive // we update our costs automatically because we have to give the machine the factor and then our retail prices. We just print our suggested list for the new receipts so that we can know whether we can update the retail prices. Once they are recommended by our manager // then we have a program that will update other retail sales. So // when it comes to sell // you don't have to refer to any retail // any costs // or anything. You just have to *produce*. You just call up the number of the invoice of the car sale from the machine. Because in the first place // once you produce that list from the bins // that list has the serial number of the document you are going to use according to the customer number or the customer type. Then // after producing that list // that same number will be used to call up the invoice. It is kept in *memory* // you see.

J.W. took pride in the efficiency of the system and the logic of its procedures. Although he knew only elementary programming, he was able to catch entry errors and to monitor the other operators' tasks by using packaged programs. He found his work with the computer much more satisfying than his previous sales job with the company. J.W. explained "When I was in sales // I was just a clerk. When I switched to the computer // I had to learn more on my own. It's more responsibility."

T.D. separates its downtown car sales and management offices from its spare parts sales, repairs, and computer operations in the industrial area. The computer operations are part of the daily "backstage" functions of the company. In the absence of on-line connections to the downtown Nairobi office and the Mombasa dealership, daily sales records are mailed or hand-delivered to the industrial area. Car salespeople have little contact with computer operators and are not in competition with them. Although the activities of the computer manager and his assistants sustain and influence the decisions of the general management, there is little daily interaction between management and the computer personnel.

Figure 5-3. The work of daily inventory control is conducted on a multi-user system at the Toyota dealership.

The computer manager at T.D. is in charge of training technical assistants, computer programming, program adaptation, coordination with the external branches of the company, and internal coordination of all computer services at the warehouse. The operations superintendent assists him in these tasks. In addition to on-the-job training, the operations superintendent and the seven computer operators receive periodic training in the form of miniseminars with IBM. The computer manager keeps the personnel relatively up-to-date on basic computer skills and attempts to expand the operating capacity of the firm. In mid-1985, T.D. made plans to purchase an additional IBM 36 minicomputer and to install two more workstations. The computer manager was sent to work temporarily with the firm's Mombasa branch with an eye toward eventual on-site computerization there.

With a staff of nine in the computer division, including the two managers, T.D. achieved efficient computerization. General management viewed the computer as an important tool for effective day-to-day operations and the organization of inventory, stock and sales operations. According to the operations superintendent, the IBM system upgraded the operating efficiency of the business and simplified routine tasks. He explained: "Before we installed the computer, customers could waste thirty minutes to an hour while we searched for spare parts. Now the counterman can consult the computer for a part and have an answer in five to ten minutes." Not surprisingly, the data processing manager also provided a glowing account of the results of computerization at T.D.

5.2 M.M., DPM, T.D.:

I think we have reaped so many benefits. I wasn't here before *this* computer came. I came after the computer was installed. The sales operator was here and she knows the situation before now. Well // I can tell you that we have been breaking our sales records about three or four times since we bought this computer. Turn around times // ordering systems // efficiency have improved very much. The machine has taken up a lot of work that used to be done manually.

The computer has been integrated into T.D. as an enterprise facilitator and a technological partner in business expansion. Note that the manager refers to "this computer" and is emphatic about the fact that the changes were brought about by the choice of a specific system. Although the computer personnel cannot always perform equivalent functions, the staff is able to operate efficiently during the brief absences of the DPM and the operations superintendent. Consequently, T.D. does not have a top-heavy data processing division, and management is able to use the computer personnel interchangeably in sales and invoicing. Mechanics in the warehouse have learned to prepare their spare parts requests for rapid entry into the computer, and they too rely fully on its presence to complete their repair work. It is not unusual to see mechanics at T.D. enter spare parts requests on the computer, thereby bridging the gap between manual and clerical work at the car dealership.

Figure 5-4. Spare parts to be listed for inventory control at an automobile dealership in Nairobi.

2. *A Datsun Dealership in Nairobi (D.D.)*

On a busy thoroughfare a few miles away from T.D. is one of Nairobi's largest Datsun dealerships. In 1978, D.D. installed a CR-7 minicomputer with fifteen workstations. The CR-7 is the only computer of its type assembled in Kenya. It is distributed by the Comp-Rite Company, the local dealership for Apple microcomputers. Reports on the effectiveness of the CR-7 system vary. Users at one of the computer training centers that I studied reported breakdowns and lack of flexibility with the system. The senior programmer at D.D., however, explained that this system was chosen in preference to imported minicomputers because of the appropriateness of the software and the effectiveness of the on-line communications system. Few modifications in the system have been required since its installation. The data processing manager and senior programmer proudly reported to me that the memory capacity was expanded from 64K to 128K. The senior programmer described D.D.'s computer system.

5.3 G.K., senior programmer, D.D.:
We are very satisfied with the reliability of the CR-7 system. We don't keep calling engineers now and then. Its down time // the time it's not running // is very, very little. We have a very efficient system. If something happens to our selling system // you see // we are in a bit of a problem because it updates the master file there and then. And even the invoicing // everything is done while the customer is waiting there. We have never encountered a collapse of it actually // because we have a lot of back-up // a lot of security. So when customers are waiting // let's say for an invoice // and there's a power failure // we put on the generator immediately, and the system can print a copy of the invoice that was in the machine.

Part of the appeal of the computer system at D.D. is the rapidity with which it assures that business transactions can be made. The senior programmer's description of this process resembles the computer success story presented by the DPM at the Toyota dealership.

5.4 G.K.:
According to the people who were here before the computer was installed // the system was definitely worse. It actually took fifty minutes to search for a spare part. Now it takes five minutes. And another thing // you know // D.D. at that time had about 40,000 line items // 40,000 different line items. So we had a Cardex system of about 40,000 cards. Now to update the history of those cards // I think it was such a // it was bound to be a bit chaotic // yeah. Forty thousand cards! Currently we have over 55,000 line items. With that kind of magnitude // only a computer can keep track of all that. So // definitely // the computer system has improved on the old Cardex system we had.

Absence of "chaos" and increased efficiency are the immediate benefits of computerization at D.D. The computer installation at D.D. is larger than that at the Toyota dealership and occupies the entire upper floor of a large warehouse building. There are twice as many computer workers as there are at T.D., with twelve computer operators and two managers overseeing the fifteen workstations. The size of the computer operations at D.D. accounts for some interesting aspects of its organization. The DPM functions both in the capacity of a systems analyst and general manager but seems to lack information about the details of the computer operations. During a joint interview with D.N., the DPM, and G.K., the senior programmer, most technical questions about the operation of the computer system were handled by the programmer, whose job entails programming and technical coordination of the twelve video operators at their workstations. The senior programmer's tasks may be compared with those of the operations superintendent at T.D. The senior programmer, however, has broader training and more programming skills than the operations superintendent. He is in charge of the technical management of the computer division while the DPM is an organizational manager who provides a liaison between the computer division and other areas of the firm's operation. This distinction is important in Kenya because the interpretation of the computer management function varies across firms and is often a point of discussion and friction among data processing personnel.[3]

Both the DPM and the senior programmer are Kenyans who were recruited into the company when the computer was introduced. The remaining twelve computer workers were drawn from the existing sales and clerical staff and were reskilled through a Comp-Rite training course. The DPM's duties include overseeing inventory control, parts management, and vehicle sales. In this capacity, he prepares program notes and reports for the senior programmer. This part of the DPM's job coincides with the standard job specifications for a systems analyst, as depicted in Figure 2-1. There are, however, some significant differences. According to the DPM's own description, his daily tasks are mixed.

5.5 D.N., D.D.:

In most cases // I am in charge of everything // the system and the jobs which are done daily. I have to check them and make sure they are okay. // And also at the moment // I am designing the branch's new system.

B. J.-R.: What does that involve?

D.N.: It involves writing manuals and reports and then writing instructions so that my friend G.K. here [the senior programmer] can write the programs.

G.K.: You know D.D. is an automobile vehicle company. Basically // that's it. So that over 80% of our work is with parts management and then there's vehicle sales. Now we are starting to do some programs about vehicle statistics. That's something we couldn't have thought of in the beginning. So

> you would read in the paper that D.D. has not grown for ten years. They are trade statistics. But being an autombile company // most of our work is parts-oriented // inventory control-oriented.

A tension arises for computer experts between the content of their work as reflected in internal bureaucratic titles and their position in Nairobi's computer world. A successful balancing of this tension involves demonstrating overall competence in a specific task on the job and exhibiting broader technical expertise with computer systems. Generalized expertise contributes to a worker's reputation, status, and marketability beyond the walls of the firm. Reputation assures membership in the local computer elite, known by word of mouth to all well-informed enterprise heads looking for computer managers and expert programmers.

Automobile enterprises offer an important arena in which computer experts achieve visibility and establish reputations. These enterprises were among the first to computerize, and their products are directly linked to the image of progress. Cars in Kenya constitute a growing sector of the economy. They operate as signs of consumerism and affluence and as real sources of income. Two car dealerships have actually subdivided into three data bureaus servicing Nairobi's batch computing needs. Automobile dealerships exhibit the interplay between external capital and expertise and local-level training initiatives in the computerization process. Rather than finding an electronic cottage run by grass-roots entrepreneurs, I observed expanding automobile enterprises relying on imported materials and technology but run with local staff, innovations, and adaptations. As the senior programmer at D.D. put it: "Nissan sends us the prices but they don't get us the programs. Everything is done here." It is apparent that these types of enterprises are an integral part of a continuum extending from large multinational subsidiaries to small, local businesses in Kenya.

3. A Retail Trading Company in Nairobi (R.T.)

The Retail Trading Company (R.T.), is a firm specializing in providing wholesale goods to general stores throughout Kenya. R.T. is supported by the Industrial and Commercial Development Corporation (ICDC), a financial institution and investment corporation set up by the government to assist local traders.[4] Established before Kenyan independence, R.T. has a long commercial history.

Located on a busy commercial street near the center of Nairobi, R.T.'s offices in a narrow, two-story building are surrounded by book publishers, printers, and other retail outlets, including two computer vending companies. The computer division of R.T. consists of three small rooms on the first floor of the building. This area is separated from the administrative offices by a small corridor. Although R.T. introduced computers into its operation five years prior to my initial contact with the company, the main office is inundated with papers and record books, apparently duplicating many of the computer files. A secretary in the main office prepares a large volume of correspondence on an electric typewriter rather than a word processor. The computer is used exclusively for stocks, inventories, ledgers, and payroll.

In 1980, the company installed an NCR 8200 minicomputer with four workstations. The NCR 8200 requires preparation of data by keypunch operations. Using this system, four workers were engaged in data preparation assisted by the computer operator. The entire process was overseen by a computer controller, who served as a systems analyst. At the time of my first interview in 1985, the controller had been with R.T. for three months after leaving a position as computer controller at D.D. Although still on probation, he was planning major changes in R.T.'s computer operations. The controller was not reluctant to voice his concerns about the shortcomings of the computer system to the management. In the absence of a programmer, the system required the controller to make notes on programming changes and to send them across town to the NCR office for implementation. Apparently, the computer vendor encouraged this practice by providing programming services for a fixed monthly fee. P.M., the computer controller, felt that the keypunch system required a larger support staff in order to work efficiently. Accounting entries lagged behind, and much of the bookkeeping work was done manually and entered into the computer weeks later. The computer was a source of symbolic order in the busy workplace. In this transitional situation, P.M. remarked that his joint tasks of supervision and systems analysis should qualify him for the title of DPM rather than controller. Acquiring such a title would secure him a substantial salary increase as well. P.M. outlined the daily computer operations for me.

5.6 P.M., computer controller, R.T.:

We have the NCR 8200 with four working stations. Right

now in our system // we normally use the batch forms for general ledgers // the computer input forms for payroll, and we have direct *capture* of information from receipts on stocks and inventory. I have one computer operator and four data preparation workers. I am also in the process of acquiring an additional computer operator and having internal promotions. In this case // I would like each section at least have a supervisor. So computing and data preparation will each have a supervisor.

B. J.-R.: Will the workers have new titles?

P.M.: That's right.

The controller's dissatisfaction with the work environment revolved around what he saw as an inefficient division of labor in computer operations and a shortage of personnel. By reorganizing tasks and titles, not only would he increase the efficiency of the operation, he would also assure promotions for himself and his staff.

5.7 P.M.: I am in charge of the entire computer operations of the company, and I do some work on systems // although right now our applications are being maintained by a software house // that is NCR. We do the systems design here. Then I give them the programming. Then I do the systems analysis here.

B. J.-R.: So that's why you say you have two jobs as computer controller and systems analyst?

P.M.: Normally // an upcoming department // like computers in this company // may have such problems, but eventually as we come up // we will employ people to hold these different positions // but this is a bit smaller. Just as I said // when D.D. was installing the computer // I was the only person there. I was the operator of everything. Then gradually they brought in different people to man different positions. I was the pioneer to the successful upcoming installation at D.D. In fact // I was proud that I was a pioneer.

A little over a year after my initial interview with P.M., I encountered him returning materials to the NCR dealership. I inquired about the progress of his plans and expressed an interest in visiting R.T. again. P.M. informed me that he had just installed a new computer system and that many of his plans for reorganization were already in progress. In a formal interview, he presented a summary of the changes that had taken place over the past year.

5.8 P.M.: We just purchased the Wang VS/55 system. We had about six vendors who brought in their quotations. The second quote is the one we took. Hardware and software costs for other machines might have been as good // but they didn't have demos in the country. Our main aim was to have a demonstration machine.

B. J.-R.: Why did you choose the Wang VS/55 in the end?

P.M.: We find it a flexible machine which could easily meet our requirements. It is also quite a friendly user machine.

B. J.-R.: How many workers will you have?

P.M.: We will have *an expansion*. We are looking at getting two new data entry operators in addition to the current ones // one computer operator // whom we are now training // and one programmer who will also develop new systems.

B. J.-R.: As for yourself // what is your title now?

P.M.: I am the computer controller. But when we have all the new changes in // I think that the company will do something about the title.

Within a year, P.M. accomplished most of his organizational goals. Although it was too soon to see whether the new system would remedy the backlog of data preparation, it was clear that the office environment operated more smoothly. In 1980, R.T. had computerized using available equipment and a bureaucratic ideology that characterizes mainframe computing. A senior supervisor took on many tasks with most of the workers functioning as lower level keypunch and data operation personnel. The new computer system permitted the upgrading and diversification of work tasks. By 1986, the computer controller had achieved both bureaucratic and technological mastery over his environment and had succeeded in restructuring staff responsibilities in order to accomplish his organizational and personal goals.

Initially, the computer was less integrated into the operations of R.T. than into the firm operations at T.D. The initial NCR computer was often viewed as an obstacle by the data processing personnel, who termed the operating system "old and obsolete." In 1985, P.M. concluded: "If management takes a full interest and participates in the development of the computer department // its future will be quite impressive. Now the computer department is just a 'holding company,' and we cannot expect a lot of improvement until a few changes take place." R.T. was an example of moderately successful computerization with a system that needed further updating and expansion to meet the enterprise's needs. In 1986, these changes occurred.

P.M. employed adroit strategies to reach his unique position in Nairobi's world of computer experts. He had been recruited from D.D. based on his reputation as both a technical expert and potential manager. In fact, the computer controller figured among those individuals most often mentioned by consultants and vendors as a highly competent computer expert. He made plans to improve his skills in the area of management.

5.9 P.M.: I had part of my training at the Institute of Computer Science at Chiromo [University of Nairobi]. I did my practicals at Caltex and IBM. I had part of my training in Uganda. Right now // I am also reading for a diploma in modern management // general management // after which I would like to go in for a course to allow me membership in the Institute of Data Processing. That means that I would be recognized worldwide in the same field where that body is an international institute.

Through his computer expertise, P.M. found a route to upward mobility. He was able to combine immediate work goals with plans for enhancing his reputation as part of Nairobi's computer elite.

4. A Life Insurance Company in Nairobi (L.C.)

Located in an imposing building on Kenyatta Avenue, one of Nairobi's main commercial thoroughfares, L.C. was founded in 1946 by a group of Indian and British businessmen. The company handles life insurance claims and deals with mortgage and rental insurance. In 1981, L.C. installed an IBM 34 system. Prior to the computer's installation, the firm used the Data Centre to process claims, accounts, payroll, data, mortgages, rents, insurance forms, and reinsurance agreements. The assistant computer manager explained that the volume of work increased so much that the firm found it cheaper and more efficient to run its own computer center.

At the time of my first interview in 1985, the computer manager at L.C. had just been recruited by a government agency. The DPM and assistant computer manager occupied acting positions awaiting confirmation at the end of the transition period. All of the other computer employees had been with the firm since the installation of the IBM system. A.M., the assistant computer manager, explained:

5.10 A.M., assistant computer manager:
The situation is like this. We had the manager here. Now he moved to another place in a parastatal. So his assistant is now the DPM. I am supposed to be the assistant computer manager. These are acting positions. We haven't yet been confirmed.

In addition to the DPM and the assistant manager, who also serves as a systems analyst, L.C. has six permanent data-entry workers and one temporary intern from the Kenya Polytechnic Institute. Six workstations are located in the computer division on its own floor of the building, and one terminal has been placed in the life insurance section. The processing of claims, estimates, and policies has been computerized. The computer division works closely with the firm's accountants, who occupy another floor of the large building in which the firm is housed. The DPM and the assistant manager coordinate their efforts in developing new computer programs. During the set-up phase, they received some assistance from the Data Centre in establishing the basic programs for claims and accounts, but by 1985 they had modified these programs internally without further help from the data bureau.

At L.C., the computer was not used for word processing, and the secretarial staff operated independently from the computer division. Although the DPM and the assistant manager shared overlapping technical and administrative tasks, the rest of the computer users were engaged solely in data-entry work. The sales staff coordinated with the computer division, but there was no overlap in personnel or tasks across sales, claims, billing, and data processing operations. The assistant manager's responsibilities resembled those of the senior programmer at the Datsun dealership. A.M. summarized his duties for me.

5.11 A.M.: Right now we // I do the programming, the systems work, and supervision of the data control department. So I basically do the work of a DPM. And // maybe // when the manager is not around // I can represent him at the company meetings. So // well // It's a small department actually, and there are only two of us who do all of the technical work // myself and the computer manager. And there is a student from Kenya Polytechnic working as our intern.

The links between computer management and general operations were more tenuous at L.C. than they were at the automobile dealerships. The computer experts were hired expressly to run and develop the programs and were not promoted internally from the sales and accounting staff. A.M. cited technical aspects of the insurance business as one of the major sources of difficulty in his programming work.

5.12 A.M.: The operating systems that we have right now cover payroll, life business, general business, claims paid, and claims outstanding. We have also introduced a system for loans, mortgages, and rents. The company has some rents outside. We don't do any statistical analysis and we don't do word processing. But we do have some very technical systems. The things we call insurance printings for reinsurance. We have to see to it that the user department informs us. We sit down with them and they have to inform us exactly what to do // otherwise // uh // somebody like me or the DPM // who is a non-professional in the insurance business // they sometimes find it difficult to program. They keep changing the insurance printings every year, and we have to overhaul the whole system. These are forms for reinsurance.

This process resembles the entry and modification of new price lists at the car dealerships, but the programming is more complex. The computer personnel must learn about insurance procedures while developing the new programs. The daily computer entry work, however, is tedious and repetitive. A.M. qualified:

5.13 A.M.: Data entry is going on all the time. In an insurance company // data entry is a full-time job. Data from various sections keep on coming // like claims reporting and paid claims. So that is basically what is going on all the time. Our reports run on a regular basis.

The assistant manager explained that the staff of the computer division was so overloaded by new data to be entered that additional hiring of more workers was a top priority. He projected the addition of two new terminals and several computer operators to meet the company's expanding needs. Since each insurance policy is prepared for an individual client, constant modification and updating are necessary. In terms of its volume of business and its technical programming and data processing capacity, L.C. appears to be expanding. In spite of the autonomy and specialized training of its internal management, the firm's computer center necessarily operates in close coordination with all other sections of the business. There is, however, no recycling of workers between the computer division and the rest of the company. The managers hold degrees in computer science and have selected their interns from local colleges and training centers where computer skills are taught. Thus, while the computer is technically well integrated into the firm, the skills required for general business differ from those used in computer operations.

Complaints about the inadequacy of the computer system and about staff size are less in evidence at the insurance company than they were at R.T. in 1985. Three critical factors contribute to the success of the computer system at L.C.: (1) the size and the qualifications of the computer management staff; (2) the flexibility of the new minicomputer system, which does not require data bureau support; and (3) the division of labor between management and support staff. Although one could identify broader uses for the computer at L.C. – such as word processing for claims and correspondence – the staff appears to be clear on the most important uses of computerization for the firm. On the whole, L.C. displays a full integration of the computer into the company's work functions and a restrained, although efficient, use of the computer staff. When problems appear with computerization, they surface in the mild form of requests for increased personnel to deal with existing tasks or in terms of dissatisfaction with the software. A.M. concluded the interview with his hope that the company would hire more programmers to provide technical assistance and expand the pool of data-entry workers. L.C. clearly has a high volume of repetitive data processing. A.M. was unable to provide an account of how these documents were handled by the Data Centre before 1981 in the manner of the data processors at the car dealership.[5] Nevertheless, a full transition had been made to reliance on the in-house computer system for all of the company's major record-keeping needs.

5. A Small Charge Card Company in Nairobi (S.C.)

Situated in a modern building in a commercial area of Nairobi, S.C. was a new enterprise when I first visited its headquarters in 1985. Formed in December 1984, the company went into full operation in May 1985. S.C. has developed a new charge card modeled on the American Express and Diners Club systems. The card is restricted to use in Kenya. S.C.'s parent firm is the Southern Credit Group, a financial institution that is associated with an insurance company and provides a variety of investment services throughout Kenya.

Cardholders are charged a refundable deposit of 750 shillings (approximately U.S. $48.00) and an annual subscription fee of 360 shillings (approximately U.S. $23.00). With the assistance of a computer consultant from the Institute of Computer Science at the University of Nairobi, the charge card firm decided to install a Kaypro 10 microcomputer to handle accounts, billings, correspondence, and word processing. A single computer operator handles these responsibilities. Eventually, S.C. hopes to have between 3,000 and 4,000 regular subscribers. The Tabflow Company, representative for Kaypro computers in Kenya, designed a special accounting and billing program for S.C. According to the computer operator, installation and initial computer training were relatively rapid and simple. M.M., the operator, explained that she took a short training course which prepared her adequately for the daily routine.

5.14 M.M., computer operator, S.C.:
I am the only operator. I took a course from Tabflow. My course involved operating but no programming. We use the computer for data entry, and we use the word processing facility for lists of cardholders and establishments. It keeps our records. We do spreadsheets for our parent company. It's already keyed in and you just put in the figures.

Although the company was in just its third month of computer use when I interviewed the public relations manager, S.C.'s basic response to the computer installation was positive. The manager confirmed that the volume of work currently handled by S.C. would be impossible to maintain without computerization. He qualified:

5.15 K.K., public relations manager, S.C.:
The major advantage of the computer is rapid and easy processing. Computers are too expensive for most small companies //

so they do not invest in them. They need volume to justify their investment. For us // the computer is essential because of the rapid turnover of business. We need to process all charge card purchases within the week in which they are made. We do not go through the banking system for each purchase. Local people cannot use our cards abroad. We need the rapid processing speed provided by the computer.

S.C. has a staff of eight people, including the general manager, a public relations and marketing manager, a computer operator, an accountant, an accounts clerk, two receptionists, and a secretary. According to K.K., the computer system is perfectly suited to deal with the company's present volume of work. S.C., however, is far from reaching its full quota of subscribers, and management anticipates steady expansion as the number of cardholders increases. The company plans to purchase a Kaypro 4 microcomputer to supplement the Kaypro 10 and to hire an additional computer operator. In addition, S.C.'s management intends to collaborate with the Southern Credit Group, which includes insurance and finance companies, for the purchase of a more sophisticated multi-user system. K.K. explained the current situation of the business and his projections for its growth with relation to the computer system.

5.16 K.K.:

The volume of work is just right for the number of staff that we have now. As soon as it goes up a bit // we'll need more equipment. We would not develop a whole data division. The computer operator would train one more person for a Kaypro 4. For the time being, this system meets our needs. The entire accounting system is on the machine. It's enough for the time being while we are small. When we grow bigger // no. Later we'll need a larger machine.

Although the public relations manager at S.C. was optimistic about the computer's role in the expansion of the business, several potential problems surfaced in our discussion. At one point, the manager described coping with expansion by purchasing a Kaypro 4 computer and hiring an additional worker. Later, he mentioned the possibility of installing a minicomputer system and added: "In that case, we would eventually hook up to the parent company's computer." The manager was aware that the computing capacity of the Kaypro 10 might prove insufficient when S.C. meets the baseline goal of 3,000 subscribers. The plan to install more microcomputers may create problems of labor duplication and lack of integration of the overall computer system. If, on the other hand, the projected computer system proves to be adequate, other problems could arise with piecemeal equipment purchases. Although S.C. is one of the smallest firms studied in terms of its personnel and organizational structure, its volume of business transactions is high, and these transactions are complex. This firm represents a case in which the computer system is appropriate for the first stage of the company's development and its *immediate* data processing needs but not particularly well suited to business expansion.

Training of computer personnel, in this case one individual, has been conducted through the computer supplier and was reported to be effective. The projections for computer staff in the future were, however, ambiguous, and plans for expansion could result in problems of adjustment and internal coordination. Consequently, there was apprehension about computerization at S.C. and an optimistic but guarded attitude toward the new technology. The attitudes of S.C.'s management contrast with the perspectives of managers in small firms that adopted integrated multi-user systems (such as the IBM 34 and IBM 36) for which expansion involves the addition of new programs and terminals but not a radical change in the computing system. It is interesting to contrast S.C. with a larger charge card company that has taken a different path toward the computerization of its accounting and billing activities.

6. A National Charge Card Company in Nairobi (N.C.C.)

The fourth floor of one of Nairobi's tallest modern buildings houses the offices of the National Charge Card Company. When I interviewed the computer consultant, S.A., the company was in the process of installing an NCR 9100 minicomputer system with three Decision Mate V microcomputers connected to it for accounting and record-keeping purposes. Accounts, billings, general ledgers, and mailings were scheduled to be handled by the new computer.

N.C.C. employs a total of 30 workers and is the nation's oldest and largest local charge card company. The company has been in existence since 1975, and by 1985 the company had 5,200 subscribers with a growth rate of 10% per fiscal quarter. Although N.C.C. has special arrangements with an international charge card company, the

N.C.C. card is valid only in Kenya. The target audience for the card consists of individuals already holding other international charge cards.

From the date of establishment until 1984, N.C.C. used the Data Centre to meet its accounting, payroll, and billing needs. In September of 1984, N.C.C. hired an independent, local computer consultant who holds a diploma from the Institute of Computer Science at the University of Nairobi. The consultant monitored the installation of the new NCR 9100 system. He explained these changes to me.

5.17 S.A., computer consultant, N.C.C.:
N.C.C. in its original form has been in existence for ten years. It has changed management at least three times since. In the beginning // N.C.C. was computerized using the services of the Data Centre. Now we are switching onto the NCR system // actually *this minute*. The model is 9100 // using an I-MOS 5 operating system. In the business // we are also supported by three microcomputers // NCR DMV's using an ICPM/MS-DOS operating system. We are currently improving our services. Our target is to manage to overtake American Express, but as you know // we have to guard ourselves. N.C.C. has suffered a great deal from previous management, but now the business is up front.

S.A. associated improvements in the business structure and profits of N.C.C. with plans for the introduction of the new computer system. He saw these changes primarily in the areas of billing and daily record-keeping. S.A. described this process to me.

5.18 S.A.:
Primarily for the business // it takes two days turnaround to get a bill to us from all over the country, and // uh // the record is held on the system for a period of sixty days. In between // you get a monthly statement, and for bills not paid you get a reminder up to sixty days. Mostly // we use packaged programs. Specialized programs are only written for cases where we have no alternative. Our system has been specially prepared by NCR. However // the implementation and associated programs on the micros have all been done by ourselves // or rather myself.

The new system constituted a definite improvement over the old batch computing approach by allowing N.C.C. to have more control over data entry and processing. S.A. explained: "We had little control over scheduling work from the outside." Some vendors, according to S.A., could impede the progress of a firm's computerization by advising that the wrong system be installed and by failing to provide educational support on the system. In comparing his company's decisions with the approaches of other firms, S.A. remarked on the appropriateness of the new NCR system.

5.19 S.A.:

Another company // I don't know the name // has a Kaypro 10 with a ten megabite system. I know everything about how that systems works. It could work out if the micros are hooked up in a multi-user system // but // they've got an uphill task // *I must say*.

Although the present computer staff is being utilized as the company switches to its own system, the consultant foresaw problems with the hiring of additional, well trained computer workers. Initially, the computer department was part of the accounting division and was projected to become autonomous when it expanded. S.A. outlined these changes.

5.20 S.A.:

We have four data-entry staff // one supervisor and one assistant supervisor. The assistant is actually a programmer. Presently // we have not yet established a computer department. It is part of the accounts and will grow up into a department in a relatively short period. We have left the titles that way because we still have to start doing a lot of in-house development. All right? And our bread and butter is on the present system. So // system implementation has been done by myself and is subsequently being monitored. No development work is going on in that line until we are satisfied with NCR's program.

Two months after my initial contact with N.C.C., an article appeared in *East African Computer News* (1, 2, October 1985e:13) describing the firm's successful computerization in glowing terms. N.C.C.'s computerization was newsworthy because of its size and visibility as a growing local company. The article states:

5.21 S.A.:

Using the 9100, NCR (Kenya) developed a system for N.C.C. [pseudonym] to maintain and monitor accounts for their members and establishments as well as overseas franchises. Recent tests of the custom-built software were extremely successful.

N.C.C.'s computer success story grew out of an internally run feasibility study and a six-month transition to the new system under the supervision of a computer consultant hired to work within the company. The success story was publicized immediately both within the firm and in the mass media. In 1985, the transitional process of building a computer department was underway with considerable advanced planning at N.C.C. As in the case of the two car dealerships, the two charge card companies chose quite different strategies of computerization and experienced contrasting results in terms of the organization of office work. Size and the volume of transactions were undoubtedly factors contributing to the variation in computer strategies adopted by these companies.

The six case studies point to significant contrasts across business types, sizes, and needs with respect to computerization. These cases highlight important differences in the stages and types of decision making used in adopting computers. In all cases except one, computers were installed after considerable business experience. In four of the cases, computer installation followed the use of an external batch computing system. Computer consultants from either the vending company or an independent agency were used, but the consultants were employed in very different ways. The functions to which computers were put varied, but in most instances maximal use of computers was not made because of staffing problems, internal management decisions, competing interpretations of the role of the computer as an organizational and recordkeeping device, and gender stratification in word processing.

Implications of the Kenyan Data

In Chapter Two, six rhetorical strategies were outlined in order to characterize the confrontation stage of computer use. Workers use technology as a justification for the successes and shortcomings of their daily routines. I have described these ways of talking about computers as strategies of technological mediation. These strategies may be applied to analyze the activities of computer users in the six Kenya firms. We may compare these strategies to Harold Garfinkel's description of the interpretation of a patient's medial record folders by psychiatrists in a university clinic. Garfinkel (1967:201) states:[6]

> Thus the *folder contents much less than revealing an order of interaction, presuppose an understanding of that order for a correct reading.* The understanding of that order is not one, however, that strives for theoretical clarity, but is one that is appropriate to a reader's pragmatic interest in that order .... The possibility of understanding is based on a shared, practical, and entitled understanding of common tasks between writer and reader.

In the case of computer adoption, the program constitutes the text that is interpreted through program modification, data entry, and the social interactions surrounding computer use. The pragmatic interest in the order of reading derives from two sources: (1) the organizational objectives that computers are employed to accomplish and (2) the workers' relationships to the technology. The status of the computer expert is equivalent to what Garfinkel (1967:201-202) has referred to as the entitlement to read a text. Entitlement to read does not always correspond with bureaucratic and technical titles, which, as we have seen, vary for similar activities from firm to firm. In some cases, the data processing manager has little to do with programming or the close reading of computer texts. In other cases, the systems analyst in charge of making programming notes performs general managerial tasks, and is, thus, both a reader of the computer text and a monitor of the firm's organizational structure. Entitlement provides a grounds for reading a text in terms of the organizational interests that emerge under specific circumstances.

In the Kenyan cases, computer managers and workers reiterate four positive attributes of computerization: (1) rapidity; (2) improved access to records through the computer memory and systems operations; (3) increased responsibility for computer workers; and (4) a more efficient division of labor based upon changes in organizational hierarchy and work tasks. These four attributes are organizational ideals expressed through rhetorical strategies such as references to the technological power of the machine, descriptions of the mechanized structure of work, and technological accounts. These accounts describe computer-related successes and reflect the adjustments of managers and workers to the presence of computers in the office. As an object of power, the computer has transformative capabilities that are manipulated, used, or misused in various ways in each enterprise.

S.A., the computer consultant at N.C.C., described some of the consequences of computer adoption in terms of what he called "misuse."

5.22 S.A.:

There is going to be a big threat of misuse of computers. We are going to have the situation that is in *even Europe // even America* where you've got people very well placed in the computer field // without the professional aptitude or ability // forget about training. People with good training do not necessarily keep the best systems because of the tremendous developments happening now. I know people // managers and supervisors // who have totally a "*cake*" system running. It takes years to develop a solid computer expert. Training alone is not enough.

The precarious relationship of expertise to entitlement and work tasks is at the heart of many debates about office automation (cf. Perrolle, 1987:155). Does computer work draw on the logic of the factory or the logic of the professions in its insertion into the organization of an enterprise? Harry Braverman (1974:329-330) sees the equation of computer operations with workers' tasks as a way of placing a ceiling on workers' capabilities and prospects for vertical mobility by mechanizing and degrading their work. According to Braverman (1974:329): "[T]he concentration of knowledge and control in a very small portion of the hierarchy became the key here, as with automatic machines in the factory, to control over the process." In contrast to Braverman's pessimistic predictions concerning worker control and deskilling, other scholars have painted a glowing picture of new management strategies in the office of the future while noting that computerization represents a complex "technical intervention" into the organization of the workplace (cf. Kling and Iacono, 1988: 233). The Kenyan case studies exhibit a range of alternatives, including the restratification of workers, the upgrading of their tasks, idealization of the computer, and a concomitant isolation of the new technical managers from both data-entry workers and general management. The computer's attributes and the rhetorical strategies that interviewees use to present them seldom touch upon the complexities of integrating computers into the work organization of firms. Instead, computer users' descriptions focus on the effectiveness of the results of computerization at the domination stage.

The question of which technology is appropriate for a firm is another salient issue (cf. Harper, 1984:116-117). Studies of appropriate technology and technology transfer often overlook the relationship of technology to the organizational structure of an enterprise. In certain cases, for example the Kenyan automobile dealerships, data processing decisions are closely linked to all operational functions of the enterprise. I shall refer to this linkage as "computer integration" within the firm. In addition to technical coordination, computer integration entails a close working relationship between general management and computer personnel.

In the automobile dealerships and the life insurance company described in this chapter, data management is a subsection of general management, and the DPM works in an executive as well as a technical capacity. In other cases, there is a large gap of responsibility between the computer department and management. Consequently, the computer division is isolated as a technical subsector of the enterprise. The majority of employees process data in a mechanical fashion, and they have marginal connections to activities in the rest of the enterprise. This process of technical linkage is reflected on the lower levels of the data processing hierarchy. In this case, data-entry personnel serve only one function – entering materials into the computer. They are not called upon to participate in other sales, clerical, or decision-making duties. Data processing overload is handled by the hiring or projected employment of more lower level computer workers with the possibility of redundancy in these tasks. Although this development is characteristic of the new mechanical solidarity, it is also a holdover from older, labor-intensive work strategies.

Redundancy, as a way of handling data processing overload, however, is not necessarily a negative development for enterprises making the transition to computerization. From management's perspective, one could argue that the duplication of data-entry work is an important ingredient in improving the efficiency of work operations at a low cost. On the other hand, my data suggest that for Kenyan cases in which the recycling of employees is an enterprise strategy, there is actually little duplication of work during the transition phase, and workers' duties increase. Employees who have been recycled from clerical or sales positions to jobs that involve computerization bring their previous competencies and skills with them and are able to expand their responsibilities. Thus, they can cooperate effectively with experienced computer personnel as skilled workers whose tasks have been upgraded by the addition of computer skills. This process is evident in the recycling of workers at T.D. and the plans for computer retraining at N.C.C.

Another important variable in the computerization process is the firm's prior exposure to computer use. Managers in firms that previously had batch computing systems know the uses to which they wish to put computers. If batch computing has been employed to handle payroll, ledgers, and inventory, a similar system will probably be developed for in-house computing. This process of adaptation, however, is not always the most obvious or internally effective solution to a firm's data processing problems. For example, T.D. continued to use the batch computing services of the Data Centre for its payroll system and adopted the new minicomputer to deal with inventories, stock control, and invoicing. The computerization of these activities, previously handled manually, increased the speed of business transactions according to the management. Thus, familiarity with batch computing enabled the firm to make decisions about which business activities were to be computerized internally without duplicating the data processing functions of the batch system.

Early critics of automation in the office use batch computing as a test case (cf. Crompton and Jones, 1984:44). They argue that tight data preparation deadlines, repetitive busy-work, and procedures of error control involved in batch computing contribute to deskilling the work of the clerk. Since these data-entry workers have no control over the computer's operating system, they are unable to modify their work procedures or to handle periodic problems with data processing. In four of the six Kenyan case studies, the firms moved from external batch computing to internal control over their own systems. Subsequently, computer workers were reskilled. In one case, R.T., a transitional system was employed from 1980 to 1985 in which programming was done by the vendor, but keypunch operations and systems management were handled by the firm. R.T. moved away from this system to one that provided greater autonomy for the workers. Hence, in these cases, the firms used batch computing as a transitional method of data processing that eventually generated a new organization of work through in-house computing. The "batch computing" thesis of deskilling computer work does not apply in these cases. Instead, these companies have opted for a gradual and selective computerization strategy that routinizes work tasks in a different way.

Selective computerization is linked to internal decisions about the importance and role of automated data processing within a firm.

In the insurance and automobile enterprises, computer operations are closely tied to sales and accounting activities, but the computer departments have developed an autonomous internal management. In these cases, the DPM supervises the computer workers in an organizational as well as a technical capacity. At R.T., the wholesale trading company, initially the managerial scope of the computer supervisor was limited. P.M. held the title of computer controller and was a technical rather than an organizational manager. Consequently, the controller complained about his job description and disagreed with management about his responsibilities. His problems were not merely terminological. In Kenya, a DPM generally earns the equivalent of U.S. $1,500 to U.S. $2,000 per month while controllers, systems analysts, and programmers earn half of that salary. When managerial duties are included in the analyst's or the programmer's duties, problems with defining the scope and level of responsibility of the job arise. These problems affect the computer expert's income and reputation.

Interviewees presented me with the following ideal hierarchy of computer-related positions (cf. Figure 2-1):

(1) Data Processing Manager (including technical and managerial functions) or Supervisor
(2) Systems Operations Manager, Systems Analyst, Computer Controller, or Assistant Supervisor
(3) Operations Superintendent
(4) Programmer
(5) Data-Entry Staff and Computer Operators
(6) Word Processors (typists and clerical staff)

The practical definition of these positions and their associated responsibilities varies across cases. The data processing manager and systems analyst often perform the same technical tasks in different businesses. If the DPM is involved with administration rather than technical management, analysts and programmers assume the technical duties associated with the DPM title. In most cases, if an operations superintendent is present, he or she is in charge of the daily supervision of workers, much like a plant foreman. The position of operations superintendent, however, was present in only two of the six Kenyan cases. These duties were usually included in the tasks of the data processing manager or the senior programmer.

The title "programmer" in the Kenyan context has a special meaning.

As a result of a lack of standardization in the certification process, programmers may range from individuals with a rudimentary knowledge of computer programming to experienced engineers capable of organizing complex systems. Secondary school graduates with some exposure to programming on microcomputers were hired as beginning programmers by some firms. In all six cases discussed here, packaged computer programs for accounting and inventory had been provided externally by vendors or other sources and directly adopted, or slightly modified, by the firms. Despite claims made by systems analysts and programmers that firms' programs were constantly "updated," the firms rarely developed specialized programs. Programming, in this context, involved minimal skills and was often devalued in terms of salary scale.

In four of the cases, between three and seven individuals were involved with data entry. In one case, S.C., one individual handled all data processing needs and in another case, D.D., there were twelve computer operators. This variation is an artifact of the type of computer system adopted and the size and goals of the enterprise. S.C. used a microcomputer while D.D. employed a large minicomputer with fifteen workstations. The other enterprises used minicomputers that permitted the expansion of full-time computer personnel through the addition of workstations. Statements of the management in all six cases projected expansions of the computer division on the data-entry level rather than at higher levels of computer work through either recycling or outside hiring.

All of the enterprises except R.T. had the capacity to employ the computers for word processing of correspondence, invoices, and internal memoranda, but they elected to use it minimally, or not at all. Invoices at the car dealerships were preprinted, so little word processing was involved. Data for claims were processed by computer at the insurance company, but the system was not used for correspondence or in-house memoranda. Thus, the female clerical staff was separate from the male computer staff. Women used the typewriter as their primary tool. In none of the other cases had a full recycling of clerical staff into computer staff occurred. This finding has two important implications: (1) computerization tends to remain restricted and compartmentalized in Kenyan enterprises; and (2) there is an organizational and fiscal reluctance to upgrade clerical staff by including computer-related duties in their existing jobs. This development coincides with a general tendency in Kenya to avoid the use of computers in cases where simpler machines such as manual or electric typewriters serve similar functions. This approach reinforces gender stratification in the workplace.

Although patterns of restratification may be found in computerized offices worldwide, Kenyan offices are influenced by the dual legacies of the colonial experience and local cultures. The tendency to form rigid hierarchies that separate managers and workers is inherited from the colonial experience. Labor-intensive, communal work patterns are part of the African experience. When the two cultural forms are combined, the result is a rigid authority structure with a proliferation of workers at the lower end of the scale.

Computerization poses special managerial challenges to enterprises because it introduces the possibility of a semiautonomous computer management, or technocratic hierarchy. The computer technocracy possesses the technological power to transform not only the authority structure and internal social organization of the enterprise but also its fiscal priorities. Consequently, the role of computer managers is either narrowly restricted to computer departments or downgraded to a technical-managerial level in order to minimize potential conflicts with a firm's general management. When the computer division surpasses these limitations, it may break away and form an autonomous enterprise.

Up to 1984, this pattern of fragmentation, in which semiautonomous computing divisions break away from parent businesses, characterized the Kenyan situation, as the link between data bureaus and car dealerships illustrates. Nevertheless, the rapid increase of microcomputers in Kenya between 1984 and 1986, with three new distributorships established in a five month period, suggests other possibilities. The uses of microcomputers can easily be restricted by management. They require less personnel at both managerial and lower levels. Consequently, a small enterprise such as S.C., can adopt a microcomputer with little organizational challenge to the enterprise. Although the computing power of an integrated network of microcomputers may nearly equal that of a minicomputer system, the technological power of the computer division in such an enterprise is lower. Creating a balance between technical and managerial activities is one of the major management problems for computerized enterprises in Kenya. This balance may be achieved through general managerial control of the computerization process and strict limitation of the uses of computers within an enterprise. Therefore, it comes as no surprise that computers *are not used to their full technological capacity*. To do so would increase the technological power of computer managers and challenge the established organizational hierarchy of the enterprise.

A pertinent test case for this proposition arises in an instance where computer use was not successfully implemented. In 1984, I introduced a Kaypro 4 microcomputer into the work operations of the Nairobi Handicraft Cooperative. I initially proposed to use the handicraft cooperative as an experimental case in which the process of computerization could be observed from its inception. Arrangements were made to trace the use of the computer over a period of one year and to record the results of this technological intervention.

The Computer Experiment

The Nairobi Handicraft Industrial Cooperative Society is located near the Gikomba market in the Pumwani section of Nairobi. The cooperative began in 1968 with a small group of Kamba carvers and now has 700 members and 300 nonmember apprentices on its rosters. All members are encouraged to give 30% of their annual carving output to the cooperative to sell, and they are supposed to receive a 15% commission.[7] One of the major tasks of the manager and bookkeeper at Nairobi Handicraft is recording the carver's monthly output and commissions. At the end of each month, the bookkeeper and his staff conduct an inventory of carvings deposited in the cooperative showroom. This inventory is conducted by physical count and is recorded by hand. The bookkeeper maintains these ledgers in a handwritten report and updates the report several times a month.

From 1981 through 1983, I conducted research among the Kamba carvers at Nairobi Handicraft as part of a comparative study of tourist art (cf. Jules-Rosette, 1984:105-141). My familiarity with the operations of the handicraft cooperative led me to believe that it would be an ideal setting in which to test hypotheses about computer adoption. I presented my ideas to the members of the cooperative's Management Committee, and they responded enthusiastically by granting me permission to install a microcomputer in the front office on a temporary basis with the understanding that I would observe the operations and assist with each phase of the computer's use. Since my research grant included the purchase of a Kaypro 4 computer to be used to record field data, with the management's approval, I decided that the entry of inventory data from the cooperative's main showroom and storehouse would be a productive beginning. Recording the inventory would be beneficial to both the carvers and to the research project. I was informed that there were approximately 500 items in stock, a far cry from the 55,000 inventory items at the Datsun dealership. Working with the bookkeeper and a research assistant from the cooperative, I began the task of entering a list of the 500 stock items onto floppy disks.

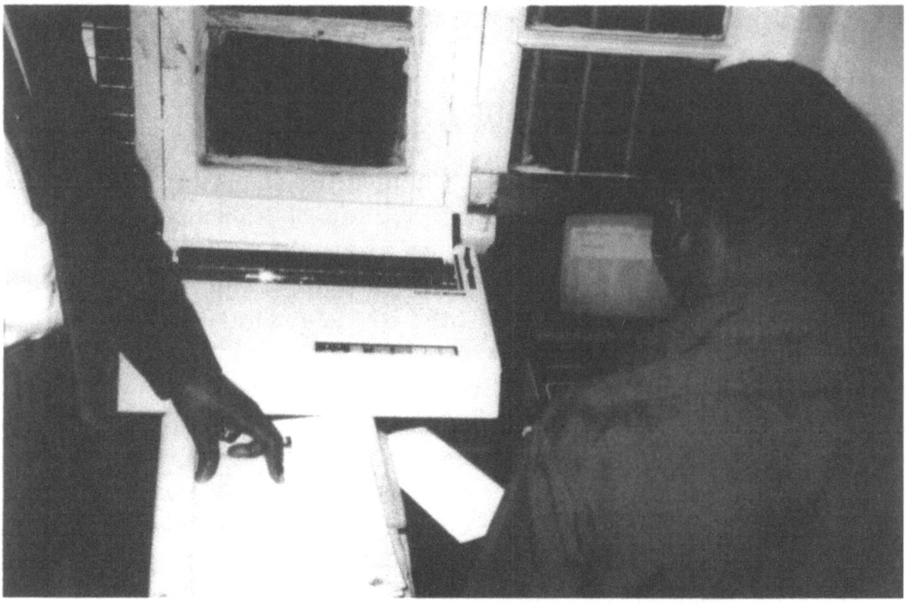

Figure 5-5. An assistant enters inventory data during the computer experiment at Nairobi Handicraft.

We did not have an inventory or stock control program, so we started by transferring inventory data onto the computer in the form of a listing. The first task was teaching the research assistant, a young carver at the cooperative, and the bookkeeper how to employ basic commands for entering data, how to use the keyboard, and how to recognize simple problems with the computer. This training session gave me an opportunity to observe the impact of first encounters with the computer in a setting over which I had some methodological control.[8] After the first week of computer use, the bookkeeper and the research assistant were able to make sales entries without much difficulty. They worked on this project together, both under my supervision and independently. A sample of a set of their entries on sales and commissions appears in Figure 5-6. In order to itemize carvings according to the bookkeeper's method, it was necessary to list each carver's submission and the amount that he received for it. This procedure uncovered inconsistencies in the cooperative's bookkeeping records.

The cooperative buzzed with gossip about the new computer. Some carvers thought that the computer was an elaborate form of cash

Kamba
Members Payments

M. No	Name	Sales of Carvings	Commission	Net
501	James Musembi	20 Shillings	3 Shillings	17
620	Shadrack Mutiso	150	22.50	127.50
373	Sakay Kioko	60	9	51
526	Daniel Mwivandi	100	10	90
721	Kamuti Kinai	30	4.50	25.50
519	Francis Kilinga	40	6.00	34.00
739	Bernard Nzau	150	22.50	127.50
641	Ngati Ndisya	60	9.00	51.00
452	Joseph Wolile	610	60.00	549.00
623	Seth Kithome	290	43.50	246.50
706	Joseph Mukula	90	13.50	76.50
507	Bonface	28	4.20	23.80
282	John	10	1.50	8.50
606	Titus	20	3.00	17.00
2	Daud	30	3.00	27.00
417	Peter	275	41.75	233.75
44	Jackson Musolo	65	9.75	55.25
262	Jones Wambua	25	3.75	21.25
215	Peter Mwangi	33	3.30	29.80
739	Bernard Nzau	150	22.50	127.50
295	Joseph Kamau	500	75.00	425.00
452	Joseph Wolile	80	8.00	72.00
507	Bonface Kamende	15	2.25	12.75
501	James Musembi	60	9.00	51.00

Figure 5-6. Members' Payment Printout, Nairobi Handicraft.

register. Others were certain that it would solve all organizational and inventory problems when its use had been mastered. The two cooperative members working with the computer showed absolutely no reluctance to use the machine. From the moment that they assisted me with unpacking the machine, the bookkeeper and the research assistant enthusiastically followed the instruction manuals. They were not afraid of making minor mistakes when I was present.

During the second week of the experiment, while he was entering commissions, the assistant noted discrepancies in the amounts paid to different carvers. Placing the list of members' payments on the computer revealed that some carvers received 10% commissions or less rather than the standard 15% commission. The research assistant brough this matter to the attention of the bookkeeper and the co-operativ secretary, who then discussed the sliding scale of commissions with the entire Management Committee. By recording the sales and the commissions on the computer and making several printouts, information which had formerly been kept on the bookkeeper's hand-written ledger became public. The computer had already begun to broaden the channels of communication within the cooperative and to alter aspects of its social organization. As a recordkeeping device, the computer had brought to light information that may have been common knowledge but that was rarely discussed by cooperative members. A text that only the bookkeeper had been entitled to read now officially entered the public domain.

The bookkeeper was asked to give the Management Committee an account of the sliding scale of commission rates. After he did so, he was placed under observation by the committee, which became especially interested in scrutinizing daily computer printouts. Ultimately, the committee accepted the bookkeeper's explanation and even promoted him. But committee members modified their views of the requirements for acceptable reporting, accounting, and record-keeping on the computer.

From August 1984 until July 1985, a Peace Corps volunteer, K.C., hired as a management assistant for the Kenya Craft Cooperative Union with which Nairobi Handicrafts is affiliated, assumed the responsibility for overseeing the computer training of the bookkeeper and members of the cooperative's management staff. As part of this reskilling process, he was to install a specially developed inventory system and send me monthly progress reports on the use of the computer. K.C., however, encountered a number of problems in monitoring the research project during my absence. As a volunteer assigned to the Union, K.C. was regularly at Nairobi Handicraft and was well accepted by its Management Committee, but he had no influence over their decisions.

Between September and November, K.C. took a computer training course at Tabflow. Nevertheless, he had difficulties putting into effect the final inventory and bookkeeping system that was projected as part of the computer experiment. Some of these problems resulted from resistance on the part of the Management Committee. Although the Management Committee claimed to be willing to cooperate with K.C. and wanted the computer in the front office, K.C. claimed that no cooperative member was ready to assist him on a regular basis with a comprehensive inventory review. Members of the cooperative observed K.C. at work on the computer with great curiosity. Despite considerable praise of his activities, they were reluctant to participate in the task of computerization. During the course of an in-depth interview on the computer experiment, K.C. gave the following description of his experiences (cf. excerpt 1.3).

5.23 K.C. computer volunteer, Nairobi Handicraft:
They are more into instant gratification today. They are not into investment in tomorrow. And that leads us into the problem of the computer. These guys saw a free asset coming to them, and it was the computer. Therefore they said, "Yes // bring the computer" // instant gratification. It gave them prestige. Our Society now has a computer. We are computer-run.

B. J.-R.: Now did they say to people after my departure that they were actually computer-run // that they had been computerized?

K.C.: No // they said we have a computer and then they let the other people figure out all the other innuendoes or whatever other associations they made about the situation. Now instant gratification was fulfilled by number one // prestige // and two // the actual requirement of the obtainment of the computer. Now they did not see // these guys did not understand // a long-term system. Because first thing they would not do is to protect the asset that they just acquired. I tried desperately to get them to take out a contract // a service contract // on the computer. It would have only cost them 4,000 shillings [approximately

U.S. $251] to have this piece of equipment totally protected for a year, with total service and regular check-ups. They would not do it because they said it would cost too much money. We had the money to do it. // Okay // another thing // going back to the point of not seeing investment // investing time, energy, and the effort into something to set up a system which would bring them over the long-run a good rate of return // or increased revenue from expanding their sales. No one there would bother learning how to use the computer . . . .

B. J-R.: So to sum up // you feel that the computer was almost a prestige symbol and people were interested in that. But they were not interested in really acquiring knowledge of the computer. They were interested in having it sit there so it was a symbol that they had it in their office.

K.C.: The computer was definitely a prestige thing because they would not let anyone else take the computer anywhere. Because I wanted to bring the computer to the Union where the data was accessible // there's facilities to use it. Because facilities at the Society were not conducive to do serious work with the computer because of the cramped quarters. And the constant traffic in and out of that office. And then when you turned the computer on // you instantly got a crowd // so that everyone stands there and you can't do serious work when you have // you know // people staring over your shoulders and people constantly asking you questions about this, that, and the other. So I suggested that I take the computer over to the Union where we had the facilities to set it down, use it like an open book, do calculations, do our programming and the rest of it. The answer was "no." Until I reminded them I was in charge of the computer. So then I was able to do that. And so I was beginning to plug in correspondence, bookkeeping, and other items.

This account of the computer experiment was not corroborated by the members of the Management Committee. On the contrary, they felt that K.C. should have been responsible for computerizing their business activities single-handedly without direction or assistance from cooperative members. Exactly what was involved in computerization was unclear to the management in spite of long discussions with them. The original discovery of the bookkeeping errors was considered to be a harmless side-effect of computerization. The committee wished to continue the experiment after this incident with the hopes that: (1) computerization could proceed with minimal input of data that would reflect on the cooperative's organizational activities and could, perhaps, be relegated to the level of word processing; and (2) that the volunteer would figure out what information would be most appropriate and which areas to computerize without substantially changing or influencing the cooperative's daily business activities. In the end, the Management Committee was adamant that the bookkeeping procedures should not be altered.

These interpretations of the functions of the computer led to an eventual decision to remove it temporarily from the cooperative until the Management Committee decided to allocate the manpower and resources for computerization. Nairobi Handicraft represents an enterprise that has not computerized because of insufficient personnel and mixed leadership messages regarding the computer's use. I believe that the need and capacity to computerize exist within Nairobi Handicraft, but that the decision to commit resources to the development of a computer staff, consisting of one or more permanent workers, was not made. This problem reflects in microcosm other difficulties of Nairobi Handicraft in developing a long-term bookkeeping and accounting staff. The turnover of the bookkeeping staff has historically been rapid due to the tensions between the Management Committee, consisting of clan elders who are carvers, and the hired managerial staff of bookkeepers and salespeople who are not carvers or cooperative members.

The computer experiment resonates with some of the problems of computer adoption in other Kenyan cases. Sufficient capital, manpower, and managerial commitment are essential in order to facilitate the computerization process. Although adequate business volume was present at the cooperative, there was little long-term commitment to the process of computerization. We have seen that in certain cases where computerization has occurred, there has been a tenuous integration of the computer into enterprise activities. Nevertheless, even in these cases, the need for computers and the staff to operate them has been recognized in the form of a managerial commitment.

At Nairobi Handicraft, the computer was a temporary donation. Although the management expressed an interest in purchasing a computer if the experiment worked, they remained uncertain about the benefits of the computer throughout the experiment. This ambivalence toward new technologies tends to surface in firms where computers have been viable business tools, but problems have arisen with the systems chosen and the organization of the computer staff. R.T. and S.C. are firms where ambivalence was evident during the early stages of computer adoption. The Kenyan ministries described in Chapter Three exhibited similar difficulties during the initial stages of computer adoption.

The ability of firms to computerize is closely linked to the training process and the availability of local personnel. Public and private training institutions provide education in programming and systems analysis, but the credentials from many of the private institutions are not standardized. Moreover, clerks and data-entry workers are reskilled within firms with few opportunities for vertical mobility. In spite of the gap between titles and job functions, computer workers without programming skills have little hope for advancement. Education places a ceiling on job opportunities for Kenya's computer professionals.

The Educational Ceiling

In five of the six cases presented in this chapter, computer professionals were recruited outside of the firms. These computer experts see their principal tasks as data management, systems analysis, and programming. Women do not figure in this group of interviewees but are instead part of the clerical work force in the firms. Only in the case of S.C. was a woman in charge of computer operations, and her tasks included many clerical functions. Internal stratification separates computer professionals from operators and men from women in these firms. A diploma in computer science is less important than a combination of training and demonstrated skills in the area of programming. The level of this training creates an educational ceiling on internal advancement in firms.

To the extent that a secondary labor force of clerical workers is created by computerization, the deskilling process characterizes computer adoption and mastery. Based upon an analysis of British firms, Crompton and Jones (1984:53) argue that computerization "'deskills' tasks, enhances the level of functional specialization, and centralizes control within the organization."[9] Achieving the attributes of rapidity and efficiency means aligning work tasks to the new organizational structure. In the case studies, we have seen that increased worker autonomy is one consequence of these changes, but so too is a tendency to expand the number of data-entry workers.

The educational ceiling separates technical from general management. If the firm's old hierarchy is to be maintained, computer operations are often insulated from administrative functions. Rather than adopting the goals of the office of the future with a terminal on every desk, Kenyan firms isolate computer operations by placing them on a separate floor or in a remote wing of the firm with communications across departments supervised by a DPM or systems analyst. Consequently, for many workers in the firm, computers become the invisible agents of organizational change. This insulation was evident at the Datsun dealership, the Retail Trading Company, and the Life Insurance Company. Combined with education and internal stratification, the symbolic and physical isolation of computers in firms increases the authority and aura of power of their users.

Figure 5-7. A computer training session at a commercial college in Nairobi.

Conclusions

The interviews presented in this chapter may be treated as micro-narratives that provide accounts of how technology works. The fit between social activities and narratives is inexact. This gray area between accounts and actions leads researchers to question the truth-value of what their informants tell them and subjects to be curious about the presentation of findings revealed in research reports. Narrative space, the gap between accounts and actions, is of semiotic interest, for it is on this pivotal point that the interpretation of meaning turns. Lyotard (1979:7) claims that postmodern culture is characterized by a "crisis of narratives." This crisis is due, in part, to the intervention of technology in the structure of the text.

Computer mastery has been defined in the narrative program in terms of a moment at which one subject dominates another ($S_1 \longrightarrow S_2$), leading to a final attribution in which the valued object, in this case the computer, transforms the subject. In each case, the Kenyan computer managers and programmers strive to master their systems and to bring the computer's functions into alignment with the firm's objectives. This process of alignment is difficult because a new layer of workers, comprised of both outside experts and reskilled clerks, has changed the face of the organization. The broad technological goals of these workers coincide with the organization's economic and social objectives. Yet, particularly in the case of computer experts, displaying mastery of the computer system is a personal goal. Once again, a narrative space emerges in the discourse of these workers as they strive to fit into the organization through their entitled reading of computer texts.

The Kenyan case studies demonstrate that job titles and task content are not necessarily synonymous. Titles, however, are linked symbolically to the requirements of the new technologies. The firm's management may manipulate the ways in which these titles are applied and the ceilings placed on job and career advancement. Computerization requires restructuring of tasks with reference to the new technology and the redefinition of job titles and positions. Technology-centered work is governed by the procedures surrounding the machines. Although tasks are functionally specific, e.g., accounting, entering inventory data, or word processing, many of these functions are interchangeable and are controlled not so much by general management as they are by the technology itself. Consequently, increasing the range of a worker's activities, including the series of tasks to be performed or the number of workers supervised, is not a direct source of upward mobility in a firm. The polarization between skilled and less skilled computer jobs creates a structural gap that makes it difficult for a clerk in a data-entry post to advance to a technical managerial position without extensive post-entry training. Time on the job and knowledge of the organization do not necessarily lead to advancement because positions and titles are technologically frozen even though the range of the worker's supervisory responsibility increases.

Barker and Downing (1985:150) take a pessimistic view of the situation of worker control in the automated office when they state:

> This process of stripping workers of control is dependent upon changes in the means of production – the introduction of new forms of technology into the workplace as part of the trend towards the real subordination of labor to capital in the form of machinery. It is within this overall context that we must view the introduction of word processors and other micro-electronically based office equipment into offices – not as part of a technology which is autonomous, driving itself forward by its own momentum, but related to and influenced by and crucially part of capital's strategy to continually reproduce itself.

Other theorists argue that capitalism is unable to reproduce itself symbolically (cf. Baudrillard, 1981:16-17).[10] On the symbolic level, a process of recoding takes place in the computer mechanized office. This recoding involves a redefinition of skills, job content, and procedures in relationship to the new technologies. A technological structure of accountability is introduced whereby members of an enterprise reference the technical features of their work as sufficient evidence for organizational policies and decisions. Technological references frame the accounts that workers give of their tasks and serve as descriptive devices for representing the activities and goals of the organization. From this symbolic perspective, the deskilling of work is actually a complex phenomenon that refers not only to the routinization of tasks but also to the workers' strategies for interpreting and recoding skills and performance.

In all of the enterprises studied, the primary function of computers was recordkeeping. Used in this way, the computer becomes a "third eye" – a powerful device that records the very activities that it facilitates. This record, which we consider a text, can be recalled at any point. As the operations superintendent at T.D. remarked, everything "is kept in *memory*." Producing an exact record of work transactions has both benefits and drawbacks. Although efficiency is increased, the record easily becomes public. The accessibility of the computer record was one of the factors that inhibited computerization at Nairobi Handicraft.

Recordkeeping involves a group of workers whose task it is to report rather than to produce. These workers do not sell products, repair cars, or prepare insurance claims. Although they are labeled "productive workers" in Ivory Coast, computer operators record the transactions that other workers make. Thus, a gap emerges between computer workers and the general staff, especially in Kenya where the ideal is to isolate the computer's functions rather than to place a computer on every desk. The immediate results of this isolation involve the creation of a dual firm, the electronic enterprise and the general firm. The global results consist of the emergence of new technicians of the sacred, who comprise an isolated and elite network of computer experts.

In the six case studies, workers reported an increased sense of responsibility, autonomy, and control through the use of computers. Especially at the level of systems management and programming, workers were able to take charge of their jobs rather than merely respond to orders. Their newly found autonomy, however, was worrisome to general management. The six case studies demonstrate ways in which computer adoption has altered the tasks, goals, and organizational structure of the enterprises. Bureaucratic positions, entitled use of the technology, and job training within the enterprises offer important keys to understanding the use and interpretation of computers in each setting. These individual efforts and successes take place in the context of a national ambivalence toward technological change. The following chapter examines the ways in which individual successes in the computer world are shaped by education in Kenya and Ivory Coast.

Appendix B: Computerizing and Computer-Related Enterprises in Nairobi, Kenya

The information contained in items 2 to 4 is drawn from Scott, 1982:41, 42, 44, 46, and 47 with permission.

1. Selected Computer Firms and Schools in Nairobi

Aga Khan Academy
Apricot/Apple Promotions and Systems
Business Computer Systems (IBM)
Business Machines Kenya, Ltd. (Olivetti)
Comp-Rite (Apple)
Comproducts
Computaccount
Computech E.A.
Computer Applications (Wang and Osborne)
Computer Bureau
Computer Consultants
Computerland (IBM)
Computer Software
Computer Stationery Suppliers
Computer Systems
Computer Training Centre
Data Centre
Data Equipment (Burroughs)
Data Recording Bureau
Graffins College
International Computers (ICL)
National Cash Registers (NCR)
Rank Xerox
Starehe Boys Centre
Tabflow Systems (Kaypro)
University of Nairobi
Wilkins Electronics Agency

2. Selected Public Agencies and Large Private Companies Using Computers

AGIP Oil
Bank of Credit and Commerce
CALTEX
Central Bureau of Statistics
E.A. Airways
E.A. Customs & Excise
E.A. Oxygen
E.A. Posts & Telecommunications
E.A. Power & Lighting Company
E.A. Railways & Harbours
E.A. Tanning
Government Computer Services Centre
Hughes
Kenya Breweries
Kenya Bus Service
Kenya Commercial Bank
Kenya National Examinations Council
Kenya Oil Company
Kenya Ports Authority
Marshalls
Ministry of Agriculture and Livestock Development
Ministry of Finance and Planning
Mombasa Municipality
Nairobi City Council
Shell & B.P. Oil
Total Oil
Treasury of Kenya

3. Selected Parastatals Using Computers

Agricultural Finance Corporation
Industrial and Commercial Development Corporation
Kenya Agricultural Research Institute
Kenya Cooperative Creameries
Kenya Farmers Association
Kenya Meat Commission
Kenya Planters Cooperative Union (Coffee and Tea)
Kenya Tea Development Authority
National Cereals Board

(To be Cont'd)

(Appendix B Cont'd)

4. Selected Private Corporations Using Computers

Amazon Motors
Associated Battery Manufacturers
Cadbury-Schweppes
Ciba-Geigy
Consolidated Holdings
Construction Engineers & Builders
Coopers & Lybrand
Cooper Motors
Datsun
Diners Club of Kenya
E.A. Engineering Consultants
E.A. Retail Traders
FAIF Management
House of Manji
Indamar Insurance
International University
Kicomi
Mobil
Nzoia Sugar
Orbit Chemicals
Ossica Security
Phoenix Insurance
Price Waterhouse
Printpak
Provincial Insurance
Securicor
Senator Charge Card
Sever Inn Sea Lodge
South Nyanza Sugar
Toyota
World Christian Handbook

Chapter Six
Socializing Workers into the New Mechanical Solidarity

> **It takes years to develop a solid computer expert. Training alone is not enough.**
> **Computer Consultant, Nairobi, Kenya, 1985**

Philosophies of Computer Education in Kenya and Ivory Coast

Under the shadow of the 1978 Nora/Minc report on computerization in France, the French government established the Centre Mondial d'Informatique et Ressources Humaines (CMIRH) with a goal of advancing computer education in the country and in French-speaking West Africa. Seymour Papert, MIT professor of artificial intelligence and inventor of the innovative educational programming language LOGO, was invited to France to collaborate on the project. With the assistance of Nicholas Negroponte, Papert set up a computer literacy experiment in Sénégal (cf. Roszak, 1986:73). In spite of Papert's example in West Africa, many African countries have been cautious about introducing the computer into the curricula of primary and secondary schools (cf. Papagiannis *et al.*, 1987:67-69). Administrative concerns with universal literacy, upgrading educational standards, and remaining within heavily constrained educational budgets have made computer literacy a dubious educational goal for African nations. Nevertheless, individual schools in Kenya and Ivory Coast have pursued computer literacy and training programs.

Three basic approaches to computer education predominate in Kenya and Ivory Coast: (1) employment preparation for automated offices and industry; (2) computer literacy and training in elementary programming in secondary schools and institutes; and (3) high-level engineering and computer education in undergraduate and postgraduate university programs. These three foci of computer education depend upon the clarity of government planning goals for new technologies.

In Kenya, where the computer contract is fragile, training is fragmented and diverse. Although the University of Nairobi as well as several schools and institutes offer curricula extending from high-level training to data entry, the objectives of computer education have not been crystallized by government planners and policymakers. In Ivory Coast, the goals and standards for the training of engineers are directly tied to the outlines presented in the first and second national informatics plans. As part of the leap into the postindustrial era, planners in Ivory Coast hope to nationalize the entire computer industry.

Beyond policymaking, computer education is sustained by broader philosophical interests and agendas. Traditionally, much learning in Africa took place through oral transmission and imitation (cf. Vansina, 1985:47-48). Village ceremonies, initiations, and storytelling sessions were essential to the richness of the child's education. The new media culture – including radio, television, and computers – returns the audience to oral and tactile forms of learning and communication. In Kenyan classrooms, once rigidly structured by the colonial educational format, the computer has become a welcome intruder that harks back to a former time. Teachers and students group around the machine, point to the computer, and even touch it in the atmosphere of an open classroom. At a computer college in Ivory Coast, the instructor, Blagué Valet, explained that "everything is done *orally*" (cf. excerpt 4.8). Orality detaches a text from the page and places it on the screen for all to see and to discuss.

Walter Ong (1982:103-104) summarizes the effects of writing on speech and oral storytelling.

> The distancing which writing effects develops a new kind of precision in verbalization by removing it from the rich but chaotic existential context of much oral utterance. Oral performances can be impressive in their magniloquence and communal wisdom, whether they are lengthy, as in formal narrative, or brief and apophthegmatic, as in proverbs. Yet wisdom has to do with a total and relatively infrangible social context. Orally managed language and thought is not noted for analytic precision.

Although oral language may lack analytic precision, it is by talking about computers that mastery of them begins. The question is not whether computer programming requires algorithmic logic, associated with abstract thought and calculation, but instead how programming is learned through interaction. As we shall see in Chapter Seven, the oral, storytelling context is an important component of the computer classroom. This interactive context is also built into Papert's LOGO experiments and other cases in which computers have been used as evocative learning devices (cf. Papert, 1980:177-189; Papagiannis *et al.*, 1987:30-31).

In Kenya approximately fifteen schools and colleges offer computer training.[1] At the secondary school level, all institutions providing computer training are private, with the exception of the Starehe Boys Centre that initiated an innovative computer training program in 1981. The Kenya Polytechnic Institute offers a diploma in computer science, and the Institute of Computer Science at the University of Nairobi has an advanced program leading to a postgraduate degree in computer science (P. Dip. C.S.). Other teacher training institutions, including the Kenya Technical Teachers College and Kenyatta University College, teach courses in computer use. These institutions are followed by commercial colleges and secretarial schools that offer courses in basic programming along with data and text processing.

The commercial colleges appear to be on the rise, and the type of training that they offer varies. Nevertheless, graduates of these colleges are able to obtain positions as programmers and computer experts. Computer training is an area that has not been subject to national planning in Kenya. Consequently, the proliferation of training institutions is an indication of lax regulation and ambivalent planning rather than a byproduct of a concerted national effort to focus on computer studies. This assessment is not intended as a condemnation. Instead, it is an evaluation of the present status of computerization in Kenya. With reference to the model presented in the narrative program, the situation in Kenya remains at the pivotal transition point between confrontation and domination. This stage of computerization suggests a possible blueprint for future planning.

The Association of Computer Professionals (A.C.P.) in Great Britain has developed examinations for a certificate in computer programming and a diploma in computer systems design. Students at various schools and training centers in Kenya are able to take A.C.P. examinations locally and to present these certificates as qualifications for entry into the job market. In addition, the National Computer Centre in Great Britain has developed a packaged course for basic training in systems analysis. According to Robert Scott (1984:18), this course has been administered to "several thousand students in more than six different countries."

In conjunction with the Kenya Polytechnic Institute, the Kenya National Examinations Council (KNEC) has developed a two-year diploma course in computer science, including systems analysis, data processing, data management, and business applications of the computer. The training received by students who take these examinations varies, and so do the local standards for evaluating examination results. Consequently, several employers that I interviewed insisted on the importance of their firm's IQ tests, letters of reference, and, above all, a previous track record of on-the-job experience.

Among computer specialists in Nairobi, an unspoken hierarchy has emerged for evaluating training qualifications. At the top of the hierarchy are the University of Nairobi, the Kenya Polytechnic Institute, and private secondary schools.[2] All of these institutions involve training programs that last for at least two years and require the students to master the fundamentals of computer programming in BASIC or COBOL languages. These programs are followed in employer preference by short-term courses that have been sponsored by vendors such as NCR, ICL, and Apple (Comp-Rite). These courses cater to individuals who already have some computer knowledge and wish to master specific systems, programs, or applications of newly imported software. Although these programs offer certificates of attendance and completion, potential employers often refer to these documents as worthless unless they are considered in the context of the applicant's job experience and references.

According to employers in many private firms, the rapidly growing commercial computer colleges fall at the bottom of the training hierarchy. This statement is not meant to imply that all of these institutions are of the same caliber, or that they are inadequate. On the contrary, the colleges suffer from "bad press" precisely because they are commercial ventures responding to changing economic demands. Although the commercial colleges administer IQ and placement tests to potential students, the popular belief among some employers is that their standards are not rigorous. To paraphrase one computer expert, practically anyone who can pay for a course can take it. On the other hand, the administrators of the commercial colleges vigorously fight such stereotypes by limiting course enrollment to qualified students, providing closely supervised instruction, and offering job placement assistance.

The commercial colleges organize six-month, twelve-month, and eighteen-month courses. Students range from secondary school leavers who have not passed A-level qualifying exams to employees seeking computer training while remaining on the job. The colleges openly court the private business sector. They have devised business-oriented curricula, emphasizing commercial applications of computers. They also encourage their more talented students to continue with twelve-month and eighteen-month courses in payroll programs, accounting, and other business specializations. In spite of these measures, however, the commercial colleges continue to be looked down upon and treated cautiously by employers.

The commercial colleges highlight the challenge of computer training in Kenya. As already indicated, the need for computer literacy is not yet extensive enough to incorporate it as a goal of general education in Kenya.[3] Moreover, the credentialing system for computer training is still unstable and lacks uniformity. Imported exams outweigh local tests and diplomas on the job market in the opinion of students and enterprise managers. Certificates vary in meaning and credibility, extending from attendance slips issued by computer vendors to valid diplomas. The picture is one of a computer educational system in crisis, and this impression mirrors the employment situation except for one crucial point. The need for higher level computer professionals perceived by the business world does not appear to be met by the educational institutions. Managers describe the process of raiding experienced computer professionals from other companies. In the words of one computer manager, if one wants to hire a systems analyst or senior programmer, it is best to "check out the network" to find out who is really competent and successful, then hire that person. A local IBM executive confirmed this view in a 1985 interview when he asserted that reputations in Nairobi's computer world are common knowledge. "We know who's really good and who's bad." Thus, at the higher levels, Kenya's computer world remains closed. Whether the skilled labor shortage is an illusion, an artifact of training, or a product of current business practices depends upon the potential for expansion in the local computer industry and computer-related enterprises.

This computer educational crisis is not restricted to African countries such as Kenya. Analyzing the problem of computer education in the United States, Judith Perrolle (1987:117) argues: "In areas of particularly rapid technological change, businesses complain that students are inappropriately trained by schools. They often sponsor on-the-job training programs for workers and professional reeducation programs." Perrolle further notes that some of these problems with training and reskilling arise because of disagreements about what education for employment in the automated office and industry means. Should workers have a higher level of competence or merely different skills?

Ivory Coast's *Plan Informatique, National, 1986-1990* (p. 42) clearly outlines two major goals for educational advancement in computer-related fields: (1) to enrich the possible resources for teaching informatics and computer science in relationship to specialized fields and (2) to educate the society at large about the benefits of computerization, algorithmic thinking, and problem-solving enhanced by computers. In other words, the plan argues that computer education will uplift the "mentality" of the national population. These broad objectives are supported by the specific economic goal of placing a majority of Ivorian nationals in high-level engineering posts, intermediary positions, and data processing jobs by 1990. The implications of these objectives were discussed in Chapter Four. In terms of a comparative overview of computer training in Kenya and Ivory Coast, these planning goals indicate that Ivorian computer education is directly linked to an employment and development strategy in which the computer industry is projected to be at the vanguard of economic growth. This assumption is absent in Kenya where technology is kept at arm's length from education.

INSET, the university of the future, is specifically designed to accomplish these aims. In addition to establishing INSET, the government has specified the credentials necessary for computer professionals in Ivory Coast. The *Plan Informatique, 1981-1985* (pp. 97-98) outlines the educational criteria for computer specialists. Passing the baccalaureate examination (BAC) after secondary school and two years of professional training are necessary for programmer-analysts. The BAC and four to five years of training are necessary for research engineers, principal engineers, and project heads. Although this training may be obtained in INSET's nine industrial and commercial schools, alternative routes to training through study abroad are also encouraged. Therefore, an equivalency may be established for computer degrees and certificates obtained in Europe, the United States, or other African countries. Interviews with heads of the SSII, computer vendors, and other company employers indicate that these formal educational standards are taken seriously in Abidjan's business milieus.[4] J.R., a research engineer at one of the computer consulting firms in Abidjan, was optimistic about the prospects for training in Ivory Coast.

6.1 J.R., research engineer, S.G.:
Well // at the level of training we have everything that's necessary. There are a few difficulties, but that's the fate of all countries. The government plans foresee the production of a certain number of computer experts to accomplish the informatics projects of the government ministries. So // the manpower // as you say in English // exists. And little by little // thanks to the director of INSET // that will help the private sector. And // little by little // we have the possibility of raising the standards for success // if you will. So // at the level of training // our prospects are good.

R.M., the French director of another consulting firm, was cautiously enthusiastic about the quality of locally trained engineers and computer experts. He compared INSET to both the Institut Africain d'Informatique in Gabon and to the IUT (Institut Universitaire de Technologie) in France.

6.2 R.M., director, S.L.:
Okay // there is a very // very good school here // ISI // the Institut Supérieur d'Informatique // which within the program of INSET is one of the best computer schools in Africa. The first // I would say // is the IAI in Gabon at Libreville. // Good // I think that in the order of things // the IAI may be the best school, but ISI within INSET is the second. They give excellent training. That's certain! Our interns from ISI are in general very, very well trained. They are well developed as engineers, and they have training and a type of diploma comparable to that of the IUT in France.

Although the Institute of Computer Science in Nairobi was praised by employers, both its graduates and the companies hiring them often qualified that the training was overly "theoretical." Complaints about the commercial colleges were even more strident. An article on computer training in *East African Computer News* (1, 1, May 1985b: 12) provides one perspective on the problem.

6.3 *How Good Are the Training Facilities?*
The Computer Association of East Africa which was very active in the early and mid-seventies and which could have assisted on this issue seems to have gone into cold storage.
As a result, each training institution is left on its own to

determine what to teach. The success or failure of each training institution therefore depends on how well that institution has judged the requirements of industry. In the final analysis, it is the requirements of East African industry which should dictate the type of training being given.

These opinions epitomize some of the crucial differences between the Kenyan and Ivorian approaches to computer education. A major problem in educational planning is whether goals for instruction in new technologies should be set by the government, industry, special interest groups, or the educational institutions. In Ivory Coast, the government has assumed the role of educational goal-setting while industry and the private sector respond to the training situation. Although the demands of business influence educational priorities, business does not establish guidelines and standards. Ambivalence toward the regulation of computerization in Kenya has led to a plea for industry to take the lead. In fact, hiring policies are used to sort experts from neophytes in Kenya's computer world, but these educational standards are mixed and implicit. Although similar hiring practices are present in Ivory Coast, agreement on the credentials required for engineers and experts exists among employers and appears to coincide with official policies of computer training. Ivory Coast's problem centers around the ability of the emerging computer industry and computer-related enterprises to absorb the new experts and not around the clarity or quality of educational standards.

Computer Training and Employment in Comparative Perspective

Vocational education was a source of debate during the industrial revolution (cf. Perrolle, 1987:117). Should the government finance training for industrial workers? With the advent of new technologies, computer education has become the focus of similar concerns. In the United States, the argument has been advanced that schools should consider computer literacy a basic skill in the manner of reading and mathematical computation (cf. Mehan, 1985:2-3). Projections about the purposes that computer education will serve, however, are often ambiguous. Computer Assisted Instruction (CAI) teaches the logic necessary for general programming and basic familiarity with new technologies. These skills will become increasingly necessary in both the office of the future and in industry worldwide. More advanced training in computers and informatics poses a different set of problems. As the Ivory Coast's planning policy suggests, advanced informatics training is tied directly to projected changes in the composition of the labor force and the organization of the workplace.

In an analysis of the educational implications of high technology in the United States, Rumberger and Levin (1984) state that while jobs will continue to expand for computer technicians and programmers during the 1980s, the overall job growth of 22% will be exceeded by the number of new jobs for unskilled workers.[5] Earlier reports from the Organization of Economic Cooperation and Development (OECD) projected a 3% increase in computer-related jobs for nine European countries between 1950 and 1975 (cf. Gill, 1985:58). Although estimates of job growth in the computer field are contradictory, figures indicate that a manpower shortage is not likely in these fields in Europe and the United States.

If the example of the United States is an indication of future trends, African countries such as Ivory Coast, with progressive computer policies, may eventually experience an overproduction of engineers and highly trained computer specialists. In their assessment of the prospects for computer training in Third World countries, Papagiannis *et al.* (1987:70) assert that "a plan to increase computers throughout a national educational system would border on folly." Instead, they recommend that computer training programs be kept small and focus on the upper secondary grades. Their approach assumes that most secondary school graduates with computer training enter the job market to fill the economic demand for lower level programmers, computer operators, and data-entry personnel. Specialized training in science and engineering is not considered in these recommendations.

There is currently no survey available on industry requirements for high-level programmers, systems analysts, and data processing managers in Kenya. Robert Scott (1984:18) claims: "There are also several hundreds of Kenyans employed as Data Processing Managers, Systems Programming and Operations Personnel that have qualifications in data processing apart from their on-the-job experience." His statement appears impressionistic. My interview data suggest a shortage of such personnel, but assessing the extent to which such a shortage is critical relative to high-level positions in the computer field requires further empirical study. After perusing several reports, I was unable to extrapolate clear projections about the relationship between high-tech education and employment in Kenya and Ivory Coast.

In an ideal situation, the job market for computer specialists would expand proportionately with training opportunities. More jobs would mean more computer training institutes. Policy planners would make predictions and design projects accordingly. This ideal world, however, does not take into account the intervention of political agendas and ideologies. Nor does it accommodate the aspirations of new computer professionals who are prepared for a transformed, postindustrial job market, regardless of whether it exists. Educational planning is based on informed predictions about the fit between training and employment. Individual aspirations, however, are part of the social and cultural imaginary – the images of the workplace of the future.

One version of the ideal fit between education and employment in contemporary society is described by Nora and Minc (1981:136) in their report on computerization in France.

> In an ideal world of fully informed "wise men," organization would coincide with spontaneity: a perfect market society, in which education and information would make each person conscious of the collective constraints, and a fully planned society, in which the center would receive from each unit at the base correct messages concerning its order of preferences and would have the same structure and the same attitude. Information and participation advance together.

Another version of this ideal is presented in Ivory Coast's plan to nationalize its computer industry and computer-related jobs. The goal of reaching an employment figure of 99% of all computer operators as Ivorian nationals and 55% of all high-level engineers carries with it an educational agenda. The explicit part of this agenda involves INSET in the training of engineers and computer experts. The unstated goals surround the training of computer operators, who will either receive on-the-job preparation or vocational training. The plans for this type of training have not been mandated by the Ivorian government. Although three secondary schools and several commercial institutions provide training for computer operators and data processing personnel, discussions of these workers are strangely absent from the educational portion of the master plans. This absence suggests educational as well as employment stratification in the new computerized society. Stratification is evidenced by the slight consideration given to the potential fate of workers once industry has become automated. It is assumed that somehow the training of higher level computer experts and engineers will filter down to workers who will assume an appropriate place in the emerging postindustrial social order. The ideological burden of change will be handled by increasing and popularizing the information available about computers.

The Ivorian model for computer education and employment goes beyond simple technological determinism, in which a technology alone motivates educational change. Instead, this approach emphasizes that technological change should lead to the goals of increasing economic autonomy by reinforcing national development and building a national elite. Computer training, thus, becomes a source of educational and cultural capital used to achieve these ends. The government is the major actor in this drama. It facilitates change and sets educational goals but does not regulate employment and professional standards. At this point, professional associations, unions, and employers enter the picture. The Secretary General of the Association Professionelle des Informaticiens de Côte d'Ivoire (APICI) explained the educational role of the association for its members and the general public.

6.4 Z.F., Secretary General, APICI:
The APICI is a professional association that is essentially concerned with bringing together computer experts and organizing the profession in order to exchange information and to assure higher professional standards. So // we hold seminars, lectures, and debates for our members. The APICI also participates in the promotion of computerization and its popularization for the general public. So // we hold meetings, lectures, and seminars which are open to the public and we bring in specialists from France // the United States // *and even here.*

In contrast, GIMI (Groupement Interprofessionel pour le Management et l'Informatique) operates as a syndicate and lobby representing the interests of the private sector.[6] The educational interests of GIMI center around monitoring the level of training received by engineers and computer managers entering the private sector work force. The official members of GIMI are private corporations rather than individuals. One proponent of GIMI outlined the organization's goals for me.

6.5 GIMI participant:

> We have created a group known as GIMI that brings together all of the computer consulting companies and vendors that have paid their dues. The aim of this group is to bring a little bit of order into the profession, to raise standards, and to facilitate relations with the state on a certain number of things. This group has existed for two years. GIMI comes from the private companies // which is not the case with the APICI. Because it comes from the companies // GIMI is not an arm of the government. They don't have to systematically say "yes" to everything that the government says // while the APICI is an association that could not live without the state. Please understand // I don't have anything against associations linked to the government. But perhaps that is not the best way to defend a profession.

This account of GIMI suggests that the private sector influences computer education by intervening in areas of employment not closely monitored by the state and in areas where conflicts of interest between the government and business might arise. Thus, although the government takes a strong lead in the planning of higher levels of computer education, employment training is an area that is less well defined and is subject to influences from the computer profession and private enterprise. As emphasized in Chapter Three, computer professionals – labeled as new technicians of the sacred – have a vested interest in keeping their profession closed and mysterious. Both the Secretary General of the APICI and the GIMI participant spoke of raising standards and creating order in the profession. Nevertheless, private enterprise is also interested in the availability and quality of workers at all levels of computer specialization. Thus, educational closure at the top of the profession is counterbalanced by the need to expand the number of workers prepared to enter computer-related fields at the base of the pyramid.

As we have seen, the government's role as the central actor in the computerization drama is not sustained in the Kenyan setting. Although a government-sponsored computer commission was appointed in 1986, as of late 1987 meeting reports and planning documents had not been made public. Computer education has proceeded in a piecemeal fashion, with the University of Nairobi taking the lead at the postgraduate level and Kenya Polytechnic playing a major role in the training of programmers, systems analysts, and computer operators. Mixed leadership messages, ambiguous national policies, and fluctuating goals relative to the profession have left a void. Secondary schools such as the Starehe Boys Centre have started to fill the gap by providing programmers for government offices and private enterprises. Nevertheless, there is no steady source of computer operators apart from on-the-job training. Many private enterprises seek secondary school graduates or students who have left school after their A-level examinations to fill in the base of the pyramid of computer workers.

Although there may be a need for computer engineers, this need remains unvoiced and is absent from large-scale government planning. Consequently, Kenya's success stories in computer education are based on individual efforts which, although they are innovative, lack coordination. Professional associations have not formed to articulate the demands of computer experts or companies. Instead, the informal process of credentialing is based upon job experience and reputation. In spite of crucial differences in computer education and professionalization between Kenya and Ivory Coast, there is a similarity across the settings. The computer world in both cases is rigidly stratified, and the education of computer operators and data-entry workers is not subject to national planning in either case. Comparisons with Western Europe and the United States reveal similar stratification practices. Heather Menzies (1981:61) states: "Given the well publicized shortages of systems analysts, data processing and other informatics professionals, it might be more appropriate to describe the 'unemployment' problem as a mobility problem." In both Kenya and Ivory Coast, this problem has three components: (1) the training of higher level computer professionals; (2) the barriers to upward mobility for computer operators and data-entry personnel; and (3) lack of coordinated training in the areas of computer literacy and elementary computer programming. In Ivory Coast, some of the basic computer services have been assumed by the rapidly growing SSII, while the anxiety over training and lack of regulation in Kenya has led to the rise of commercial computer colleges.

The problem of computer education may be diagrammed in terms of the relationship of computer education policies to schools and to ideologies of computerization (cf. Figure 6-1). Training is part of a cycle of symbolic and economic exchange. National educational policies are played out and recoded in the curricula of schools and training institutes. Where professional computer associations exist, as they do in Ivory Coast and Tanzania, these groups modify, reinterpret, and even upgrade educational standards for the job market. The efforts of schools and associations are supported by different ideologies and goals of computerization. In Ivory Coast, the ideology of computerization is tied to the education of a national elite composed, in part, of highly trained engineers. On the contrary, in Kenya the emerging group of technocrats and computer specialists operates as a closed and elite network whose resources have not been considered part of the official agenda for national development.

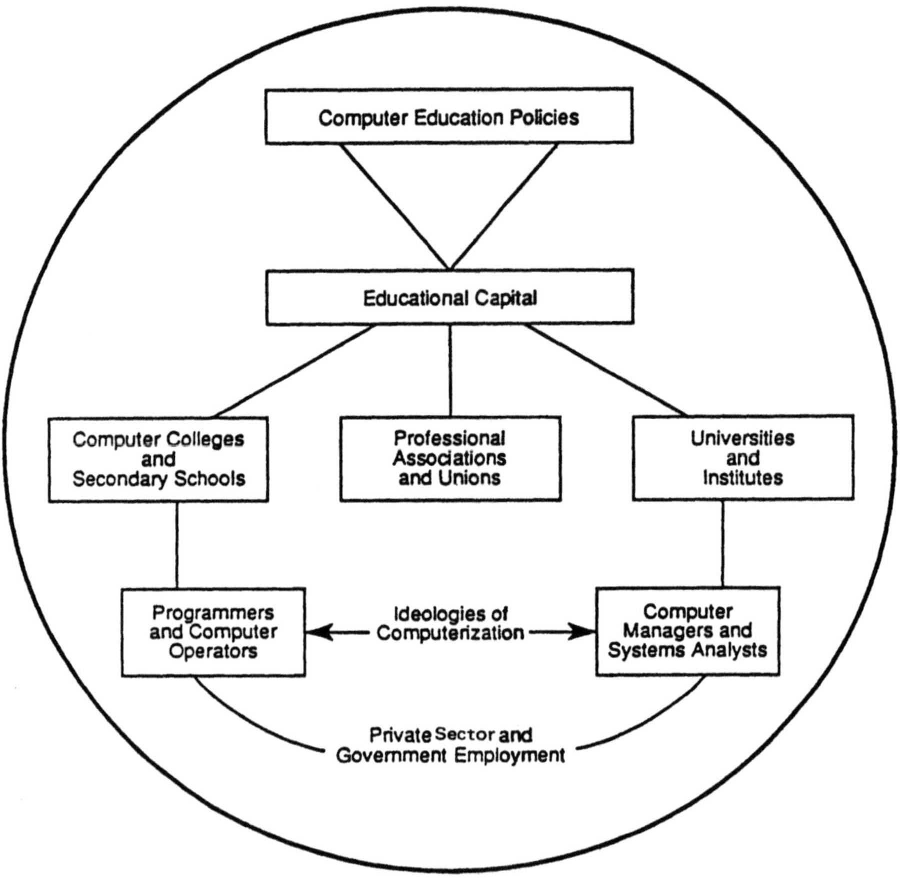

Figure 6-1. The Cycle of Computer Training.
Computer education policies frame the process of computer training and the cycle of education and employment.

The major areas of employment that have been computerized in Kenya include the automobile industry which has branched out into data bureaus, banking, finance, insurance, the retail trade, government ministries, education, and, of course, computer vending and consulting companies. Computerization in the retail trade has been sparse and limited to large-scale enterprises. The hotel and tourist industry and the airlines just began to install computers in 1984. Thus, individual pockets of the public and private sector have started to computerize without coordinated efforts. For many companies, on-the-job training and reskilling provide most of the data-entry workers. In contrast, banking, finance, and government have led the push toward computerization in Ivory Coast. The educational component of the national planning effort has resulted in coordination between government and business. Although the training of non-specialist computer workers is not planned, educational criteria and guidelines have been set for computer experts.

New computer jobs transform the character of work through the intervention of the machine. I have defined the new mechanical solidarity as the relationship between computers and people in the workplace. This relationship is not the love of the machine and its systems described by Turkle (1984:213-214) in her analysis of computer hackers in the United States. Rather, it is a view of the computer as an integral feature of office procedures, conversation, and hierarchy. With the advent of microcomputers in Africa, computer consultants have equated the new mechanical solidarity with the increasing competence and autonomy of individuals to complete work tasks that were previously difficult, dispersed, and hierarchically controlled (cf. Ingle, Berge, and Hamilton, 1983:114-118).

A labor force that benefits from the new mechanical solidarity creates an increasing need for computer training. Ideally, workers must be able to complete tasks and solve problems with the hardware and software available to them. Beyond the day-to-day mastery of work tasks, they need to be able to adapt systems and software to the local environment. For example, imported accounting packages have to be adapted to African monetary exchange and taxation standards. Although some program modification is relatively simple, much of it has to be conducted in-house. In view of this situation, one may ask what sort of local training is available to meet employment needs and what forms of certification are acceptable to employers.

These questions are not new in studies of education in Africa.

Philip Foster (1965) found that early technical and vocational education programs encountered many problems in newly independent Ghana because job training was focused on individual preparation with little attention to the structural possibilities for employment in the Ghanaian labor market. Foster (1977:358) further suggests that although a "real demand" may exist for highly trained personnel, these individuals are often underutilized or diverted into areas where their skills are of little use. In the case of Kenyan computer training, the problem of matching training and labor market demands is acute. The perceived shortage of highly trained programmers may be an empirical reality, but it is also an artifact of the underutilization of reorientation that Foster describes. To the extent that data management training is used as a stepping stone to general management, Foster's hypothesis about the diversion of job training to alternative occupational careers may be correct.

It has been argued that schools reflect and perpetuate existing systems of social inequality (cf. Bowles and Gintis, 1976:208-209). Access to training and certification is controlled as a way of creating a professional elite (cf. Weber, 1946:240-242). According to this argument, economic capital is protected by the educational system, which, in turn, develops its own form of cultural capital – the certificates and credentials of higher education.

Pierre Bourdieu and Jean-Claude Passeron (1970:90-102) present a model of education as a source of the reproduction of cultural and class structures in a society. One consequence of this process of educational reproduction is the appearance of what Randall Collins (1974:419-422) has referred to as status cultures that pose barriers of access to outsiders. The computer world may be viewed as a status culture in which familiarity with complex machines confers technological power to those who possess the knowledge. In Kenya, one shortcoming of the reproduction argument for education arises from the fact that old social structures are challenged by the creation of a new technocratic class.

In Kenyan businesses, the challenge of technological power enters into direct confrontation with nontechnical bureaucratic authority. Within the business sector, two status cultures are operative: the bureaucratic and the technocratic. When the criteria for credentialing of computer users and experts remain in flux, entrepreneurs can control technological power within their enterprises and limit the scope of authority and mobility of computer users. The deskilling of word processors in the United States and Western Europe represents an instance of such control. It is in the context of the credentialing crisis surrounding technology that the situation of computer training in Kenya may be brought more sharply into focus.

Public Training Institutions

The University of Nairobi is the major public institution that provides computer training in Kenya. An article in *East African Computer News* (1, 4, December 1985-January, 1986:20) describes the University of Nairobi's Institute of Computer Science as "one of those rare computer installations that boast of excellence in two spheres – operations and education." In 1969, the Computing Centre was established as a department of the Faculty of Engineering at the University. The first computer, an ICL 1902A, a mainframe installation, was set up in 1970 to assist with instruction in computer programming and to provide computer support services for the university's administration and researchers. Administrative records were placed on the computer, and the Centre became essential to the daily operations of the campus. By 1976, the teaching function of the Centre had expanded. It was moved to the science wing of the university at Chiromo, and plans were made for the installation of a larger mainframe computer that was eventually donated by the British government.

In May of 1979, the new ICL 2950/20, the largest mainframe computer in Kenya, was installed, and administrators planned to establish a formal curriculum in computer science. The University Computing Centre became the Institute of Computer Science, and by the end of 1981, UNESCO's Regional Office for Science and Technology for Africa (ROSTA) submitted a recommendation that UNESCO support a postgraduate diploma in computer science. The Institute's curriculum was designed to provide advanced training in computer systems and engineering for the East and Central African regions. UNESCO has continued to support the Institute by renewing annual contracts for teaching materials and computer equipment between 1983 and 1988. The Institute currently offers basic courses in computer science and information processing to over 1,200 undergraduates in commerce, engineering, agriculture, the physical and social sciences, and other departments during their second and third years of university studies. Students begin by learning to program in BASIC on Sinclair Spectrum, Olivetti, and BBC desk-top computers. They also have on-line connections to the university's ICL 2950/20 computer. More advanced programming courses are given in COBOL and FORTRAN along with training in systems analysis techniques.

In addition to the undergraduate courses, the Institute, as already stated, offers a postgraduate diploma in computer science open to the graduates of the University of Nairobi who have successfully completed one undergraduate computer programming course and to other individuals who have worked as computer programmers for at least six months outside of the university. This sequence covers one academic year spread over four terms. The courses include programming methodology, mathematical programming, software systems, business applications of computers, data base management, and computer assisted engineering design. According to Robert Scott, Professor of Computer Science and former head of the Institute, an emphasis is placed on computer education for industry and government. The Institute's placement rate for successful graduates is high. The catalogue describes the goal of the course of study as preparation of "graduates of commerce, pure and applied sciences, social sciences, and engineering for senior positions in data processing installations and computer applications departments in government, industry, and the private sector." In contrast to the program at INSET in Ivory Coast, the Institute's program is not designed to nationalize a cadre of computer experts. Direct plans linking higher education and national employment appear to be absent.

Through special seminars, reports, and faculty activities, the Institute has incorporated practical applications of computing into its program. In spite of these efforts, the Institute's image in the private sector carries a highly academic aura. Perhaps without the benefit of direct contact with the Institute, some enterprise heads and most of the computer vendors whom I interviewed stated that the Institute's training was "overly theoretical" or "too mathematical."

A 1985 article on computer training published in the *East African Computer News* reports industry's impression that university computer graduates are "inexperienced" and "tend to require another three years' training before they can become useful."[7] Unfortunately, the article does not qualify whether the computer manager cited is referring to undergraduates with basic computer courses or to postgraduates.

This problem points to an area of ambiguity in the credentialing process. University students who have taken an introductory computer course may present themselves as experts in programming when they apply for jobs. Since there is a general mistrust of computer credentials in Kenyan industry, this sort of representation perpetuates stereotypes about local institutions, even those of the highest caliber. Hence, industrial concerns and small firms cling to their selection standards and do not attempt to mesh with the training capabilities of academic institutions. This credentialing crisis is further exacerbated by the fact that computer technology is still relatively new in Kenya, and the examination process for computer specialists is not well established.

In 1983, Kenya Polytechnic, a public institution, began a course within the Department of Mathematics, Statistics, and Computing. Depending on whether a student has full or part-time status, the diploma course is designed to take two to two-and-a-half years. Kenya Polytechnic focuses on vocational training for students who are already employed and adults who wish to retool themselves for the labor market. There are currently about 200 students enrolled for the computer studies diploma. Over 100 students have already graduated from the program. The course of study includes programming in BASIC and COBOL, mathematical computing, an introduction to systems analysis, computer technology, data base management, business applications of computing, and applied statistics and quantitative techniques. For students not already employed in jobs using computers and information processing, a three to five month industrial internship is arranged. Mr. Oyomno, head of the Mathematics, Statistics, and Computing Department, described the history of the institute's computer curriculum for me.

6.6 Mr. Oyomno, Department Head, Kenya Polytechnic:
The diploma in computer studies at Kenya Polytechnic started in 1983. The idea about the course was arrived at as early as 1980. So between 1981 and 1983 // the curriculum was developed. A draft copy of the curriculum was ready by 1981 and was submitted to the Kenya Institute of Education [KIE] for approval. The approval came in early 1983 and the first intake of about twenty students was effected in May 1983. The course is lasting about two or two-and-a-half years // depending on whether the students are taking it full time or part time.

Oyomno continued to outline in detail the subject matter covered in the two-year course.

6.7 Mr. Oyomno:

The course essentially comprises six or seven subject areas. We give them a bit of basic mathematics // also some mathematics relevant for computing // Subject "0" mathematics. After that // they have the Subject "0" mathematics program. First we give them an introduction to programming languages // BASIC or COBOL. If they have taken BASIC the first year // the second year they take COBOL and also programming concepts. And we give them a bit of systems analysis. Then we introduce them to computer technology. Ideally // we are looking at the hardware parts of computers // going as low as chips. Then later on we look at how you can interface computers. How do you connect computers to printers, telephone lines, or any other physical device you want to control? The aim of computer technology is to familiarize them with the architecture, organization, and physical assembly of computers. The next area is conventional data processing. So we look at typical data processing systems. Then we take them to data base management // which is a new development // and introduce them to a few data base management systems available like D-Base II and other systems. And the next subject is called computer applications. Here we are looking at business and industrial organizations and how computers can be used to increase productivity. The next subject is called quantitative techniques // which is essentially applied statistics. We thought that this was necessary for people who are going into scientific applications // those who are going into research *like you*. And the last element of the course is the *human element*. You can be an expert working with machines but still // whenever you are working // you have to interact with people. So we have developed something called general studies. Here what we are looking at is communication skills // typical // normal communication levels // like how do you relate to your bosses // how do you relate to your peers // how do you relate to your juniors. We give them skills of how to make a project proposal and things like that. These are essentially the areas we are covering within the two years.

At the end of the first year, the computer studies students at Kenya Polytechnic take a preliminary examination, and at the end of the second year they take a standardized examination that certifies them for the diploma in computer studies. After the first year, unemployed students are placed in internship programs with business and industry. Oyomno was enthusiastic about the high quality of students trained in his program. He stated, "Employers want to absorb our students immediately after the internship, but they are encouraged to complete their diplomas first. We tell them that salaries will increase with the diploma." Kenya Polytechnic arranges job interviews with employers in government and industry as well as with the major computer vendors and distributors in Kenya. Oyomno reported positive responses from employers and a high success rate among graduates hired from his school. Managers whom I interviewed praised Kenya Polytechnic's computer studies program as "relevant" and "adequate to the needs" of their firms. They qualified, however, that an employee who had already been working as a computer programmer would consider the first year of studies to be redundant.

When the head of the Central Informatics Service of INSET in Ivory Coast praised that institute's highly trained graduates, he referred to students with advanced training in mathematics, engineering, and computer software construction. These graduates were hired as student engineers and were expected to be able to run entire companies on their own after a few years of practical experience. Oyomno, however, is describing a different type of graduate. His students are prepared to become systems analysts, programmers, and computer operators with skills relevant to the business world. They are not seen as part of a national intellectual and technocratic elite groomed by a computer institute and supported by the government. In the Ivorian case, computer training leads to professionalization. On the other hand, in Kenya, computer training in institutes and colleges produces educated workers whose leadership role in business and industry must be tested as the new technologies in the country continue to evolve. As already stated, some institutions, such as Kenyatta University College and the Kenya Technical Teachers College, offer computer studies as part of degree or certificate programs in business, education, and mathematics. Because these institutions do not confer degrees in computer studies, many company managers are unaware of their programs and generally do not recruit computer employees from them.

Secondary Schools

The secondary schools offering computer training are a potential "gold mine" for the private sector. Graduates with a basic introduction to computer programming are hired at entry-level wages and understand that they will have to be retrained by the firm. The objective of the secondary school programs is computer literacy and an introduction to programming, mostly in BASIC. Some of these training programs do not differ greatly from the commercial colleges. A crucial distinction, however, lies in the dedication of the teachers and the schools to introducing programs in computer education as a supplement to the standard academic curriculum.

The oldest secondary school computer course is that of the Starehe Boys Centre, a school that was established for disadvantaged and orphaned boys with above average scholastic aptitude. In 1980, David Wilson, an Australian educational volunteer and computer expert, organized an experimental computer course at Starehe in conjunction with the University of Nairobi and the Kenya Polytechnic. In 1982, Wilson received a donation from Comp-Rite (Apple) of a CR-7 locally assembled minicomputer and twenty-one Apple IIE's. Later he acquired eight BBC microcomputers and eight Sinclair Spectrums. Computer studies became one of the most popular electives at Starehe. Students approach their computer assignments with energy and dedication. They even give up recess and lunch breaks to work in the computer room. Pupils are introduced to programming in BASIC through self-directed study sheets. They start learning programming immediately, and only after they have mastered basic skills are they encouraged to make applications of the computer to the rest of their studies. Motivating Wilson's educational concept was the goal of making graduates competent to enter the job market as computer programmers and operators upon completing their A-level secondary school examinations. In 1984, Wilson's post was taken over by Glyn Turner, who left a computer training position in England to join the Starehe Centre. Turner continues to write and update the self-study programs for the classes and offer private instruction to pupils who wish to advance more rapidly. His hope is that computer literacy will one day be included as part of the national secondary school curriculum in Kenya (cf. Friedland, 1985:23).[8] The Starehe computer program begins at the primary school level and continues as a secondary school elective. The teachers at the Starehe Boys Centre see their computer education project as a potential model for a national curriculum in computer literacy. The Centre has been an innovator in using its graduates as instructors in special evening and summer classes for students and teachers. Returning graduates have assisted in introducing a special course in COBOL on the CR-7 computer.

Figure 6-2. Starehe Boys Centre, Nairobi, Kenya. Secondary school students at Starehe Boys Centre participate in a computer training session.

In my discussions with the managers of private businesses, the Starehe Boys Centre received uniformly positive evaluations from employers. Scott (1984:59) notes that the first group of six graduates trained in Wilson's program were employed without difficulty and that the second group of eight graduates received high marks from their employers as successful workers with advancement potential. Representatives from both the Kenyan government and industry have supported the continuation of the Starehe scheme, and the Institute of Computer Studies at the University of Nairobi has urged that the Starehe model be used as a point of departure for implementing a national computer training policy. Perhaps employers were surprised by the level of competence possessed by young Starehe graduates who were skilled without the benefit of computer diplomas and extensive job training. These graduates were just "ready enough" to be hired in computerizing firms at a price that struggling enterprises could easily meet.

In a feature article on computer training, the *East African Computer News* (1, 1, 1985b:12) referred to the Starehe Boys Centre as an "exception" in the Kenyan computer training process because "all its graduates were easily absorbed by industry." Although it is difficult to evaluate whether this program has really been exceptional in relation to others, interviews with Starehe graduates suggest that they consider themselves to be thoroughly prepared for entry-level positions in computer programming.

Brian Wray worked with David Wilson during the early days at the Starehe Centre and then joined the Aga Khan Academy as head of its educational computing project. In 1982, the Aga Khan Education Foundation sponsored a pilot study of computers in developing nations.[9] The portion of the program sponsored by the Aga Khan Academy in Nairobi was intended to improve the quality of secondary education while simultaneously studying the effects of microcomputers in the classroom (cf. Papiagiannis, 1985). Wray adopted the philosophy expounded by Papert for U.S. and African computer literacy programs. Rather than viewing the computer as a potential key to new employment opportunities, Wray conceived of it as an all-purpose educational tool. In a 1985 interview, Wray stated: "We don't teach students how to format a disk. We show them how to learn with the computer." Students begin by using the LOGO program for graphics, lessons, and computer games. They work in small groups around the computers and use the computer as a tutor for exercises in English composition, mathematics, and social studies. Full competence in computer programming is not a goal at the Aga Khan Academy.

The Aga Khan project began with two BBC microcomputers, two Sinclairs, and four Apple IIE's. Subsequently, two more Apple IIE's were purchased by the school, and there is a projected donation of several more. The Aga Khan experiment represents the beginning of a general computer assisted learning program in Kenyan schools. Because the Aga Khan focus is purely educational rather than "pre-professional," the model that Wray has established may well be adopted by some Kenyan public schools in the future despite university support for the more vocationally oriented Starehe model. Other private schools in Kenya are beginning to use computers for educational stimulation in response to the Aga Khan example.

An article appearing in *East African Computer News* (1, 2, October, 1985:23) places computer training at the secondary schools in a positive light.

> 6.8 *Learning by Computer*
> One can envision a rosy future for computer education at the Aga Khan Academy and at Starehe Boys Centre. Both institutions talk of expansion. Starehe anticipates offering courses in advanced programming languages once it can get hold of some IBM PCs. Aga Khan's Brian Wray is right now researching the next phase of the project which, following a thorough evaluation of the Pilot Project, will seed computer education in other Kenyan schools early next year.

These schools project the hope of major educational changes in new technologies for Kenya.

Commercial Colleges

The commercial colleges bring to the forefront conflicts and differences of opinion involved in the credentialing crisis. As with similar two-year commercial institutions in the United States and Europe, these colleges have mushroomed. They capitalize on the attraction and sense of novelty surrounding the new computer industry and hold forth the promise of glamorous, highly paid, white collar jobs. In Kenya, some of these computer training programs have developed within secretarial colleges that are shifting their focus to train word processors and computer operators. Other institutions such as the Computer Training Centre were started uniquely as computer and data processing schools.

It is easy to find fault with the commercial colleges because of their newness, their problems with accreditation, and their links with secretarial training programs rather than education in science and technology. The colleges are on the margins of the educational system. Scott (1984:18) claims that some of the colleges have been closely scrutinized by the Computer Association of East Africa because they allegedly "exploit the uninformed school leaver by providing a training that is not relevant to the needs of industry." Standards, however, vary across the colleges, and their educational goals differ. Some colleges focus on basic computer literacy and word processing while others offer longer and more in-depth programs, including systems analysis, computer technology, and data base management. One representative of a large computer distributorship in Nairobi stated that all of the college courses "are too long." He felt that they offer only the fundamentals of programming, usually in BASIC, over a protracted period and that curricula are "padded" from several weeks to year-long courses in order to increase the revenue of the colleges. There is, however, a history of rivalry between vendor-sponsored courses and the commercial colleges because they draw on an overlapping market of individuals who are new to computers and are seeking on-the-job training in a relatively short time.

Representatives of industry and some of the computer professionals interviewed were overtly hostile toward the colleges. One computer programmer thought that the colleges deceived neophyte students by giving them simple computer exercises that provided insufficient introduction to programming. A representative of industry referred to the colleges' practices as "daylight robbery" in the May 1985 edition of *East African Computer News*.[10] A balanced assessment of the colleges, however, should take into account their purported educational objectives in relationship to credentials and results.

A student may actually believe that a basic introduction to computers is adequate to qualify for a job that requires far more technical skill, or the student may need the computer certificate for continuation on the job, whether or not these skills are actually used at work. Confusion generated by popular images and the demand for computer training works to the advantage of the colleges in terms of their enrollment potential but to their detriment with regard to the impression that they make with industry and the private sector. Profiles of two commercial colleges help to clarify this problem.

I conducted interviews at the Computer Training Centre and Graffins College. The Computer Training Centre, established in 1974 by Mr. and Mrs. N.G. Fazal, is one of Nairobi's oldest commercial training institutions. In January 1986, the Centre moved from its cramped quarters on Kenyatta Avenue to the new and centrally located Kimathi House. The Centre occupies the seventh floor in the ultramodern building and has an enrollment of approximately 70 students. The enrollment has nearly doubled since the Centre moved to the new premises. The director, Mrs. Fazal, explains the Centre's goal as career training in the areas of computers and information processing. The Centre offers short certificate courses in computer fundamentals and programming and a diploma in computer studies after a year of courses. In addition, students take word processing courses from which they receive a certificate for the job market.

The Centre admits a cross-section of students. The major requirements are completion of A-level certification or the equivalent and passing an aptitude test devised by the Centre. Mrs. Fazal hopes to meet the needs of large numbers of students for whom computer training abroad or through vendor-sponsored courses would be too costly. Indeed, one segment of the Centre's population consists of school leavers. The Centre provides computer training for employees in private industry and business. Students are prepared to take the A.C.P. (Association of Computer Professionals) examinations in computer programming, systems design, and word processing. Apart from the overseas certificates, the Centre awards successful students with certificates of completion and letters of recommendation. Although the Centre does not have a formal job placement program, Mrs. Fazal receives many requests for referrals and matches students with local employers whenever possible. During an in-depth interview, Mrs. Fazal described the Centre's goals, courses, and examination requirements to me.

6.9 Mrs. Fazal, Director, Computer Training Centre:
Our students are now able to sit for overseas examinations. The examining board is A.C.P. // the Association of Computer Professionals. We train students in the certificate and the diploma course. The certificate course is five months // half-a-day // three hours a day. And the diploma course is four months after that. It's a very // very intensive course. Students work a lot on the computer field. They not only cover programming, but they cover programming, operating, software packages // the in thing of the day! Right. Everything they cover. And they cover systems analysis, commercial study, and so on. It's a very intensive concentrated course.

Mrs. Fazal explained that although her students would soon be able to take local examinations approved by the Kenya National Examination Council, many of them preferred to compete for the overseas computer diplomas.

6.10 Mrs. Fazal:

They've also been approved for KNEC examination // the Kenya National Examinations Council exams // which we hope will start as soon as we get the intake for that. It seems that the students go more for overseas examinations. Our students have just sat for our first examination in July. We are going to have examinations three times a year // July, September, February // every year. The examinations are conducted by the British Consul. So students sit for examinations at the British Consul. The papers come straight to the British Consul and are sent back by them. The A.C.P. course certificate is five months. They have to attend three hours a day. And the diploma course is four more months. In U.K. // there are many schools also giving the same type of examinations // because A.C.P. is an examining body. If you compare these courses, the certificate course in London costs 2,300 pounds. And plus // they have to pay their airfares and accommodations // all separate. And here is the same thing // they do it for 9,000 shillings [U.S.$566] // ummm // you can't even *compare* // you know. So it pays our country a lot of foreign exchange if you really look at it. And students end up getting a college examination because they end up sitting for all these examinations now.

A reliable external evaluation of the Centre's admission procedures is difficult to provide. Students come to the college to meet a variety of needs ranging from a short-term introduction to computers to full-scale job training. Although the Centre operates as a commercial institution, these incentives do not compromise its educational goals. The Computer Training Centre offers an introduction to programming in BASIC, RPGII, and COBOL. Students are able to work with a Wang 2200-L multi-user system with six terminals. In addition, the Centre has recently installed eight Commodore microcomputers. During their introductory course, students spend a minimum of thirty-two hours at the Wang computer under the supervision of special tutors. Instruction is divided between hands-on tutorials on the computers, formal lectures, and videotaped demonstrations on programming and data management. Students who are interested in the microcomputers are given additional practice time to work on the Commodores. Before a classroom observation session began, Mrs. Fazal toured the Centre's facilities with me and described its computer systems.

6.11 Mrs. Fazal:
The students enjoy a lot of practical facilities. We've got a multi-user Wang system, and we have installed different types of Commodore systems. We've got the plus four. We've got the 54. And we've got the new Commodore which has come out on the market // the C-128D. The students have a chance to use both // micros and minis // and they use software packages, data base, word processing, spreadsheets, graphics, everything on it. So // by the time they're finished // they don't only know programming // they know a lot more than just programming.

Figure 6-3. Computer Training Centre, Nairobi, Kenya. Students receive individual instruction during practice sessions at the Computer Training Centre in Nairobi, Kenya.

During a morning of instruction at the Centre, tutors were available to all students with special questions. I observed one tutor present for every three students working on the terminals. Students worked at their own pace and presented a final project at the end of the course. For the six-month and year courses, the final projects entailed setting up fairly complex payroll and accounting programs as well as providing sample computerized accounts, ledgers, and taxation schedules. Instructors assert that students who have completed the longer course should be capable of setting up such programs independently within a firm. The focus was on business applications of computers and on training students to use and modify existing packaged programs. Many of the final projects were thorough and impressive. The results suggest that the students who have successfully completed the longer course should have the skills and background to enter the job market as beginning computer professionals. Mrs. Fazal, however, cautioned against the overconfidence of some beginning students. "We do not mislead the students. Those who have just completed the six-month course need more training before they can apply for jobs." Although Mrs. Fazal emphasized her standards of high quality in computer education, she was candid about the differences in her students' performance.

6.12 Mrs. Fazal:

We've got a collection of students // like we have students who are company sponsored, do the short course, and go back to the company. We've got school leavers // some of them // here just to do the course for the certificate diploma. // Okay. // And others who finish the course and look for a job. When it comes to placement, we have companies that call and say // "Could we have your good students?" So we place them. We send them for interviews and they get in. But you cannot *definitely* be able to accommodate all your students in jobs // you know. It doesn't look like that in anything // you know? But those which we can place, we try to place // the good ones. I always try to take them a bit higher than what the syllabus requires // right? Our aim is not only the overseas examinations. My aim is really that my students be absorbed. And so far // the people who have taken my students are very happy. And whenever they have a vacancy // they always call me back and say // "Come on // send me someone now." So you can tell they're happy with the ones you've sent // you know?

Mrs. Fazal emphatically stated, "Where the student is trained makes a difference." She attributed her success in placing students with employers to the quality of training at the Centre and to the instructors' efforts in taking students beyond basic examination requirements.

Practical exercises with the latest equipment were a key to Mrs. Fazal's success. Although some computer experts in Nairobi claimed that the type of training received at the commercial colleges is limited, Mrs. Fazal emphasized the strict requirements and high standards of her courses. She stressed that these courses were intended to prepare students for further university education and employment rather than substituting for advanced study and degrees in computer science. From her perspective, no misrepresentation was involved.

Several other institutions offer computer curricula similar to the program at the Computer Training Centre. Among them is Graffins College, which began as a secretarial and business school. Graffins College is located in an unassuming, narrow building on the fringes of Nairobi's downtown commercial district. The college initiated its program in computer studies in March of 1986. Students currently complete a six-month course that covers an introduction to the computer, programming in BASIC, and a final programming project. The course includes 360 hours of instruction and practical work. Students are encouraged to use the computers on their own during free periods. The qualifications for entry are the same as those outlined by the Computer Training Centre. Although Graffins College does not mention an aptitude test as an entry requirement in its catalogue, Nishit Chauhan, head of the Computer Department, assured me that students must pass such a test with a score of 75% or above before they can begin computer training. Chauhan, who was trained in computer science in Britain, and his teaching partner Majid Jaffer, make computer facilities available to the Secretarial Department of Graffins College for word processing courses and to individuals specializing in accounting and sales management. Chauhan outlined the new program and its admissions requirements for me.

6.13 Nishit Chauhan, head of Computer Department, Graffins College: Well // we started this year // March of 1986 with the Computer Department. So we are still very new. We have a six-month course here. And // uh // you might have gathered from our brochure that we are offering overseas examinations. We are offering certificate level programming // diploma level programming // the overseas examination after the six month course. Of course // before we take any student into our program // we make sure that he or she passes an aptitude test // which is very important outside. So on enrollment // on inquiry rather //

we have the student taking up an aptitude test. After that // we mark the aptitude test and // uh // only those who get 75% or above are admitted. Because our computer students must have good logic. You cannot have just anybody to do computers. It's unlike the regular courses.

The organization of the courses at Graffins College resembles those at the Computer Training Centre. Although the diploma course, lasting a full year, had not yet gone into effect at the time of my interview, the shorter certificate course follows the same model as that used by the Computer Training Centre. Chauhan described the organization of the short certificate course.

6.14 Nishit Chauhan:

Our course is basically split up. Our certificate course is split up into three sections // fundamentals, programming, and practicals // accounting and all that. We introduce our students to the general applications where computers are used. And that is one meaning of practical. The other is // they use the machines practically. Because programming is not only theoretical // you see. The practical side is the better half of it. You see // what we do // our courses last three hours a day // isn't it? They work three hours every day and we split it up into one-and-a-half // one-and-a-half. They've got one-and-a-half hours lecture // theory. And the next one-and-a-half hours is practical. They go and apply the theory. There is a final individual project they work through. The final project is a programming exercise they work through. The minimum time is fifty hours // half the time for practical machine time and half for theory. They are programming in BASIC.

In Chauhan's opinion, students completing the six-month course were well prepared for the job market. Although he did not yet have information documenting their performance on the job, Chauhan emphasized that inquiries from employers were abundant.

6.15 Nishit Chauhan:

After they finish the six-month course // the chances are they are *very good* programmers because here we give a lot of machine time to our students. And // side by side with that // they have learned about a lot of applications. So I think that knowledge // they can mainly put it into a firm // probably running accounting

> packages or word processing. So // once you finish // you stand a chance getting a job as a programmer or someone who will run a microcomputer. We do have inquiries from firms to give them a particular combination of students. And // then // of course // after a student finishes // we give them letters of recommendation so that the finding of jobs is much easier. But we are very new. We haven't actually started the diploma class yet.

Graffins College follows the pattern of other commercial training institutions in Nairobi. Students are given an introduction to the fundamentals of computer use and programming with the expectation that they will move immediately into jobs where the business applications of computer technology are necessary. Hands-on training is essential to accomplish this goal. Graffins College's facilities include five Apple IIE's, two Epson Taxi personal computers, and two printers. Students learn programming in BASIC and D-Base II. Eventually, the college hopes to expand its computer offerings and facilities. In January 1987, Graffins inaugurated its twelve-month diploma course that prepares students for the A.C.P. certificate in computer programming and the diploma in computer systems design. As with the Computer Training Centre, Graffins relies on overseas examination as a major source of credentialing for its students. Whether students with a preliminary exposure to programming in BASIC will be able to pass those examinations remains to be seen.

The Graffins program, which had sixteen entering students in 1986, is still in its early phases. I visited the program just after the first examination period. The student's spirits and hopes seemed high. They were comfortable with microcomputers. The students will be computer literate and will be able to use the microcomputers in their future jobs. Whether or not their competency lives up to their job expectations, the Graffins graduates will acquire a competitive edge in the job market within a society where computer use is new and growing rapidly.

Profiles of Starehe Graduates

One test of the viability of an institution's training programs consists of the careers of its graduates. Starehe Centre graduates figured prominently among the computer personnel interviewed at data bureaus, automobile dealerships, and small enterprises. One graduate was among the programmers interviewed at the Toyota dealership. Others were employed in insurance companies, data bureaus, and government ministries.[11] One American computer consultant that I interviewed ironically remarked: "One day // these Starehe kids will run the government."

A.I., Senior Programmer. From 1977 until 1981, A.I. attended secondary school at the Starehe Boys Centre. He was among David Wilson's first pupils in the new computer program. A.I. was 25 years old at the time of his interview. While at Starehe, he concentrated on business studies and took introductory programming courses in BASIC and COBOL, a computer language with business applications. A.I. said that he became intrigued with computers at Starehe and was encouraged by David Wilson's enthusiasm. During his last year at Starehe, A.I. began working as a summer intern at the Data Centre and was hired as a programmer trainee two days after his graduation in 1981. During his internship, he was introduced to FORTRAN and RPG II. He began to learn more about writing programs and indicated his interest in the computer field. A.I. explained the value of his training at the Starehe Centre.

6.16 A.I., senior programmer, Data Centre:
I attended the Starehe Boys Centre from 1977 to 1981. After graduation // I came to the Data Centre // right after graduation // actually a difference of about two days. I had worked here for about four weeks. This was on school holidays. And // after that // you know // they started to instruct me. And actually I didn't go looking for a job anywhere else. I was kept here.

B. J.-R.: Looking back at your training at Starehe // what were some of the things that you had to begin to learn here?

A.I.: You know // initially when I was at Starehe // what we had was tier courses. That way // we did not slip behind. I did not take my A-level preparation // because instead of A-levels // I was getting computer courses. Then //

it was really // No // I should say // How can I say it? Just probably // just slightly better than going to Polytech // you see. But now that the whole thing has been changed to allow everybody at least to get the computer exposure. It has become like a curriculum in the whole school. It's not as adequate because it's less focused than when David Wilson was there. But as far as secondary schools are concerned, I think it's still on top of the rest. At Starehe // it was really more of a general introduction // because we covered so much in such a short time that we just touched on it // and then moved on to the next. When I came over here // I had to go into more specialized training because // Since I've joined this company I've gone for four trainings outside // namely special aspects of computers and programming. Mainly again // here the main thing has been more job training and experience than anything else.

B. J.-R.: What type of courses did you take?

A.I.: One course was COBOL programming writing. And another one was on-line systems // which was really an ICL type of thing // which was meant for development on a new machine. And the other two were on Data General machines.

Because he started as an intern and learned quickly, A.I. was spared the ubiquitous IQ test when he was hired by the Data Centre and was allowed to develop skills in his area of interest. He wrote programs first under supervision and then independently, and he improved his ability to program in COBOL. This project was not enough. A.I. continued to perfect his skills by taking vendor-sponsored courses in new software and equipment at the ICL training centre during his spare time. Now, after five years at the Data Centre, A.I. is a senior programmer in charge of setting up new on-line systems for Data Centre clients and organizing two-week training courses for new computer users who employ the Data Centre's services.

A.I. wishes to make progress in both the technical and public relations aspects of his work. He says that he views his job as a challenge that has spurred him to learn more about computers and telecommunications. Eventually, A.I. would like to open a computer consulting firm of his own with a focus on telecommunications. But that is still a dream. At the age of 25, A.I., a Starehe graduate, is well on his way as a high-level programmer at a major Kenyan computer firm. I asked A.I. about the content of his present job and what his future ambitions were.

6.17 A.I.: Mostly // the job is writing programs. But besides that // now I have come to a position where I can go to our clients and find out what they really want. And from them // I may now prepare my data analysis for the programs and everything else related to the job. And then I come here // maybe confer with my director // and then from there go out and implement whatever they wanted me to do.

B.J.-R.: How do you view your own future // your own career?

A.I.: It looks really positive. That's true. But // uh // maybe my own ambition is to sit in my own office and do my own thing // like this eventually.

A.I.'s ambition to own a computer business is not unusual among the young professionals whom I interviewed. He sees his training and opportunities as the keys to eventual autonomy and leadership in Nairobi's computer world. For A.I., his present job as a senior programmer is the training ground for future independence.

A.O., Programmer Trainee. In May of 1986, A.O. was hired as a programmer trainee at the Data Centre. When I interviewed him, he had been on the job for three months. A.O. was hired to assist with the Data Centre's transition from an ICL 2950/20 system to a new Data General "supercomputer." Most of his work consisted of converting and rewriting programs for the new computer. In addition, A.O. continued to perfect his skills with programming in COBOL. A.O. viewed his job at the Data Centre as an important opportunity to expand his training with computers.

After graduating from Starehe Boys Centre in 1984, A.O. spent a year working for Comp-Rite, first as a computer operator in the data-entry division and then as a control clerk in charge of coding and data entry for Comp-Rite's computer bureau. This job led A.O. to seek a better position where he would be able to obtain more experience with programming. Mastering programming in BASIC and COBOL became his goal. A.O. described his training and interests to me.

Figure 6-4. Young Computer Professionals in Kenya.

6.18 A.O., programmer trainee, Data Centre:

I have been at the Data Centre since May 8 of this year as a trainee programmer. I have been exposed to computers since 1983. You remember I saw you in 1984 when I took the computer course at Starehe. I had to stop to study for my A-levels. Then // after that // I started commercial work with computers. And // uh // I tell you // I've got a *burning interest* in computers. I just don't think I know enough. I feel every time // I must keep learning // keep training about computers. I not only do things like learning systems. I also like to teach about them. I helped teach a course at Comp-Rite. I also help to teach COBOL programming at Starehe in my spare time.

A.O. does not have a computer at home. If he did, he would certainly become a 24-hour hacker. Instead, he leaves the Data Centre at five o'clock in the evening and heads directly for Starehe where he helps to tutor students in programming and works on his new pet project, a computerized system for the Starehe library. A.O. proudly showed me the program that he had helped to develop for the Starehe library system. For him, recreation is more work with computers – teaching, learning, and assisting others. He summed up the situation: "Computers are very important. If you interact with computers // you increase your knowledge. Knowledge of computers is a very great advantage." Early in our interview, A.O. outlined his daily routine for me.

6.19 A.O.:

So // how I spend my time is this. On Mondays // I come in at eight o'clock and work until lunch at twelve-thirty // writing programs. I come back at two o'clock and work until five // Monday through Friday. After five o'clock // I go to Starehe. And on weekends // especially Saturdays // and during the holidays // I can go there several times. They have the computer room open up to eight-thirty. Yeah! And I stay until they close. I just wish I could get more computer time. All my interests are in computers // teaching, software development, and programming.

For A.O., constant work with computers is the route to the dream of establishing his own computer company. He believes that mastering the programming languages with which he works would be the first step toward realizing his ambitions. Eventually, A.O. would like to become more involved with systems design and data management. He also hopes to teach computer courses on the side both at Starehe and Kenya Polytechnic. While A.I. has moved more into management and public relations, A.O., a young hacker, is increasingly absorbed in the technical intricacies of computers. He added:

6.20 A.O.:

I would like to spend a lot of time doing software development because I want to know it. That's why I go to Starehe to work on computer programming. And there is another place. There is a small computer bureau // very // very small. The owner was just doing word processing. But // uh // he hopes to buy an IBM compatible. On that // he could run high-level programs like COBOL. So // I'm spending my full time on having a variety of things at the moment. And you asked me what I will do in the long-run. It is my wish that I do some management. Yeah! Notably systems design for the computer systems market. Yeah! I hope I can be able to organize a group of people to achieve a computer service. I want to organize data and manage a company. But that's for later. I don't wish for it now because I don't know that much.

Both Starehe graduates entered the job market with a competitive edge. They had solid computer training, but they knew that they needed more preparation on the job. They were both driven by a fascination with new technological horizons that they had to conquer, and both improved themselves through outside training and professional activities. For them, entering the computer world was not accompanied by the expectation of overnight success, but their new careers held the promise of a brighter and very different future after graduating from secondary school. As computers become more widespread in Kenya, these Starehe graduates may be the information entrepreneurs of the future. Their concerns are less with credentials than with an ability to work innovatively in a new information environment.

Although it has been asserted that high technology has made limited inroads in Africa (cf. Perrolle, 1987:208), technology's effect on social mobility is great. Within small enterprises, this process of change involves a delicate balance between computer management and administrative management. Can this influence of technological power in Africa extend beyond the workplace to society at large? Evidence indicates that there are moves in this direction in the West. Whether African nations, where computers and other high technologies are just beginning to appear, are seedbeds for the rise of a new African technology and a new technocratic elite will become clearer over the next decade. Suspicion of foreign experts may give rise to a new type of local information broker and entrepreneur.

Some critics question the relevance of computer training for the needs of industry in the Kenyan context.[12] They recommend that consistent training standards and credentialing policies be developed. Such policies are an essential piece of the larger puzzle created by new technologies in Africa. Nevertheless, a broader question remains. When training and examination criteria have been standardized, what larger educational, employment, and leadership objectives will be met? Although statistical forecasts will not be attempted here in response to these questions, a partial answer may be found in a cross-cultural analysis of the relationship of technology and belief systems to the training process.

The Conversion Experience in Computer Education

Socializing workers into the new mechanical solidarity involves the subjective aspects of working with computers in addition to the process of formal training. When they experience mastery of a technology even in a limited way, new computer users begin to change their attitudes. This attitudinal shift is characteristic of the domination stage of computer use. Mastery alters the ways in which computers users experience and describe their jobs. Part of the technological conversion experience is the worker's sense that increased knowledge leads to greater control of tasks and skills. A.O. expressed this relationship between knowledge and control when he stated: "If you interact with computers // you increase your knowledge." This increased knowledge should not be confused with the information at the computer user's fingertips by virtue of access to a data base or a computer file. Rather, it is an understanding of how computer operations and systems function in the workplace. In Chapter Two, I emphasized that this knowledge is displayed by talking confidently about the machine. Beyond talk, the computer user's heightened sense of responsibility is at issue. As the operations superintendent at a

Nairobi automobile dealership put it, "When I switched to the computer // I had to learn more on my own. It's more responsibility."

Skill and responsibility create the referential illusion of power and control. Several computer managers whom I interviewed in both Kenya and Ivory Coast began descriptions of their jobs by asserting that they were "in charge of everything." By "everything," they meant the machines *and* the people. This expanded sense of control and responsibility is part of the subjective experience of technological conversion. Since the technological structure of accountability creates an equivalence between the organization's operations and the computer's functions, responsibility for both computers and people leads to a sense of global power that cannot be achieved solely by the management of an organization. Aspiring toward titles of project heads and data processing managers, computer operators, programmers, and systems analysts look forward to the day when they will be "in charge of everything." In her interviews with children learning to use computers and with experienced hackers, Turkle (1984:218-220) remarks upon the emergence of a similar self-perception of expanded personal control. She even uses the term rite of passage when describing the hacker's initiation into computer culture (Turkle, 1984:231):

> Rituals of initiation bring novices, defined as being on the "outside" of a culture, and subject them to experiences designed to alter their identity. The novices emerge on the other side as different people, full members of the group. In some societies, rituals of initiation take the form of large collective festivals. In the hacker culture, the principal relationship is dyadic, between the individual and the computer. So the rites tend to be more private, although they are often followed by some collective expression: a Chinese banquet, a pancake breakfast. Certain experiences of an individual with a computer are invested by the group with the consequence of a passage.

Turkle's analysis deals primarily with environments in which human-computer interactions are dyadic and are not couched in the larger structure of work organizations. The expanded sense of control, power, and responsibility that I observed was a consequence of recoding job content and titles in the workplace. In work settings, initiatory experiences are often difficult to isolate, and they may appear contradictory. In evaluating research on the continuous-process form of production in automated chemical plants, Shoshana Zuboff (1988:53) remarks: "Skill and effort are no longer inextricably linked. The operator must put forward a minimum of strenuous physical exertion, but the most critical skills he or she brings to the process still depend on the subtle and inarticulate know-how that a body accumulates with years of experience." Thus, when workers perceive their jobs as requiring greater skill and more technological knowledge, what they are actually referring to is "know-how" rather than skill as such (cf. Greimas and Courtés, 1979:231-232).

In the firms that I studied, know-how became significant when it could be displayed in the organization through job titles and supervisory control. If such knowledge was not recognized, individuals often went to great lengths to make it evident by advocating changes in computer systems, representing absentee managers at board meetings, and increasing the supervisory responsibilities of their departments. When reskilling threatened to downgrade jobs, computer managers and operators responded by reconfiguring the knowledge and skills necessary to exhibit their new status. The administrator at Kenya Polytechnic defined this aspect of computer training as learning to deal with the "human element."

The subjective aspect of computer education involves acquiring skill with the technology and displaying that know-how in the work setting. The technological conversion experience is the realization that knowledge acquires meaning with reference to the machine. Working with the computer is an index event that generates the user's sense of expanded control. In the workplace, "learning," as Zuboff (1988: 395) puts it, "is the new form of labor." The following chapter will explore an interpretation of interacting with computers in instructional and work settings as a subjective and tactile experience for computer users.

Part III
Simulating Postmodernity

Chapter Seven
Indigenizing the Computer: Social and Interpretive Practices Surrounding New Technologies

> Learning is no longer a separate activity that occurs before one enters the workplace or in remote classroom settings. Nor is it an activity preserved for a managerial group. The behaviors that define learning and the behaviors that define being productive are one and the same (Shoshana Zuboff, 1988:395).

The Computer as a Tutor

Introduction of computers means that learning is no longer isolated within educational institutions. The computer alters conceptions of learning, labor, and productivity. In this chapter, I shall explore these assumptions via the microanalysis of human-computer interactions in educational and work settings in which the computer's roles as a second subject and a valued object are paramount. The narrative program for computer adoption will be employed to examine specific cases of computer adoption in Kenya.[1]

The domination stage of the narrative program for computers is characterized by the manipulation and mastery of various computer programs and lesson plans.[2] This analysis draws upon data from educational and work settings in order to explore how mastery of the computer is achieved and to examine the relationship between human-machine and human-human interaction. In contrast to discourse at the confrontation stage of computer use, in which potential applications of the computer are discussed in the subjunctive mode, the discourse of domination involves the mastery of computer languages and programs at school and work.

One young senior programmer in Nairobi characterized the domination stage as "cutting down the processing time from the time the document is prepared to the time they've got actual results to affect . . . the decisions" (cf. excerpt 1.5). This statement, however, merely provides a general description of the domination stage. Based upon transcriptions of videotaped learning and work sessions in Kenya, the features of the discourse of domination may be delineated. I have divided the videotaped data into image-discourse sets, that is, collections of dialogues and images transcribed from each recorded segment. Both nonverbal behavior, or images, and discourse are relevant to this discussion. These image-discourse sets are supplemented by ethnographic data from each setting in order to provide a sociosemiotic analysis of the processes of computer learning and use. In these interactions, the computer functions as a tutor shaping and modifying human actions.

By employing the narrative program as a point of departure for analysis, it is possible to examine contrasting types of texts and interactions that appear at each stage. Both naturally occurring conversations and written texts, arising at the same stage of the narrative program, share similar characteristics. For example, the discourse of confrontation, which consists of subjunctive accounts of computer successes and potential machine errors analyzed in Chapter Two, includes personal testimonies, media accounts, and written manuals and commentaries. In contrast, the discourse of domination and mastery encompasses both human-machine communications and human interactions specifically oriented toward computer language and the execution of programs. Public policy statements and final printouts are the products appearing at the attribution stage. This application of the narrative program has the virtue of facilitating the interpretation of verbal exchanges, written texts, media accounts, and computer texts within a single analytic model. These textual formats are outlined in Figure 7-1.

Two interesting methodological considerations arise in this type of analysis: (1) the problem of the relationship of time to text and narrative and (2) the categorical distinction between human interpersonal interactions and human-machine interactions. In one sense, a text may be viewed as a preconstituted whole. In another sense, however, a spoken text unfolds over time. In spite of elaborate conventions of transcription, it is difficult to reproduce a textual version of spoken discourse with the timing and immediacy of naturally occurring speech. The linear format of print is simply not adequate to this task. Within this discussion, a computer text is defined as a graphically produced representation including both programming commands and data entries in the format in which they appear on

Narrative Stage:	Confrontation	Domination	Attribution
Text Format:	Training manuals	Computer languages (e.g., BASIC, COBOL, FORTRAN)	Printouts
	National government plans Government and official reports	Packaged computer applications for business (e.g., VISICALC, LOTUS 1, 2, 3, TABLEUR)	Ledgers, stock controls, inventory, and budgets.
	Grant proposals	Lesson Plans	Completed lessons
	Subjunctive media accounts	Computer programs for education (e.g., APPLE ORCHARD)	

Figure 7-1. Texts and the Narrative Program for Computer Use

the computer screen. The computer transforms these initial entries into terminal representations. The texts discussed here include both lesson plans and prepackaged business programs.

The computer text appears rapidly on the screen in a matter of seconds. Capturing this process analytically is difficult. *Durée*, or lived time, intersects with linear time, and a clash of models of experience occurs.[3] In videotape analysis, using a footage counter adds another dimension of digital timing. Thus, in this exploration of human-machine interaction, we are faced with time as perceived by the human subjects; time as counted by lines of computer text; time as counted by videotape footage; and the segmentation of human-computer interactions in terms of "real" time as measured by clocking the exchanges. This distinction points to the complexity of texts as they unfold and leads to a reconsideration of problems of time and narrative. My approach to time and narrative differs from a psychological analysis of cognitive processes such as that presented in the investigation by David Rumelhart (1984:1-20) where reading time is clocked word by word in milliseconds. In contrast, my analysis focuses on longer, naturally occurring sequences of human-machine interaction that encompass reading, inputting data, waiting time, and social interaction.

The categorical distinction between human-human and human-machine interactions raises a second set of problems. In her study of computer hackers in the United States, Turkle (1984:212-213) describes the unconscious flow of thought from the human user to the machine. This flow of thought is accompanied by a psychological disengagement from human interactions and a total absorption in exchanges with the machine. The result is a trance-like state, or computer "trance-formation," in which the rapid exchanges appear to be produced by a single subjectivity. David Sudnow (1983:65) notes a similar "electric intensity" experienced by children during their first encounters with computers and video games. Turkle (1984:212) describes this type of reaction as perceived by an experienced programmer.

> Some people don't program straight from their mind. They still have to consciously think about all the intermediate steps between a thought and its expression on a computer in a computer language. I have basically assimilated the process to the point that the computer is like an extension of my mind. Maybe of my body. I see it but I don't consciously think about using it. I think about the design, not implementation. Once I know in my mind exactly what I want to do, I can express it on a computer without much further conscious effort.

The image-discourse sets drawn from the Kenyan data show that this trance-formation, or absorption in computer communications, is reinforced by the rapidity of the human-machine exchanges, most of which do not exceed eight seconds for the student subjects whom I observed in Kenya. Yet, the computer remains a second subject in this exchange and has the ability to guide, transform, and correct the communications of its human user.

Computers and the Semiotics of Invention

Umberto Eco (1976:245) defines semiotic invention as a "mode of production whereby the producer of the sign-function chooses a new material continuum not yet segmented for that purpose." The expression of ideas and human communications in the algorithmic format of a computer program is a case of semiotic invention. The reproduction of the program and the entry of data might be viewed as what Eco terms "a moderate invention." He states that a moderate invention "occurs when one projects directly from a perceptual re-presentation into an expression-continuum" (Eco, 1976:252). This projection of perception onto the computer screen creates a new form of expression with unique content. Moderate inventions may be interpreted with respect to the complexes of socially organized activities within which they emerge (cf. Collins, 1986:80-85). Thus, invention may be examined in the domains of cognition and expression and from a sociosemiotic perspective in which expression is viewed as a means of constructing a universe of discourse.

In applying the concept of moderate invention to the learning of a computer program, it is possible to examine how neophyte users move from perception to expression within a new format. For computer users, this process involves a transitional moment during which an idea is transformed and then mapped onto the computer language. Thus, we may expand Eco's model to include four steps: (1) perception; (2) transformation; (3) mapping; and (4) expression. The machine intervenes in this exchange as a respondent and as a second communicating subject. Mapping entails the full deployment of the computer program and leads to a terminal expression in the form of the text on the screen and the computer printout.

As an extension of the algorithmic format of computer programs, we may note the transfer of this algorithmic process to human transactions associated with computer use. Greimas and Courtés (1979: 12) define the concept of algorithm in semiotic terms as "the prescription of a fixed order in the execution of a number of explicit instructions aiming at the solution of a certain type of problem." They see algorithmic rules in texts as an expression of the subject's "knowing-how-to-do" (*savoir-faire*) and as a reflection of the narrative process of "doing," or the execution of an act. We have already seen the importance of know-how in defining the work associated with the computer-related job titles discussed in Chapters Two and Five.

Different levels of mastery may be achieved during the processes of transformation, mapping, and expression. Thus, for example, student users master the commands in an educational program and achieve smooth and rapid communications with the computer. Such mastery, however, does not necessarily entail an understanding of the content of the program and the background cultural knowledge that the programming content presupposes. Hence, persons involved in data entry need not understand the details and purposes of the material that they transfer onto the computer. Although this lack of knowledge may serve as a source of organizational control and order in the workplace, the absence of content comprehension clearly has negative effects in educational settings.

Theodore Roszak (1986:68) describes simulation as one of the computer's "most impressive technical capacities." He sees simulation, however, as a potential source of problems in educational settings. According to Roszak, the computer's capacity to represent a simulated, or hypothetical, universe creates an increasing hiatus between two orders of reality: the technological and the social. At the confrontation stage of the narrative program, vendors, instructors, computer programmers, and managers present the computer's models as infallible. For Roszak, this infallibility is a byproduct of the ways in which users eventually encounter and manipulate the computer's simulation capabilities. In describing the social effects of the computer's similation capacities, Roszak (1986:68) states:

> Its simulations will always be coherent and logically consistent. Of course, what it is doing may have no relationship whatever to the world outside its program. Yet, because it offers a manageable little "reality" of its own, it may tempt its user's attention away from the messy, frustrating angularities of imperfect daily life. This is especially apt to be the case where hypothetical "facts" take the place of data for the sake of the simulation.

The social reality surrounding computer use is managed by educators in order to create specific outcomes. Two practical approaches guide learning about computers: the technical training approach and the integrated educational philosophy. Technical training emphasizes the acquisition of specific data processing and programming skills without developing the user's sensitivity to the content and the range of applications of these programs. Integrated education is based upon the use of computers as tools to accomplish specific content-oriented tasks, e.g., learning physics, solving mathematical problems, or writing essays. These contrasting approaches to computer instruction appear in schools and on-the-job training of computer users the the workplace.

Papert (1980:24-25), Dreyfus and Dreyfus (1986:146-147), and Roszak (1986:72-86) address the problem of the influence of the computer's algorithmic and logical systems on how children think. Following Seymour Papert's lead, the other authors make similar arguments for the development of computer programs that encourage creative problem solving and stimulate computer users to link logical operations to the experience of work in a particular domain, to use commonsense as well as logical reasoning, and to apply programs to "real world" learning tasks. These authors, however, shy away from a description of the repetitive and ritualized aspects of learning to use and interact with computers.

In a study of computer instruction in 21 California public schools, Hugh Mehan (1985:2-3) emphasizes that different strata of the population are exposed to highly contrasting forms of computer education. Purely technical training tracks students toward certain types of vocational jobs in information processing while integrated education leads to a fuller application of computers to a variety of fields. A segmented approach to computer instruction illustrates how social stratification influences the tracking and training of computer students. Certain students are restricted to the more vocational uses of computers while others are encouraged to employ computers as a learning tool for academic subjects. Computer literacy curricula in California schools are often oriented toward training students in programming, usually in BASIC, rather than showing them how to use computers meaningfully to accomplish educational tasks. This training orientation is also evident in computer education at the Starehe Boy's Centre in Nairobi where data for this chapter were collected.

In his model for computer adoption, Mehan points out that lower class and ethnic minority students are routinely given computer drills that limit the applications of computers and the students' possibilities for working creatively with the machines. Mehan (1985:9) summarizes:

> In short, the educational approach adopted in computer literacy curricula is atomistic and stratified. The assembly of the parts of the task (machine operations) into the whole task (machine applications) comes after students master the components of the subtask. Access to advanced courses is stratified along ethnic, gender, and socioeconomic lines.

In Kenya, a similar approach has been adopted in computer training centers and commercial colleges where the objective is to train computer operators and programmers for industry and private sector enterprises. Although these individuals eventually join the ranks of computer professionals and, therefore, occupy a privileged position in Kenyan society, they are trained to begin at the lower rungs of the professional hierarchy. The type of computer training displayed in my videotapes has certain affinities with shopfloor supervision in the workplace. Students are taught the fundamentals of programming and then pursue a lesson plan with brief supervision and correction from the instructor. This shop-floor approach, along with the segmentation of the instructional format, prepares students to enter the lower rungs of the world of information processing. They are able to map ideas onto a program, but they have only a partial understanding of the content of the data that they are manipulating.

Analysis of the Image-Discourse Sets

In the present collection of data from Starehe Boys Centre, students are taught to program in BASIC through specially devised lesson plans (cf. Appendix C). The computer classroom houses 32 Apple II microcomputers and five BBC microcomputers. Interactions with the computer have been divided into four image-discourse sets that include talk about the computer, gestures and nonverbal actions, and the computer texts appearing on the screen. The actions and dialogue within each image-discourse set have been enumerated in order to facilitate analysis. Since nonverbal behavior has been included, these image-discourse sets are more comprehensive than the discourse sets analyzed in the previous chapters.

In image-discourse set 1, a student employs one of the Apple program lesson plans to test his knowledge about historical figures and current events in the United States and Europe (cf. Figure 7-2). Image-discourse sets 2 and 3 contain instructional sequences in which teachers advise students about how to proceed with their programming tasks (cf. Figure 7-3 and Figure 7-4). This brief advice takes the form of shop-floor supervision. Exchanges between the teachers and the students are rapid, fragmented, and mechanical.

Programming in BASIC, as illustrated in image-discourse sets 1, 2, and 3, involves a linear, question-answer approach to problem solving (cf. Dreyfus and Dreyfus, 1986:132-138). Answers are either right or wrong. In contrast, LOGO, developed by Seymour Papert is based on a "discovery-learning" approach that encourages students to think through the program and devise innovative strategies to solve problems. In describing his philosophy, Papert (1980:23) states:

> Learning to be a master programmer is learning to become highly skilled at isolating and correcting "bugs," the parts that keep the program from working. The question to ask about the program is not whether it is right or wrong, but if it is fixable. If this way of looking at intellectual products were generalized to how the larger culture thinks about knowledge and its acquisition, we all might be less intimidated by our fears of "being wrong." This potential influence of the computer on changing our notion of a black and white version of our successes and failures is an example of using the computer as an "object-to-think-with."

The student in image-discourse set 1, however, programs the computer to print "that is correct" and "that is wrong." Human-machine interaction proceeds according to a narrative format in which the computer acknowledges a correct or a wrong answer. The student receives a response that offers him no opportunity to reconceptualize this knowledge process. Although it encourages technical skills, this approach to learning creates a condition for the unidimensional imposition of knowledge forms mediated by the machine.

As image-discourse set 1 unfolds, the student inputs the Apple Orchard program and data (cf. Figure 7-2). The program tests the student's knowledge of historical figures who are not part of his daily cultural context. In image 1.5, the student learns that he has answered the question about "Mozart, the composer," incorrectly. He has misspelled Mozart's name as "Mnotzat" and Mortzat." One wonders exactly what the student's perception of the knowledge and of the question is. As the student moves through the processes of perception, transformation, and mapping, he handles the responses correctly as far as the computer program is concerned. That is, the question has been understood by the computer, and communication has occurred. At what Eco terms the content-unit level, however, the student has not answered the question testing his knowledge correctly. Eco (1976:252) argues that understanding a new expression in a moderate invention requires collaboration between the sender and receiver of the message. In this case, both the student and the computer operate as message senders and receivers. The student treats the computer program in a technical manner but fails to receive and interpret the message content. Therefore, he is unable to answer the question about Mozart correctly. Nevertheless, he is able to proceed with the programming lesson without any difficulty.

The student inputs materials in just seconds. His embodied mastery of computer use and the rapidity of his exchanges with the computer appear highly competent, but his comprehension of the question asked is low. After completing the question about Mozart, the student enters questions about Adolf Hitler and John F. Kennedy. The subject's move from perception to expression on the computer is complex and relies on both communicating in computer language and the mastery of cultural background knowledge. The student is uncertain about the background knowledge to be employed and hesitates while entering the data. As the student inputs data in image 1.4, another student attempts to assist him. The computer user brushes away the other student's hand without looking at him.[4] The focus of the interaction is between S_1, the computer user, and S_2, the computer as actant. To engage in a full invention process, the student would have to master both the computer program and the cultural knowledge content that it tests.

In the student's key stroke, as in the painter's brush stroke, we observe what Jean Baudrillard (1968:66) has termed "*le gestuel d'effort.*" The embodied act takes on a cultural significance. The execution and stylization of the act, however, are based upon the human gestural relationship to the object. The technological apparatus of the computer intervenes in the communication. Baudrillard views the anthropomorphic status of the "tool-object" as emerging from the gesture of effort. In turn, we may read the text produced by this effort with relationship to the student's initial gesture of working with the computer. Gestural effort is a personal, and originally a private, act. Yet, this private act enters the domain of interaction via the objects that are manipulated.[5]

James Bunn argues that through a process of synedoche, tools

Figure 7-2

Image	Video Counter #	Line	Text	Time	Commentary
1.1 Student in front of Apple Orchard program.	000		Computer + H-ET 32K	5.58 seconds	During this interaction string between the student and the computer, the student establishes the program.
1.2 Student inputs program.	008	20 30 40	Read OS, FS Print "who was" // ; DS // Input // SS // If // correction A // SS- // FS Then Print // "that is correct" // Then Print // "that is wrong" //	1 minute 4.51 seconds	The student inputs computer commands at a rapid rate.
1.3 Student concentrates on Apple Orchard program and inputs data.	032		No text available from computer screen.	3 minutes 2.43 seconds	Although student is visible, the computer screen is not.
1.4 The student is joined by another student and pushes the other student's hand away.	071		No text available from computer screen.	2 minutes 6.63 seconds	Only the student's inputting of data is visible.
1.5 Student responds to "Mozart" question.	103		I am going to test your knowledge. Who was Mnotzat ?L That is wrong. Mozart was a composer. >L	5.84 seconds	These materials were entered under image 1.3. The student's entry process was only partially visible on the screen.

(To be Cont'd)

(Figure 7-2 Cont'd)

Image	Video Counter #	Line	Text	Time	Commentary
1.6 Student enters computer commands.	104	 10 20 30 35	5 CLS Print "I am going to test your knowledge Read DS, FS Chrs (128 + RHD) (G + Chrs (14) Who was "; DS Print Chrs (128 + RHD (G >>) + Chrs (15) Who was "; DS	2.97 seconds	Student continues to enter computer commands at a rapid rate.
1.7 Student enters computer commands shown on screen.	105	40	Input SS If SS = FS Then Print "That is correct" // else Print // "That is wrong."	4.62 seconds	Student's input activities are not visible in this image. Only the screen is visible.
1.8 Student inputs data for questions on the program.	106	60	Data Mortzat composer, Hitler, Dictator	4.47 seconds	Student's data entry is not visible. Only the screen is visible.
1.9 Student continues to input data.	108-113	70	John F. Kennedy, USA President [Additional new lines on computer screen not visible.]	19.30 seconds	Data are on the screen. Visibility of the screen begins to fade out at the end of this image.

Figure 7-2. Image-Discourse Set Number One

represent the social and cultural worlds in which they are used. Office machines offer a model for the cultural environments in which they appear. Bunn (1981:24) states:

> Seen as reflections of the end itself, the principles by which a tool is constructed may be construed as hieroglyphs, omens, signatures, symptoms, laws, or models of higher function. Our definitions therefore are perspectivist; depending upon the oblique perspective given a semiotic object, it becomes sign, tool, or model.

I shall return to and expand this conception of the tool as sign and model in a discussion of computer languages and their applications as modeling systems.

In image-discourse set 2, a British teacher supervises the lessons of several Kenyan students who are seated before a row of computer consoles (cf. Figure 7-3). The teacher enters the room in an informal manner. He responds to each student's questions with brief, indexical answers and monitors the computer screens. The students are then left to practice on their own. These brief and truncated human interactions closely resemble the subjects' transactions with the computer. According to my segmentation of the interactions in image-discourse set 2, the transmission of a simple line of computer text, e.g., one minimal communication between S_1 and S_2, falls in the same time range as a very short human interaction sequence.[6] The timing of these short interaction sequences falls between four and eight seconds. Data manipulations, students' reviewing of lesson plans, and expanded discussions of problems with the computer take much longer in this data sequence, e.g., between 30 seconds and three minutes. In the cases of reviewing lesson plans and discussing problems of comprehension, the variability is great. Nevertheless, a simple question-answer instructional sequence in this particular computer classroom roughly follows the same pattern and timing as a computer questioning sequence. This similarity is not fortuitous. It results from the modeling of human interaction in the format of human-computer exchanges. The transcript for image-discourse set 2, included in Figure 7-3, illustrates this point.

In image 2.3, the teacher instructs the student to move the cursor all the way down and leave a space. The indexical reference to the cursor as "it" demonstrates how the discourse is embedded in the student's technical activities. In image 2.6, the teacher similarly alludes to the activities in process and says, "Yes, same // same." He shows the lesson plans to the students and refers them back to the instruction

Image	Video Count #	Text	Time	Commentary
2.1 Teacher comes into view in front of computer consoles.	400	No text.	12.16 seconds	Teacher enters the room in order to supervise students as they work on computer console.
2.2 Student 1 walks past teacher and approaches computer console.	402	Student 1: Hi. Excuse me. May I have some help, sir? Teacher: Yeah.	7.96 seconds	Student greets teacher and requests assistance.
2.3 Student 1 sits down in front of computer.	403	Teacher: There you go. (Gives student lesson plan.) Student 1: How do I do this? Teacher: Move it all the way down and leave a space.	25.04 seconds	Teacher gives student a lesson plan and brief advice.
2.4 Student 2 looks up at teacher, and teacher responds with advice. Teacher gives student a lesson plan.	405	Student 2: I want to place something here. Teacher: Yes. What? (Teacher looks at computer screen.)	30.06 seconds	After clarifying the student's problem, the teacher leaves him with a lesson plan.
2.5 Teacher leaves student 2.	413	No text.	4.15 seconds	
2.6 Student 3 beckons teacher, and teacher bends over to respond to student's question.	414	Student 3: I want to place something here. Teacher: Yes, same // same.	1.50 seconds	Teacher proceeds to repeat a similar instructional sequence with student 3.
2.7 Teacher and student 3 look at the computer screen together.	415	No text.	8.76 seconds	Teacher reviews the progress of student 3.
2.8 Teacher leaves the setting.	416	Teacher (to BJ-R): Interesting, yes.	4.27 seconds	Teacher leaves the students at the computer console and talks to the researcher.

manual. The emphasis is on practical activity rather than on elaborate verbal instructions.

These instructions are indexical statements that rely on the sender's and receiver's background knowledge as a basis for filling in the content of the expression (cf. Schwartz and Jacobs, 1979:256-259). In contrast to the more elaborate discourse on the hypothetical benefits of computers at the confrontation stage, these instructions depend completely upon contextualized human activities with the machine and on the execution of specific aspects of the computer program found in the lesson plan. As already stated, the human-human interactions are not only influenced by the machine, they also begin to assume a mechanical and ritualized form. This mechanization and truncation of human interaction is characteristic of the new mechanical solidarity discussed with regard to educational settings in Chapter Six.

Image-discourse set 3 involves the use of an educational graphics program in a computer class at the Starehe Boys Centre (cf. Figure 7-4). This excerpt re-emphasizes the importance of working with the computer as an embodied act of learning. Demonstrative gestures emerge as a feature of the instructional communication between the teacher and the student. The transcript included in Figure 7-4 records the teacher's monitoring processes and the students' concentration on the computer commands as the students learn to input a graphics program. The objective of these activities is to create a colorful geometric design on the screen.

The indexical features of human-machine communication are particularly evident in this videotaped segment. During this exchange, a Kenyan teacher monitors the progress of several students working independently on the graphics program. He times their work and gives them advice about programming commands. Throughout this learning session, the computer text is not completely visible on the researcher's screen. In image 3.3, the teacher reaches over the first student's head, leans forward, and points to the text on the computer screen with his left hand. At this juncture, the teacher uses a demonstrative gesture and briefly touches a line of text on the screen.

Gestural illustrators are part of a class of nonverbal demonstratives used to accent and segment a verbal communication (cf. Ekman and Friesen, 1969:49-98). According to Ray Birdwhistell (1972:338), demonstratives are referential motions employed in the process of gestural mapping.[7] In this discussion, a distinction will be drawn be-

Figure 7-4.

	Image	Video Counter #	Text	Time	Commentary
3.1	Teacher consults and sets his watch.	000	Teacher: Just one second.	6.22 seconds	Teacher appears to be preparing to time the work of student 1 at the computer.
3.2	Student 1 inputs commands for a graphics program.	001	Text not visible in detail on the computer screen. No verbal exchange takes place.	3.76 seconds	
3.3	Teacher approaches the computer console at an angle, extends his arm over the head of student 1, points to the computer screen with four fingers of his left hand, and actually touches the computer screen briefly.	002	Teacher: Two point zero, zero (2.00). Student: What? Teacher: Then // two point zero, zero (2.00). Teacher: What is that? (Teacher looks at computer screen.) Student: I don't know. (Student shrugs shoulders.)	9.17 seconds	Student 1 has stopped inputting data in order to consult with teacher. Teacher and student exchange information about the computer commands appearing on the screen.
3.4	Teacher points to the lesson plan on the table in front of the computer console.	005	No text appears in this section.	9.27 seconds	Teacher corrects student 1 by pointing to the lesson plan.

(To be Cont'd)

(Figure 7-4 Cont'd)

Image	Video Counter #	Text	Time	Commentary
3.5 Student 1 leans forward and taps the computer screen gently with the index finger of his left hand.	006	No text is clearly visible on the computer screen. There is no verbal exchange.	4.46 seconds	Student 1 attempts to comprehend the computer commands by touching the screen. This process is considered to be a tactile grounding of knowledge about the computer.
3.6 Student 1 removes left hand from the computer screen and places it on the keyboard prior to resuming work. Student 1 inputs commands that appear on the computer screen.	007	Commands are not fully visible in detail. No verbal exchange occurs at this point.	5.16 seconds	As student 1 prepares to resume his lesson, student 2 continues to input computer commands.
3.7 Student 2 disengages his hands from the keyboard and concentrates on the lesson plan.	008	The computer screen is visible only at a distance. The lesson plan is not visible.	6.57 seconds	Student 2 leans down and concentrates on the lesson plan before resuming his work. At the end of his sequence, the camera moves away from student 2.

Figure 7-4. Image-Discourse Set Number Three

tween gestural mapping and what I have identified as tactile grounding, or touching the computer screen in an effort to clarify some aspect of the programming or data-entry procedures.

In image 3.5, the student engages in tactile grounding when he taps the computer screen gently three times with the index finger of his left hand. In contrast to the teacher's illustrative movement, tactile grounding is a private act not intended to communicate to others about the computer text. Tactile grounding appears to be an extension of the student's act of concentration.

Thus, when contrasting the gestures of images 3.3 and 3.5, it appears that the teacher's demonstrative gesture is intended to instruct while the student's private touch provides a means of establishing his personal relationship with the computer. Both the student and the teacher use their left hands when pointing to and touching the computer. In Kenya, even in mundane human interactions, use of the left hand is considered antisocial and denotes the sphere of the profane. The right hand is reserved for the polite social gestures of handshaking, greeting, and offering as well as for ritual acts. Use of the left hand when pointing to the computer suggests that the machine is, in some senses, still alien to the student's cultural environment and is part of a taboo domain. Thus, the computer is a "sacred-profane" object. The act of touching the screen breaks through the symbolic barriers between the subject and this new, alien tool-object. Yet, the student who is still in the early stages of mastering this object and is not a technician of the sacred, tenatively approaches the object with his left hand.

Tactile grounding is not gestural mapping but is instead a type of relational mapping in which the student establishes a working interaction with the computer. Both gestural and relational mapping are essential to the process of moderate invention delineated in the opening sections of this analysis. These nonverbal mapping procedures facilitate the student's expression in the format of computer language by expanding his competence and embodied experience with the computer as a new learning tool integrated into the daily work environment.

The subject's usage and perceptions of time in image-discourse set 3 are worthy of comment. The tactile grounding transaction in this segment lasts 4.46 seconds, which is approximately the same time as a short computer inputting sequence in image-discourse set 1. The teacher takes 6.22 seconds when he says to the student, "Just one second." The student-teacher interaction lasts about nine seconds in each case and roughly corresponds to the timing of interactions between the Kenyan students and the British teacher in image-discourse set 1. The Kenyan teacher and students, however, are much more involved in gesturing as a feature of their communications about the computer than is the British teacher. These gestures, including both demonstratives and tactile grounding, appear to incorporate the computer into the subjects' physical grasp and domain of comprehension.

Four categories of nonverbal behavior emerge in this transcribed segment: (1) direct engagement with the computer during the data-entry phase; (2) demonstrative gesturing; (3) tactile grounding of the computer text on the screen; and (4) disengagement, or pulling away from the computer in order to concentrate on programming and data-entry problems. In images 3.6 and 3.7, student 1 and student 2 pull away from their activities of gesturing and inputting data respectively to ponder the program and decide on a course of action. Disengagement also takes place, as will be seen in image-discourse set 4, when computer users are waiting for the machine to perform certain operations. "Down time," or waiting time, is accompanied by backchannel gestures such as holding the chin, tapping fingers, and other forms of self-involved waiting behavior (cf. Duncan, 1972: 286-288). The photograph in Figure 7-5 shows student 1 pausing to reflect upon how to proceed with his computer lesson.

Based upon the complexity of the social organization and the computer user's familiarity with the programs employed, applications of the computer to work settings differ from educational uses of the computer as a tutor. Image-discourse set 4, recorded at the T.D. automobile dealership in Nairobi, illuminates some of the differences between work-site and educational uses of computers (cf. Figure 7-6). Seated before an IBM 3400 terminal, the computer operator explains the dealership's inventory, purchasing, and sales procedures to the researcher via a series of computer displays. In Chapter Five, the computer system at T.D. was described as an enterprise facilitator and a technological partner in business expansion. The degree of computer integration that has been achieved at T.D. is reflected in the ease with which the computer operator delineates the machine's functions through words and gestures.[8]

As with the case of the Kenyan teacher in image-discourse set 3, the computer operator uses demonstrative gestures to illustrate the importance of certain program entries appearing on the screen. The computer operator employs his right hand to point to and touch the screen. His self-assured, demonstrative gestures are characteristic of the final phases of the domination stage of computer use in which technological mastery has already occurred and a clear positive value is attributed to the computer. The computer is no longer a threatening sacred-profane object. It is now a valued object manipulated by one of the new technicians of the sacred (cf. Figure 7-7).

In image 4.2, the computer operator points to the IBM screen with a ballpoint pen held in his right hand. He traces a line of computer text using the pen as an indexical extension of his body. In image 4.4, the full text becomes visible on the researcher's video screen. Twenty-four entry lines are displayed for invoices, stocktaking, and

Figure 7-5. A Student Programming Session at the Starehe Boys Centre. Student pauses to work out a programming problem during a lesson at the Starehe Boys Centre.

accounts. With this display clearly in view, the computer operator touches the screen using three fingers of his right hand. In image 4.5, he withdraws his hand, rapidly followed in image 4.6, by placing his index finger on line 10 of the 24 item display. At this point, the operator gently moves his index finger up and down over line 10, "Batch Order Stacking." This movement appears to be a cross between a demonstrative gesture and tactile grounding through which the computer is integrated fully into the operator's domain of control. While gesturing in image 4.6, the computer operator begins to explain to the researcher how invoices, accounts and purchase orders are prepared by the dealership. When he completes this verbal explanation in image 4.7, the operator, depicted in Figure 7-7, removes his hand from the screen and punches in a command for the sales menu. While waiting for this display to appear in image 4.8, the operator taps his fingers gently on the side of the keyboard and concentrates on the screen. Once again, we see an example of momentary disengagement during a computer operation.

The timing of computer entries and human-machine interaction during this segment is not representative of actual work time. The computer operator is instructing the researcher about the programs and data-entry procedures used at his company rather than inputting new data into the computer. Although it is not possible to make an accurate extrapolation of work time from this informal instructional sequence, the details of the program menus suggest that data-entry and waiting time at T.D. might be somewhat longer than it is for the students' brief programming sequences in image-discourse sets 1 and 3.

Moreover, the computer operator at T.D. works with a packaged program for invoices and accounts. His job does not entail tentative searching and trial-and-error experimentation in the manner of the students' programming efforts. Therefore, his actual work process flows more smoothly and is characterized by longer sequences of concentrated effort. This smooth flow is a feature of the competent mastery of a program application in active use at the domination stage of the narrative program for individual work with computers.

Computers and the Discourse of Domination

In Chapter Two, the discourse of confrontation was analyzed in terms of the social practices and relationships surrounding talk about computers when they are first introduced into a setting. This talk is part of the surface structure of the narrative program for computer adoption. It is the manifestation of the narrative program within day-to-day

Figure 7-6

	Image	Video Counter #	Line #	Text	Time	Commentary
4.1	Computer operator sits in front of the IBM 34 keyboard and terminal. He begins to explain the use of an inventory and sales program display to the researcher (B. J.-R.).	000		Details of the program display are not yet visible on the computer screen.	10.28 seconds	Computer operator prepares to explain the display emerging on the screen to the researcher (B. J.-R.).
4.2	Computer operator uses his right hand to point to the IBM screen with a ballpoint pen.	003		Computer text is only visible at a distance on researcher's video screen.	5.75 seconds	Computer operator uses the ballpoint pen to point to several lines of text. This demonstrative act is a gestural illustration of the information communicated.
4.3	Computer operator withdraws hand from IBM screen.	004		No text appears in detail on the video screen.	6.29 seconds	

	Image	Video Counter #	Line #	Commands	Line #	Menu Invoice	Time	Commentary
4.4	Computer operator touches the IBM screen with the first three fingers of his right hand.	006	1	REAL INVOICE ENTRY	13	D/O ENTRY	1.73 seconds	The full set of commands for invoicing, stocktaking, and sales appears on the screen. Computer operator points to the commands.
			2	- - - - - - - -	14	D/O ENTRY		
			3	RECEIVING	15	- - - - - - - -		
			4	D/O MAINTENANCE	18	- - - - - - - -		
			5	P/O MAINTENANCE	17	- - - - - - - -		
			6	STOCKTAKING FILE	18	- - - - - - - -		
			7	PRICE CHANGE DATA ENTRY	19	- - - - - - - -		
			8	ORDER PLANNING FILE UPDATE	20	- - - - - - - -		
			9	REAL ORDER STACKING	21	- - - - - - - -		
			10	BATCH ORDER STACKING	22	RAP CHECK		
			11	INVOICE RELEASE MAINTENANCE	23	BRANCH STOCK INQUIRY		
				- - - - - - - -				
			12		24	INVOICE SHEET ALIGN		

(Figure 7-6 Cont'd)

	Image	Video Counter #	Line #	Text	Time	Commentary
4.5	Computer operator withdraws hand from IBM screen.	007		The computer display appearing in image 4.4 remains on the screen.	2.81 seconds	The invoicing commands remain on the screen as the operator prepares to explain them in more detail.
4.6	Computer operator points to line 10 of the display on the screen with the index finger of his left hand. While he explains the program's use, he holds his finger on the screen for 2.2 seconds.	008		Computer display remains the same as in image 4.4. Computer operator: The account customers // We have statements // monthly statements and // sales ledgers. So we need to have the invoice and account numbers. We have account numbers. We have account numbers when we are selling to them, but at the same time we identify account customers by account codes.	7.07 seconds	
4.7	Computer operator withdraws hand from IBM screen and inputs the command to change program displays. He continues to explain the invoicing and sales process to the researcher.	010		Computer continues to display the 24 entries found in image 4.4. Computer operator: So they send a P.O. We use that code so automatically it gives them a discount, the particular discount. BJ-R: Hmm // So // Yeah // Their code would tell you how much discount they would receive.	9.29 seconds	Computer operator prepares to select a new entry in order to demonstrate to the researcher how sales and discounts work.
4.8	Computer operator waits for the program to change on the IBM screen. While he is waiting, he taps his fingers on the side of the keyboard.	013		A program for sales appears on the IBM screen. Details of this program are only partially visible on the researcher's video screen.	9.31 seconds	Computer operator keys in a new display and prepares to explain the company's sales procedures in more detail to the researcher.

Figure 7-6. Image-Discourse Set Number Four

Figure 7-7. Computer Operator at Work.
Computer operator inputs inventory and stock data at a car dealership in Nairobi, Kenya. His work procedures involve tactile grounding of the computer environment.

interaction. Confrontation is superseded by mastery, or the domination stage of computer use. As a feature of the narrative program, domination also manifests itself in terms of discourse. The activities of competent computer users afford a glimpse of the discourse emerging at this phase of the narrative program. The discourse of domination has four components: (1) mastery of the machine through the use of a programming language; (2) the user's comprehension of the content of the program; (3) the execution of the program as a work sequence; and (4) the human interactions of instruction and supervision that surround the work sequence. Hence, domination refers to both human mastery of the machine and to the hierarchical social relationships that develop in settings where computers are used.

At the confrontation stage, discourse about the computer's possibilities occurs with reference to hypothetical situations of human-machine contact. The director of marketing from a large computer company in Nairobi illustrates this point.

R.T., director of marketing, ICL:
Well // we look at which organizations there are within various size categories by segments. What computer do they have // if any? If they haven't, why not? What their plans are // etc. It gives us a feel for what's inhibiting people from moving, what their development plans are (cf. excerpt 2.7).

Confrontation stage discourse involves "wanting to know" about computers as opposed to "knowing how" to use them. The two subjects, the human user and the computer, merely exist in a state of copresence, or hypothetical encounter. By using a semiotic diagram, confrontation was represented in Chapter One as the relationship between subject$_1$ and subject$_2$: $S_1 \longleftrightarrow S_2$. The computer is only vaguely projected to become a valued object at some future point. Much of the discourse of confrontation is intended to convince people that they should use computers to solve management and bureaucratic problems. Elliptical computer success stories are presented to persuade potential and neophyte users of the computer's power and effectiveness.

The discourse at the confrontation stage frames the computer as the final solution to problems of efficiency and productivity in the workplace. This type of discourse does not separate knowledge of technical procedures from problems of content expression and comprehension arising during computer use. As we move from confrontation stage discourse to the discussion of domination and mastery, problems of articulating the procedural, expressive, and content levels emerge with more clarity as they apply to specific cases of computer learning and use. The analysis in this chapter has emphasized the complexity of problems of content comprehension in human-machine and human social interaction.

In the discourse of domination, the meaning of the computer derives from the conjunctive relationship established between the human subject and the computer on actual occasions of the machine's use. This relationship implies human mastery of the machine and does not anticipate the machine's domination over the human environment as the social setting is remodeled around the computer and its functions. Mastery has been diagrammed by the following semiotic formula: $S_1 \longrightarrow S_2$. The computer is valued for its actual rather than virtual utility. Its meaning is dependent upon the gesture of effort, or the entire process of embodied interaction with the computer. The case studies presented in Chapters Four and Five demonstrate that the domination stage takes a variety of forms. It may be partial or incomplete, as the case of the retail trading company discussed in Chapter Five illustrates. In this case, workers had mastered the old computer system, but the system was too cumbersome and slow to meet the requirements of the business. Thus, although the computer controller's relationship to the computer could accurately be diagrammed as $S_1 \longrightarrow S_2$, within the context of the company, domination and integration of the computer system had not been achieved. This situation contrasts with the outcome at T.D. where computer and enterprise functions meshed.

The human interactions that form part of the domination stage are oriented toward regulating gestures of effort through instruction and supervision. Talk at this stage requires knowledge of the computer's programs and capabilities. In sumarizing what he had learned at the Starehe Boy's Centre, A.I., the Starehe graduate and senior programmer profiled in the last chapter, stated:

Senior Programmer:

> I remember the best thing I found about it is anytime anybody is talking about anything in computers anyway // you're not really lost. You probably might just have an idea what they're really talking about (cf. excerpt 2.10).

Whether it takes the form of description, instruction, or supervision, discourse at the domination stage involves "talking confidently" about the machine, based upon direct and detailed experiences with its operations. This expert talk is cryptic and elliptical. It may be played out in the shopfloor-like training interactions appearing in image-discourse sets 1, 2, and 3 and in actual shopfloor interactions in the workplace as evidenced in image-discourse set 4. Although this discourse illustrates human control over the computer, it also describes ways in which human interactions conform to the demands of the machine. Learning does not stop during the domination stage.[9] Computer users master new and more complex operations, programs, and systems. The "burning interest" in computers that one programmer trainee expressed emerges during the domination stage as a desire to learn more and to talk more knowledgeably about the machine. This characteristic of the discourse of domination and mastery leads to a unique expression of the politics of signification in human-machine interaction.

Learning to Labor with Computers

Analyzing the role of computers in business operations, Zuboff (1988: 393) states: "The electronic text becomes a vast symbolic surrogate for the vital detail of an organization's daily life." My case studies from Kenyan automboile dealerships, retail companies, and insurance firms have underscored this point. Basic business procedures and functions are part of the computer system, and the resultant texts are not merely documents containing abstract knowledge. They are records of crucial consequence to the firm. Without payroll, inventory, and stock control data, these organizations could not function. We have seen how entitlement to read and produce these documents transforms the structure of an organization and increases the status of computer managers and operators within the bureaucratic hierarchy.

Nevertheless, researchers studying the computerization process have remarked on the increasing segmentation of work tasks (cf. Crompton and Jones, 1984:58-59), the impersonalization of work, and the widening social distance of computer managers from other employees and from customers (cf. Zuboff, 1988:382-383). Although the hypothesis of increasing distance is open to question based on accounts from computer workers, a knowledge gap emerges when computers are installed. This knowledge gap separates computer users from non-users and pushes computer users to perfect their technological expertise. Michael Heim (1987:131) refers to this knowledge gap as "system opacity," which he defines as "the fundamental disparity between the user and the engineered setup of the interface." This disparity derives from the complex and hidden technical features of the computer system.

System opacity masks the full operations of the computer from its most skilled users. I have noted that at some of the technical training institutes described in Chapters Four and Six, class sessions begin with opening up the computer. This revelatory act supposedly unveils the intricacies of the computer's operations for neophyte users. In fact, however, this act may be more symbolic than informative. It initiates computer users into the long road of learning that will ensue and reinforces system opacity by revealing to computer users just how complex the machine really is. Learning becomes more difficult than the static semiotic formula $S_1 \longrightarrow S_2$ suggests, for it entails, for it entails coming to terms with the surface operations of the machine in full knowledge that the computer's internal functions remain opaque and mysterious.

This opacity inspires a reinterpretation of Zuboff's concept of "distance" with reference to the psychological distance between the computer user and the machine. The instructors and students at the Starehe Boys Centre attempted to cope with this distance by touching and talking about the computer. Tentatively, they reached out with their left hands to embrace the computer and its knowledge – to bring the computer within their embodied sphere of action. The computer operator at T.D. sought to embrace the computer within his sphere of knowledge and action by reaching out with his right hand in order to demonstrate mastery and control. Both the students and the experienced computer operator delimited their arenas of control by focusing on specific programming applications. The students prioritized the computer program over its content. They mastered the program's functions, but the content was restructured in order to encompass only a small area of the business operations. The problem of system opacity was minimized in both cases because the students and the professional computer user did not seek to go beyond a limited set of computer operations.

By requiring mastery of data bases and programs that are public and shared, computers alter the structure of learning in organizations. This process was emphasized in Chapter Six as part of the discussion of a new cybernetic orality that removes documents from the page and places them on the screen for all to see. Although this new orality challenges hierarchies and control in learning and work environments, computers do not destroy hierarchy.[10] Instead, a modified relationship between learning and labor emerges. Improved mastery of the computer becomes a goal of the worker as well as the student.

This mastery is the basis of both job performance and entitlement. When the trainee programmer from the Starehe Boys Centre stated, "I don't know that much," he was referring to both his knowledge and his position in the firm (cf. excerpt 6.18). Knowing more about programming was, for him, more than an academic exercise. Increased knowledge would lead to job advancement and would change his position in the firm. Learning and labor are interconnected in the cycle of computer mastery. One of the consequences of mastery is a positive computer attribution ($S_1 \longleftarrow 0$), in which the computer user is confident of his or her ability to manipulate the valued object. Zuboff (1988:395) characterizes this situation when she states: "The behaviors that define learning and the behaviors that define being productive are one and the same." This statement is correct up to a point. Learning leads to skill and mastery. The display of mastery supports recognition at school and entitlement in the workplace. There are, however, ceilings on both educational and job mobility. For example, in the Ivorian case, we saw how highly trained computer engineers entered the job market as trainees and student workers. Learning provided the point of departure for job entitlement and increased status. It did not substitute for job experience or challenge the new technocratic hierarchy arising in computer firms where the management of machines and of people have become two keys to success.

The Politics of Signification in Human-Machine Interaction

The computer program is a modeling system that controls both the texts produced on the screen and the human actions involved in computer use (cf. Sebeok, 1976:23). This type of modeling system creates a second language and an ideological model of the world based upon a primary natural language. In computer programs, the secondary modeling system includes commands and functions as well as a text to which the human user responds. Constraints are placed on the text by the computer system. In turn, in order to manipulate the language, users focus their gestural and cognitive activities in the particular ways outlined in my discussions of computer trance-formation, demonstrative gesturing, and tactile grounding.

Systems analysts recognize that the processes of signification in computer communication and human social organization are homologous. The politics of signification involve a conscious manipulation of technological activities in order to reinforce social control. Human-computer interaction encompasses a variety of cognitive processes and gestural manipulations. Mastery of these computer-related procedures and manipulations leads not only to a positive valorization of the computer but also to a utopian discourse about the range of its capabilities. Thus, computers may be perceived as creating bureaucratic efficiency, social order, and harmony in the workplace.

Attributing meaning to computers based on how they actually function begins during the mastery phase of computer use and is carried over into attribution. The semiotic formula $S_1 \longleftarrow 0$, used to describe the relationship between the human user and the computer at the attribution stage of the narrative program, may be interpreted in two different ways: (1) with respect to individual involvement with the computer and (2) with regard to the organizational environment in which the computer and its users become part of a larger hierarchy of technological and social control. As illustrated by the use of demonstratives and tactile grounding, what appears to be a private gesture with the computer is simultaneously a source of public accountability and social control. This shift of interpretive levels (*débrayage*) is one of the controlling processes issuing from the presence of the computer in the work site. The computer is more than a passive valued object in this semiotic exchange. It too responds and dominates through its texts, technological requirements, and the adjustments to its presence in the workplace.

Processes of social control are further reinforced by the record-keeping capacities of computers. Individual entries appear as though they are discrete and isolated. These individual entries and texts, however, assume a different meaning when they are compiled as part of an organizational dossier (cf. Garfinkel, 1967:186-190). The computer's capacity to store information transforms serial entries into an account of the student's performance or the worker's productivity. Thus, once again, we see two forms of expression: the individual gestural act and the serialization of the act in its holistic representation as a computer text. The politics of signification in computer communication involve the transfer and manipulation of these two forms of expression.

The politics of signification ultimately relies upon a theory of terminal signs. The computer is a valued object (0) in its capacity as a partner and a controlling agent in human-machine interaction. As human subjects come to see themselves reflexively via their use of the computer, a terminal attribution of value takes place (cf. Gremias, 1970:175). With regard to the early computer-learning question about Mozart, the student's terminal narrative utterance will occur when he realizes that two types of knowledge and mastery are involved. Then he will become aware of the cultural background knowledge required to execute computer programs and operations successfully. In conclusion, the student's response to the Mozart question will be complete only when he recognizes the impact of the model of postmodernity and Western culture that the correct computer answer requires him to assimilate.

The following chapter examines the computer as a valued object at the final stage of the narrative program for computer adoption. The contractual exchanges involved in computer use will be explored as part of a sociosemiotic analysis of computer adoption. Data derived from public policies, educational settings, and the workplace in Kenya and Ivory Coast will be reviewed as background for the analysis of terminal technological signs.

Appendix C: Lesson 1 in the Starehe Boys Centre BASIC Programming Sequence

LESSON 1

(BBC : CAPS LOCK is lit) (Apple: CAPS LOCK is down)	Type in this:	Before pressing RETURN write down what you think will be displayed.	Tick...If correct, also write down what was displayed.
01	HELLO	not BASIC	
02	PRINT HELLO		
03	PRINT HE		
04	PRINT H	numeric variable	
05	PRINT "H"		
06	PRINT "HE"		
07	PRINT "HELLO"	character string	
08	PRINT "HELLO"		
09	PRINT "HELLO"		
10	PRINT "HELLO"		
(Release the CAPS LOCK;) 11	PRINT "hello"		
(Use the SHIFT key) 12	PRINT "hello"		
(Ensure the CAPS LOCK is down) 13	Let H=7 = (press RETURN) PRINT H	Always press RETURN or : after each statement.	
14	HOME (or CLS for the BBC)	Clears the screen but not memory.	
15	PRINT H+2		
16	INVERSE (Apple only) * PRINT H-8		
17	FLASH (Apple only) * PRINT H *3		
18	NORMAL (Apple only) * PRINT H *H *H	Same as H^3	
19	LET Z=H/2 PRINT Z+1		
20	PRINT TAB(15); H; SPC(5); Z	Count those spaces..from the left!	
21	NEW PRINT H, Z, H *Z	Clears the memory, but not the screen.	
22	LET A=2 LET R=3; LET C=4 PRINT A+B+C	From now on, don't expect the computer to respond with ?SYNTAX or MISTAKE.	
23	PRINT A;B;C	What does the ; do?	
24	PRINT A,B,C,	What does the , do?	

PRINT H, Z, H

* = Apple differs to BBC

LESSON 1 (continued)

25	PRINT A+1, B+1; +C-1Ø		
26	PRINT A *B+C		
27	PRINT A+B *C	There is an order of evaluation	
28	PRINT A *(B+C)		
29	PRINT A *B+C/3-A		
30	PRINT A *(B+ C)/(3-A)		
31	PRINT A, A+1, C-A		
32	LET D=B+1 LET E=C+D LET F=2+(D+E)		
33	LET D=D+1 LET E=E-D LET F=(E-1) *2 PRINT D, E, F		
34	PRINT SQR(9), INT(32.927)	Square root, integer	
35	LET X=SQR(20) PRINT INT(X), INT (2 *X)		
36	LET Y=SQR(X) PRINT TAB(X); INT (Y)		
37	NEW A=4 B=A-4 PRINT A/R	The LET statement is implied.	
38	X=0 Y=X-25 PRINT SQR(Y)		
39	PRINT A *(A^B *(INT(Y))) *X		
40	X=-90 Y=0 Z=142 PRINT SGN(X), SGN(Y), SGN(Z)	SGN returns one of three values, depending upon the sign or value of the variable.	
41	PRINT ABS(X), ABS(Y), ABS(Z)	Strips the sign	
42	PRINT ABS(Z-200), ABS(X*2)		
43	PRINT 1000000000	"Scientific notation"	
44	PRINT .0000005		
45	PRINT 7.5 *10 ^9	Raised to the power of nine.	
46	PRINT -321 *10 ^11		
47	PRINT 12345 *10 ^-4		

Chapter Eight
The Computer Contract: A Sociosemiotic Analysis of Computer Adoption

> In vain are we asked in what records this charter of our liberties is registered. It was not written on parchment, nor yet on leaves or barks of trees. It preceded the use of writing, and all the other civilized arts of life (David Hume, "Of the Original Contract", *Moral and Political Essays*, 1748).

Introduction

Contract initiates the final stage of the narrative program, the moment at which all resolutions are accomplished. In turn, an implicit fiduciary contract, or an understanding between speakers, is a prerequisite for discourse and face-to-face communication (cf. Greimas and Courtés, 1979:70-71). The concept of contract has a long philosophical history. Eighteenth century European philosophers considered government by consensual contract to be the basis of modern civil society.[1] Nineteenth century politicians attempted to achieve this utopian dream. Twentieth century idealists have decried this contractual utopia as a description of the postindustrial era, and some voices proclaim that "our allegiance to the nation-state as such ceases because its claims to having a solid contract with us have been shown to be fraudulent" (cf. Bay, 1978:42). The semiotic concept of contract, however, is concerned not with macropolitical structures but instead with the structures of discourse and narrative.

Although some semioticians deny the sociological implications of the concept of contract, in my view, its sociological meaning is evident. In qualifying their description of unilateral and reciprocal communications as the basis of contractual structures, Greimas and Courtés (1979:70) state:

> The preceding remarks may seem to be inspired by philosophical or sociological preoccupations. But this is not the case. They are developed exclusively and above all on the basis of a growing number of concrete analyses of discourses, more specifically, of narrative discourses.

I shall endeavor to demonstrate that the semiotic and sociological analyses of contract are closely related when discourse is considered in social context. Specifically, I shall examine the cases of the progressive and fragile computer contracts that characterize the narratives of computer adoption in Ivory Coast and Kenya. Before analyzing these materials and drawing the narrative of computer adoption to a conclusion, it is helpful to review the semiotic theory of contract.

Contractual Structures

Contract is the relationship of understanding and obligation established between two subjects in a narrative. Subjects enter into a contract when the story begins, and they undertake actions accordingly to accomplish their goals. Greimas (1983:226) outlines the conditions for such a contract in terms of the relationship between a contractual mandate and its acceptance as follows:

$$\frac{\text{mandate}}{\text{acceptance}} = \text{establishment of a contract.}$$

The mandate motivates the narrative subject to pursue a particular course of action. As the narrative unfolds, a contract may be fulfilled or violated in different ways several times. Acceptance of the contractual mandate is a narrative function that changes the course of events and ultimately brings the narrative to a close. A broken contract alters the relationship between the sending and receiving subjects and may change the flow of the narrative or terminate it. Greimas (1983:226) outlines the broken contract in the following formula:

$$\frac{\text{interdiction}}{\text{violation}} = \text{violation of the contract.}$$

The broken contract destroys the relationship between the sending and receiving subjects and sets up a chain of new narrative sequences. These new sequences have a causal and temporal relationship to the ruptured contract. Thus, accepting or violating a contract initiates a sequence of events, or syntagmatic chain, which includes assignments, tests, and decisions to be accomplished by the acting subjects. In the context of this study, for example, acceptance of the computer leads to confrontation with the machine and eventual mastery of, or failure to master, the new technology. Although the test of mastery determines the outcome of the final computer contract, the test itself is based upon prior contractual understandings.

By viewing the narrative program as a series of interconnected sequences in a syntagmatic chain, we are able to witness the repetition of confrontation and mastery at different phases of computer use. These repetitions are narrative redundancies in which the same type of act is replayed in different contexts of computer use. For example, the same subject (S_1) may master the computer (S_2) in contrasting ways in an educational and a work setting. Mastery of increasingly complex programs at the domination stage of the narrative program differs from initial confrontation and mastery of the computer when it is first introduced. Hence, if we apply this approach to the syntagmatic chain, tests of confrontation, domination, and mastery appear at each stage of the computer narrative rather than merely at the early stages of its introduction. At each phase, there are new obstacles to be met and overcome. The cast of sending and receiving subjects shifts across the phases of narrative development. Figure 8-1 diagrams the syntagmatic chain at different stages of the narrative program.

Syntagmatic Chain	Qualifying Test	Main Test	Attribution
acceptance	computer received by human subject	mandate	assignment of computer tasks
confrontation	computer's initial use	overcoming initial problems with system	positive uses of new technology
domination	computer's ongoing application	technological adaptation	valorization of new technology

Figure 8-1. Syntagmatic Chain of Computer Adoption.

Acceptance of the contract leads to a mandate concerning computer use and attributions of value regarding the new technology. At each point, a narrative complication arises and spurs computer users to move to the next phase of technological adaptation. Success or failure to adapt to each phase of narrative complication influences the final attribution of value to the new technology.

In this book, I have employed two variations of the computer contract as the basis of comparison: a progressive contract in which the acceptance of new technologies on the national level stimulates formal planning and a fragile contract in which an ambivalent response to computers leads to mixed attributions and an emphasis on individual initiatives in computer use without formal planning. Decisions at the national level filter down to influence education and computer adoption by firms and individuals. In the case of the progressive contract, a conjunctive contractual relationship is established between the government and the users of new technologies.[2] This conjunctive relationship implies a set of positive values surrounding new technologies. In the case of the fragile contract, a disjunctive relationship between the government and the users of new technologies leads to ambiguity and confusion in the adoption process. The contract's existence is threatened.

Roland Barthes (1974:88-92) discusses contract narratives as texts in which the entire narrative establishes a relationship of exchange between the sender and receiver. This type of exchange may be likened to Marcel Mauss' concept of the gift in which the exchange contract creates a social bond and a relationship of obligation between two parties (cf. Mauss, 1967:12-13). The object of exchange is part of a larger system of circulation, distribution, and repayment. The storyteller exchanges the narrative with the listener, who awaits a reward of something of value in return for receiving the narrative. Barthes (1974:89) outlines this exchange succinctly: "A desires B, who desires something A has; and A and B will exchange this desire, this thing, this body and this narrative. . . ." Gerald Prince (1987:61) describes this contractual narrative exchange as "a tale for a day of survival as in *Arabian Nights*, a story for a night of love as in 'Sarrasine.'" On this level, the narrative exchange in computer adoption amounts to a story in return for a new technology. The computer-related successes analyzed in Chapter Two are stories intended to persuade potential users to adopt computers. They are accounts of success that are symbolically traded for computers. Similarly, disaster stories are narratives intended to lead to the acquisition of new technologies to effect radical change. The case of the systems analyst at the Kenyan retail company aptly demonstrates this process. In return for his accounts of the failures of the old card punch minicomputer system, he ultimately received a new multi-user system. The narrative itself was exchanged for the new computer.

With regard to narrative and discourse, contract may be considered an artifact of communication. From a sociological perspective, contract involves acceptance of obligation within a normative agreement. Talcott Parsons (1937:333) argues that contract and social exchanges are based upon "the noncontractual element in contract," or the norm-setting environment in which the contract is established.[3] This sociological interpretation broadens the semiotic concept of contract in productive ways by emphasizing the social constraints placed upon narratives in a living context. Text and deed are united in their implications for social action. At this moment of union, the theoretical importance of sociosemiotics emerges as a method for interpreting the social implications of the text. Although they frame their theory of contract exclusively with respect to an analysis of narrative and discourse, Greimas and Courtés (1979:71) specify the relationship of contract to mutual exchange and obligation.

> In such a case, the contract appears, at first, as a delayed exchange: the distance which separates its conclusion from its execution is filled up by a tension which is both like a credit and a debit, like a confidence and an obligation. Yet a closer look shows that a simple operation involving the exchange of two objects of value is not a mere pragmatic activity.

The exchange is based upon communicating the symbolic value of the object – an act of valorization that influences a further course of action. This act of valorization involves interpreting the object of exchange and persuading others that this object is important. These interpretations return us to the domain of discourse where the object of contractual exchange is discussed and evaluated. At this juncture, the analysis of contract inspires an examination of the contractual exchanges that arise within a larger narrative structure (cf. Landowski, 1986:20-25).

In developing an analysis of contractual exchanges, I shall begin with Garfinkel's explanation of contract as "the definition of normal transactions between clientele and . . . agencies in terms of which agencies' services are franchised and available to clients" (Garfinkel, 1967:198). With respect to computer adoption, contractual exchanges occur in different domains and at contrasting levels. The government interacts with firms by setting or refraining from establishing computer policies. In turn, firms interact with their clients and with their own personnel. Computer technology influences the ways in which firms are structured and organized. Contractual exchanges take place in each of these domains. Figure 8-2 provides a typology of contractual exchanges for computer adoption.

Sender	Contractual Exchange	Receiver
1. Government	◄────────►	Firm
2. Firm	◄────────►	Client
3. Firm Management	◄────────►	Computer Personnel
4. Computer Technology	◄────────►	Firm

Figure 8-2. Typology of Contractual Exchanges for Computer Adoption.

Receiving a computer within a narrative sequence triggers a chain of events that has been glossed as computer adoption. This chain of events involves various tests, including learning how to use the computer and integrating the computer into the structure of an organization. A schematic presentation of this chain of events is offered in Figure 8-1. In Figure 8-2, actual computer exchanges are designated. These exchanges form the basis for the analysis of discourse sets collected in Ivory Coast and Kenya.

Contractual Exchanges in Computer Adoption

Based upon comparative data, I have specified four types of contractual exchanges operating in computer adoption. The first exchange occurs between the government and the firm. In this case, the government sets computer policies and controls access to computer hardware and software. This control occurs through import regulations and direct as well as indirect monitoring of the activities of computer vendors. Although the firms may, in turn, sell all their equipment or expertise back to the government, they are essentially the recipients of government policy, calls for bids, and other agreements that regulate the distribution of computers.

In the second type of exchange, the firm interacts with a client. This client may be an individual or another agency or firm. Firms often use the term "contract" explicitly in this instance, viz., a sales contract, a maintenance contract, or a service contract. The third type of exchange deals with intrafirm relationships. In this instance, the firm's management interacts with its personnel by designating the use of the computers and the organization of activities surrounding them. Problems with the implementation of computerization within firms emerge during this exchange. In the fourth type of exchange, the computer technology is the acting subject and the firm is the recipient of the messages. The technology requires specific changes in organizational structure, work tasks, and job titles. The process of entitlement discussed in Chapters Five and Six emerges as a feature of discourse in these exchanges.

Type 1 exchanges are characteristic of the interactions between computer vendors and the government. Since government regulations set the norms for computer sales, vendors are particularly concerned with the limitations imposed by these regulations. A crisis resulting from ambiguous or overly stringent regulations generates type 1 responses in the form of complaints from vendors and other firms about government restrictions. Type 1 exchanges emerge during the course of normal business transactions. These exchanges include negotiations of sales and service contracts as well as regulation of interactions between firms and their clients. Type 3 exchanges are byproducts of the firm's internal operations. Many of the interview excerpts and conversational exchanges with enterprise managers in Chapters Four and Five may be classified as type 3 exchanges. Type 4 exchanges focus explicitly on the strengths and weaknesses of computer technology. Computer-related success and disaster stories are variations of type 4 exchanges. Broken contracts are expressed as type 4 exchanges in which the computer technology proves to be inadequate or inappropriate in a particular context. Type 2 and type 3 exchanges are prevalent in my interviews in computer consulting firms in Ivory Coast while type 1 and type 4 exchanges characterize much more of the Kenyan data. The Ivory Coast interviews are distinguished by specific references to the term "contract" while the idea of contractual exchange remains implicit in most of the Kenyan examples.

The Progressive Computer Contract in Ivory Coast

Type 2 exchanges have been characterized as typical of much "business as usual" for computer consulting firms. In Ivory Coast, consulting companies and vendors occasionally worked together on contracts with a single client. J.R., a research engineer at the S.G. firm, described this process of cooperation in a type 2 contractual exchange.

8.1 J.R., research engineer, S.G.:
Right now // I'm working on a *contract* with a private company where two computer consulting firms are collarborating, and it is going very well. // Good. // There are also examples in Ivory Coast // it hasn't happened to us yet // where one firm calls upon another to help them with a client and that second firm has broken the contract and taken away the client.
B.J.-R.: That's pretty bad.
J.R.: I'd say you could do better // hein? ((Laughter.))

The basic type 2 exchanges are straightforward sales and maintenance contracts. J.R. described how these contracts work in his firm.

8.2 J.R.:
Generally // we say that we are the best but that we are also the most expensive. That is // euh // one of the reasons for which we have certain clients and for which we lose other contracts. // Good. // Our best references are banks. There are also financial exchange boards // such as the coffee-cocoa exchange // for which we developed a statistical system. That is // after all // at the heart of Ivory Coast business. . . . And then we have maintenance contracts that are practically obligatory. So // the maintenance contract protects the clients against all sorts of things. In particular, it guarantees repairs within twenty-four hours after a call. That's especially important for banks and also for small enterprises that do not have any other computer services.

In type 2 exchanges, equipment is provided, and services are rendered for payment. If both parties are satisfied, the exchange works well. Problems may arise, however, when the clients learn to use the new computer. Other problems, as J.R. suggests, issue from the cost of the services. Type 2 contractual exchanges designate the arena in which most competition takes place for computer consulting companies and vendors. These exchanges require adequate personnel to execute the contract. R.M. of the S.L. firm in Ivory Coast explained the impact of competition and the importance of personnel in his firm.

8.3 R.M., director, S.L.:
The principal competitors for S.L. on the Ivorian market are firms that are supported by banks or have developed out of banks and deal with relatively important contracts. // So // they work with large companies // large and medium-sized companies // and they have big contracts. . . .

R.M. continued to explain how the size of a contract influenced the shifting organizational structure of his firm. A large contract required him to hire additional consultants from France.

8.4 R.M.:
Our staff consists of anywhere between seventeen and twenty people, but that number includes our consultants who are here on special missions. If I sign a contract tomorrow that asks for five or six people to be on a single job // euh // I don't have the human means // if you will // to respond to that demand. // So // I call on S.L.-Paris.

Competition demarcates the territorial domain of type 2 contractual exchanges for computer vendors. A commercial engineer in IBM's Abidjan office explained this situation to me.

8.5 Commercial Engineer, IBM:
We are in competition with other computer vendors. We don't have contractual agreements with other vendors, but we do work with other firms that offer services complementary to ours. Distributors // euh // good // that's another matter. But complementary services include programming, training // which we do not do // and the sale of complementary products that we do not carry such as cable systems, regulators, and electrical installations. Complementary services are not in competition with our own.

Profits are the major byproduct of type 2 exchanges for computer vendors and firms. Vendors attempt to regulate these exchanges in order to assure the stable growth of their firms and their profits. The commercial engineer at IBM Abidjan explained this situation to me.

8.6 Commercial Engineer, IBM:
Voilà. Our growth is not regular. You can grow and receive high profits one year and then have nothing for two or three years. So // since our policy is to offer the same services throughout the world // we install an infrastructure that generates fixed prices and we utilize reliable local personnel. We are very prudent in our local installations in order to assure the longevity of the agency. That is our overall policy. So // in general // when IBM decides to leave a country // it's a question of sales // hein? Or else there is a truly political reason // such as what happened in South Africa // for example.

Thus, the conditions for a broken contract, or violation, in type 2 exchanges are primarily commercial. The vendor seeks to maximize profits, and if financial growth does not take place, the contract is eventually abandoned. Failure to fulfill type 2 exchanges is an endemic feature of the fragile computer contract. For example, major vendors withdrew from the Kenyan computer market in 1985 and 1986 because of falling profits caused by high duties and taxes along with a restricted arena for their sales.

Type 3 exchanges establish the prerequisites for internal interactions within firms. As a firm expands its technological capabilities, new employees are hired, and the management restructures tasks and job titles. This process may involve engaging temporary employees in computer consulting firms and other small businesses. R.M., director of the S.L. firm explained:

8.7 R.M.:
Good. // In general we hire extra employees for local contracts. These local employees are usually expatriate women whose husbands work in enterprises here in Ivory Coast. Actually // we have three women working in this capacity at S.L.-Ivory Coast. They are programmer-analysts with degrees from IUT [Institut Universitaire de Technologie] in France.

Although hiring women to fill substitute positions is only one strategy for meeting organizational demands, this policy received sharp criticism from one young woman engineer interviewed in Ivory Coast. R.B. expressed her discontent when she stated: "All of the women are at a low level // keypunchers, data-entry people, and computer operators // that sort of thing. The higher one goes, the fewer women one sees" (cf. excerpt 3.10). She would have preferred to see more women in the permanent capacities of company heads, high-level technicians, and maintenance engineers, R.B. added:

8.8 R.B., research engineer:
Rarely do you see blue collar women who come in to repair the machines. Rarely. This job is essential // but women are directed away from it in their education. . . . I personally would have really liked to be a maintenance engineer. They would call me on the telephone and tell me that a computer isn't working. I would arrive, make my tests, and say its that part which doesn't work. I would prefer to be essential like that instead of scratching out lines of a computer program.

R.B.'s experience contrasts with that of research engineers whose work entails management of their firms. Ethnographic descriptions of selected firms in Ivory Coast and Kenya have demonstrated that many of these engineers and systems analysts play the dual role of part-time managers. This division of labor is indicative of a type 3 contractual exchange within the firm. An implicit contract empowers these engineers to manage the firm, even if for short periods of time. This contractual exchange restructures lines of authority, job titles, and perceptions of work within the firm. Once again, J.R. elaborated on this situation in his dual role as research engineer and part-time manager.

8.9 J.R.:
Well // I do everything and nothing. That is to say // I supervise research and maintenance, and I replace the director while he is away. That's why you are speaking with me today // because the director is away, and I assume those responsibilities in his absence.

Type 3 exchanges permit internal mobility in firms by freezing data-entry workers at the bottom of the hierarchy. In Ivory Coast, rapid vertical mobility is expected of engineers, but it is not necessarily a reasonable aspiration for beginning programmers and data-entry personnel. Nevertheless, the flexibility of internal contractual exchanges encourages workers to believe that progressing to higher levels is not only possible but likely. Computer training, as we saw in Chapter Six, uses type 3 contractual exchanges as an incentive in recruiting students, especially in the commercial colleges. The flexibility and ambiguity of contracts within firms are also a source of workers' complaints.

Although the computer itself may be viewed as a sign of status and a symbol of change, the contract is essential to the computer's presence in the workplace. Type 1 exchanges develop in contexts of regulation. One of the most frequent type 1 exchanges is the call for bids (*appel d'offres*) in which the government requests computer services. Type 1 exchanges also occur in cases where firms or individuals have complaints about national computer policy. The extremes of type 1 and type 4 exchanges characterize a fragile computer contract in which policies toward technology are ambiguous and fluctuating.

Calls for bids involve competition and potential conflicts of interest. Vendors believe that the government and large private companies discriminate against certain smaller firms. Complaints about the bidding system surfaced repeatedly in the Ivory Coast interviews while discontent with government customs regulations was more evident in the Kenyan responses. Both complaints about bids and importation regulations are examples of strained type 1 contractual exchanges. When complaints become bitter, contracts may be in jeopardy or broken. The director of a computer vending company in Ivory Coast presented his view of this situation to me.[4]

8.10 J.L., director, computer vending company:
Sixty percent of our sales involve minicomputers. There are very few calls for bids from the government for large, mainframe systems. And when there is a call for a bid // often it specifies a French company // BULL // for example // and French computers are required.

The call for bids is the first step in a contractual exchange that entails sales and maintenance contracts and, very often, a long-term financial commitment between the vendor and the contracting agency. J.L. continued to explain a situation that he considered to be threatening and dramatic.

8.11 J.L.:
Well // just this morning I received a call for a bid from a government agency. It's useless to apply. This morning also I // well I // threw other calls for bids in the waste bin because they required French computer equipment // exactly the same obligations. . . . The situation is dramatic. It is a problem.

A sample call for a bid illustrates the dramatic problem that J.L. describes (cf. Figure 8-3). The call outlines six requirements for initiating a type 1 contractual exchange in the form of a bid. (1) The sending party, in this case the post and telecommunications office, is described. (2) Financial support conditions are outlined. (3) Conditions of participation are presented. (4) Eligibility requirements for submitting the bid are specified. (5) Procedures for bidding are delineated. (6) The deadline for the receipt of the bid at a particular location is announced. J.L. drew my attention to the second and third conditions for contractual exchange in the bidding process.

The call specifies that the contract will be financed by the Caisse Centrale de Coopération Économique (CCCE). It further stipulates that respondents must have primary business operations in France or Ivory Coast *and* that their capital must be assured by French or Ivorian sources. These conditions allow the government to retain control over the bidding process and to favor Ivorian and French firms. Although bidding is not a direct form of government regulation, such as the requirements outlined in the national informatics plans, it is a contractual means of regulating the market for computer sales. The bid is both a symbolic document and the first step in a political and legal process that has economic consequences for Ivory Coast's computer world. These consequences, of course, extend beyond the computer industry to the overall regulation of the Ivorian economy.

Not only is a call for a bid a type 1 contractual exchange, it is a qualifying test in the narrative program of computer adoption (cf. Figure 8-1). The call is the basis upon which computer companies establish their credibility, and it initiates a sequence of acts – the installation of computers, the hiring and reskilling of computer personnel, and the reorganization of work tasks. Without the accepted bid, there can be no mandate for technological adaptation in a firm. The call for a bid is fundamental to the implantation and circulation of computers. Regulation of the bidding process limits the presence and uses of computers on the national scene.

Annexe II of Ivory Coast's *Plan Informatique National, 1986-1990* is explicit about the delineation of type 1 and type 2 contractual exchanges as they pertain to interactions between the government and computer consulting firms as well as to interactions between the firms and their clients.[5] The plan designates two types of contracts as the objects of government scrutiny: (1) contracts with the obligation of results ("*contrats avec obligation de résultats*") and (2) contracts

Gallo ONT informations

AVIS D'APPEL D'OFFRES Nº 2660/DMP

1 - OBJET:

Le Ministère des Postes et Télécommunications. Direction de l'Office National des Télécommunications lance un Appel d'Offres ouvert à la concurrence pour la fourniture et l'installation de matériels pour le réseau national de transmission de données à commutation de paquets.

2 - FINANCEMENT:

Cette operation est entiérement financée par la Caisse Centrale de Coopération Économique (CCCE).

3 - CONDITIONS DE PARTICIPATION:

Ne peuvent participer à cet Appel d'Offres que les fournisseurs ayant d'une part leur siège social ou leur domicile en FRANCE ou en CÔTE D'IVOIRE, et d'autre part, dont le capital est detenu et la direction effective assurée par des ressortissants ivoiriens ou des ressortissants français.

4 - SOUMISSION:

Les fournisseurs sont obligés de soumissionner pour l'ensemble de materiél et des services objet de l'Appel d'Offres.

5 - RETRAIT DU DOSSIER D'APPEL D'OFFRES:

Les dossiers d'Appel d'Offres pourront être consultés et retirés dès diffusion de cet avis, au Service de Marchés de Télécommunications sis au 12eme étage de la Tour POSTEL 2001, porte 12-07 contre versement d'un chèque certifié ou d'un mandat-lettre à raison de 20,000 F.CFA· établi à l'ordre de l'Office National des Télécommunications à ABIDJAN.

6 - DEPÔT DES OFFRES:

La date limite de remise des Offres est fixée au Lundi 21 Septembre 1987 avant 17 H 00 délai de rigueur, au Service des Marchés des Télécommunications 12eme etage de POSTEL 2001 - Porte 12-07.

Pour tout renseignement complémentaire, téléphoner: 34-67-61 Service des Marchés pour les renseignements administratifs et au 34-66-12 ou 34-66-27 pour les renseignements techniques. Télex O.N.T.: 23 790-23 750.

Figure 8-3. Call for a bid from the National Post and Telecommunications Office, Abidjan, Ivory Coast.

providing means of support ("*contrats avec obligation de moyens*"). Contracts with an obligation of results involve specific jobs undertaken by the SSII to be completed within deadlines agreed upon by the firms and their clients. These jobs include master plans, audits, accounting schemes, and computer feasibility studies. The client receives work completed with the aid of the firm's computer system in exchange for payment. *Annexe II* (pp. 57-59) of the second national informatics plan suggests that the contractual deadlines and organizational conditions for the completion of work be specified in advance by both parties in the exchange. The plan, however, does not lay out conditions for the enforcement of result contracts. *Annexe II* (p. 59) states that "positive or negative sanctions" may be applied to result contracts depending upon the quality of the work received.

Contracts providing means of support are considered to be covered by the sanctions of the economic market.[6] This type of exchange involves the firm's furnishing of human or technological means to complete certain projects. Thus, the firm supplies the computer hardware and software and, if necessary, instructional or support staff. The second national informatics plan outlines three conditions for means contracts: (1) the responsibility for mastering the work of the client; (2) the possbility for the client to question the services or materials provided by the computer firm; and (3) the obligation of the computer firm to furnish the equipment promised regardless of the reliability of the firm's sources of supply. The plan notes that violation of the third condition is the major cause of turnover in the profession. In terms of the narrative program, responsibility for mastery is the qualifying test that allows the means contract to be executed. Provision of adequate material is the main test and the source of technological adaptation. Ability to fulfill the contract leads to its termination and a positive computer attribution.

The result and means contracts have legal, sociological, and semiotic implications. The informatics plan highlights these contracts as possible areas for regulation, control, and, if violated, legal action. The two contractual exchanges are clearly economic and rely on trust between the contracting parties. From the perspective of semiotic analysis, it is noteworthy that these contracts *per se* are spelled out and thematized in the computer planning document. These contractual exchanges are not treated as implicit and problematic concerns in Ivory Coast's computer world. Instead, government planners consider the result and means contracts to be fundamental to the process of regulating computerization in Ivory Coast. Once again, it is important to recognize that in the absence of a computer plan, the Kenyan government does not formally regulate the status of transactions between computer vendors and their clients.

With respect to my four-point typology of contractual exchanges, result and means contracts may be classified simultaneously as type 1 and type 2 exchanges. They are type 1 exchanges insofar as the government is a silent partner in the exchange. Government attention to regulating the computer vendors and the SSII places result and means contracts under special scrutiny. Yet, these exchanges also fall under the type 2 (firm-client) classification. This mixed type of contractual exchange is a byproduct of a progressive computer contract in which the government oversees and potentially intervenes in private sector economic and technological activities. This intervention does not imply total state control, even over the government's type 1 exchanges. In a description of the means contract, the second national informatics plan (*Annexe II*, p. 59) states:

8.12 *Annexe II, Plan Informatique National, 1986-1990:*

> This type of contract is very common in the government [*l'administration*] which has not yet been able to impose the obligation of result that implies a positive or negative sanction. In the private sector, it is easier to impose this obligation because the economic sanction is sovereign.

Type 1 contractual exchanges have a direct influence on type 4 exchanges, or the role of computer technologies in firms. Availability of the technology and its regulation are preconditions for how computers are managed and interpreted. Another excerpt form *Annexe II* of the second national informatics plan illustrates this point. *Annexe II* (p. 55) states:

8.13 *Annexe II, Plan Informatique National, 1986-1990:*

> With progress in electronics, computer vendors have gradually become the new industrialists. This progress, coupled with new electronic equipment for the construction of hardware, has resulted in a spectacular fall in the price of computers. A parallel development is the transfer of research activities from the vendors to specialized firms.
>
> At the same time, the increasing complexity of computer systems has created numerous partners in the process or com-

puterization: the vendors, the computer consulting firms, the software firms, the telecommunications companies, the providers of computer-related services, the logistics firms, and the firms specializing in business applications of computers.

The plan continues to specify that regulation of these computer-related firms and activities is an urgent necessity. The plan's statement concerning technological advances, however, may be interpreted as a type 4 contractual exchange in which the progress of computer technology is considered to be the basis for the existence of a variety of new enterprises. Technology not only controls the workplace, it also creates the economic opportunities and organizational reasons for its existence.

Not all of the consulting firms are content with the status of computer policies in Ivory Coast. In an interview, a director of one of the SSII warned against the dangers of overregulation. He also emphasized the ways in which "the laws of the market" serve as a stimulus for type 1 and type 2 contractual exchanges.

8.14 Director, computer consulting firm, Abidjan:
I believe that the Ivorian government is modeled on the French administrative system and unfortunately it has assumed the shortcomings of the French system. The government is a *grande dame* who weighs a great, great deal. In France // to change anything in the government is a long-distance run because she is a very heavy mistress. That is // it's a long-term task. In Ivory Coast // it's pretty much the same. But the private sector is more flexible and has to deal with competition and the laws of the market // which means // well // that if a bank has computerized its operations in a very sophisticated way // other banks are obliged to follow its example.

This discussion of the four types of contractual exchanges made under the aegis of the progressive computer contract highlights a pattern in which contracts are initiated at different stages of the narrative program and are a primary concern of the firms installing and using computers. Wherever computers are an object of discourse and a feature of the workplace, so too are the contractual exchanges that surround the distribution and use of computers. The computer world is a symbolic, social, and economic environment sustained by contractual exchanges. The circulation of computers is a defining characteristic of postmodernity. Contractual exchanges delimit and regulate this circulation. My interviews with directors and engineers in Ivorian computer consulting firms contain many descriptions and references that may be classified as type 2 and type 3 exchanges that emphasize the daily activities of buying, selling, and using computers. Type 1 exchanges are more frequent in government policy statements and comments of vendors on the impact of government regulation.

I have placed type 1 and type 4 exchanges at polar extremes in my classification because their appearance signals the success or failure of the computerization process. Type 1 exchanges from vendors are generally complaints and indications that an area of conflict exists between regulatory agencies and the private sector. In the Ivorian case, type 4 exchanges focus on computer-related successes and positive conditions for technological adaptation. On the contrary, in the Kenyan setting, type 1 exchanges are common in small firms as well as among computer vendors while type 4 exchanges include disaster stories that reflect the underutilization or inappropriate application of computer technology. This contrast underscores the conclusion that differences between the progressive and fragile computer contracts are evident in the discourse and the social and economic exchanges that constitute the narrative program for computer adoption.

The Fragile Computer Contract in Kenya

The fragile computer contract is characterized by disjunctive and violated exchanges. The narrative sequence of interdiction and violation constantly threatens the fragile contract and renders both its survival and interpretation precarious. The fragile contract is often an unstated or ambiguous agreement. Since the Kenyan government has not developed an explicit computer policy, type 1 exchanges assume the form of indirect economic and political sanctions. Although type 2 and type 3 exchanges relate to the conduct of "business as usual" in Kenya, as they do in Ivory Coast, discussions of these activities are often overshadowed by complaints about indirect government regulation and inappropriate technology. In an interview with an administrator in charge of the computer division of Price Waterhouse, which serves not only in an accounting capacity but also as a computer consulting firm in Kenya, concern about government economic policies dominated our conversation.

8.15 D.S., Price Waterhouse, Nairobi:

So // as a prompt // first of all // high cost is definitely an obstacle. As you've probably gathered // the duties and taxes are very, very high // I think the highest I've come across in any country in the world // considerably higher than other African countries. It means that // for most computers // the cost has run about 300% of the direct exchange value of most Western countries when you include taxes, duties, and import. So that would be significant. I think that the slowness in adapting to the fourth generation language approach // well // they're probably no more slow than other places // but that's moving now. So they've been constrained by traditional methods // which means long and expensive implementation of these small systems.

In this instance, government policies are described as obstacles. Although there is no direct reference to contractual regulations, taxes, duties, and importation costs set by the government are considered to hinder a smooth computer adoption process. D.S. continued to elaborate on the consequences of a fragile type 1 exchange when he described the ways in which individuals attempt to overcome these problems in Kenya.

8.16 D.S.:

But I think there's a need for some more explicit policies. Tanzania // for example // has some sort of central computer committee // I believe // and I think there are a number of African countries which have that. I mean // that // I guess // is one way where they make some decisions // and then that is // to me // I suppose // one way // but // of course // they get bogged down into a sort of unfortunate bureaucracy. I don't really favor that approach, but I think there's a need for some more explicit policies. A byproduct of this situation // is that a very, very large number of micros now don't come in through official channels // probably more than fifty percent. They come in through people's personal effects // because the duties are so prohibitive. And this must be counterproductive // because rather than getting very big duties // they get nothing. So something in-between would discourage this alternative path.

The interviewee repeatedly refers to the government as "they" rather than specifying a particular agency or set of regulations. By doing so, he distances himself from the decision-making process, and he does not propose an alternative for making the voices of computer companies and vendors heard. The example of bringing in duty-free computers indicates a violation of the unstated contract between the government and potential computer users. D.S. referred to this violation as an "alternative path," but he was reluctant to propose strategies to remedy this situation. S.M., the director of a computer vending company voiced similar complaints about the role of the government *vis-à-vis* the market for computer sales.

8.17 S.M., director, computer vending company, Nairobi:
The government has a double standard. It's understood. Government *does understand* the need for computers they will have. On the other hand // they have to keep up a front. They say to people, "Yes // yes this is not very good for our government // so we're going to charge high duties on that." But half the computers here come without payment of duty. If you look at the case where duty was lower, then the people would bring it in genuinely and they would pay duty. Here // I don't think they make any more money. And people like us get clobbered // to some extent. Because for each machine we bring in // two or three come in without payment of duty. You see if duty was lower // all of them would pay, and I think government would generate more revenues.

In contrast to the Price Waterhouse administrator, S.M. expressed the belief that a Kenyan computer vendors' lobby, similar to GIMI in Ivory Coast, would eventually relieve the burden of high duties by representing the interests of vendors and data bureaus to the government. He saw increased professionalization as a way of improving type 1 contractual exchanges. Nevertheless, S.M. lamented the fact that to date a lobby had not formed for reasons that he could not explain. As in the case of D.S., he viewed the government holistically as "they" but referred to "people like us," meaning the computer vendors. S.M. saw collective action as a strategy for changing computer regulations and thereby expanding the local market for computer sales.

Type 1 contractual exchanges are subjects for discussion in the news media. A 1984 editorial in the *Daily Nation*, one of Kenya's leading newspapers, encouraged the Minister for Education, Science, and Technology to take a more aggressive role in stimulating computer education at the university level. The author concluded the article by reiterating the complaint about the government's role in restricting computer importations and sales. He advocated more funding of computerization.

8.18 *Computers Are No Longer Luxury Items*

If that requires a larger expenditure in education, so be it. Other monies have been spent on lesser worthy causes.

At any rate, there could be some savings if irrelevant subjects are dropped from the school curriculum or are trimmed. Computer science could then fill the void created that way.

Costs could be kept down, too, by providing computers only to the more advanced classes. And computer manufacturers – not retailers – could be persuaded to donate a few. After all, it is in their best interest to stimulate demand which would otherwise remain dormant due to ignorance.

The demand and knowledge about computers would be increased, too, if the Government stopped treating them as luxury items and, therefore, charging a high import duty. That forces the computers' price up so much that only big companies can afford to own them.

And even the large firms find it expensive to keep up with the computers which they may need.

A more rational import policy would treat computers the same way other industrial imports are treated. After all, their use increases productivity.

It is no good pretending that the country can do without computers. Those companies which cling to outmoded technology in the belief that it will save jobs are likely to end up losing their market share to those that take advantage of the technological breakthroughs. And what is true of companies is equally true of countries. Only more so.

The time to take the right decisions is now. Not after all is lost (*Daily Nation*, Sunday, August 12, 1984).

The tone of this editorial is reminiscent of Ivory Coast's national informatics plans. The author advocates advances in technoscience and exhorts the government to take an active role in developing outlets for computer education. Unlike the Ivorian official documents, however, this complaint originates with a private citizen who perceives government economic regulations as a major obstacle to the progress of computerization. The author concludes with the plea that the government should respond before "all is lost." He hopes that Kenya will not miss the train of postmodernity.

Type 4 contractual exchanges designate another arena in which problems of computer adoption arise. K.C., the volunteer who described his experiences with the computer experiment at a handicraft cooperative in Nairobi, cited the management's lack of interest in a maintenance contract as one indication of the failure to computerize. In Chapter Five, I described the Management Committee's ambivalence toward the computer, which was viewed as both a symbol of progress and a stumbling block hindering traditional business practices. Explained K.C.:

8.19 K.C., computer volunteer, Nairobi Handicraft:
I tried desperately to get them to take out a contract // a service contract // on the computer. It would only have cost them 4,000 shillings [approximately U.S. $251] to have this piece of equipment totally protected for a year // with total service and regular check-ups. They would not do it because they said it would cost too much money. We had the money to do it (cf. excerpt 5.23).

In this case, the type 4 contractual exchange is not based solely on the payment for services. Instead, the interpretation of technology in the cooperative is at issue. The Management Committee regarded the computer as an object of value for its display rather than as a result of its use. The potential effects of actually employing the computer would have destroyed the equilibrium of their business operations. In this case, the computer was an inappropriate means to accomplish their business objectives. The refusal to obtain a maintenance contract is a case of interdiction/violation, or the breaking of an implicit contract regarding the computer's use.[7] As a consequence of the Management Committee's ambivalence, the computer collected dust while serving as an impressive symbolic display of the cooperative's hopes for technological change.

Type 2 exchanges encompass many of the daily transactions between computer firms and their clients. These exchanges include formal and informal arrangements concerning batch computing, payroll, inventory, and financial planning projects. Type 2 exchanges involve the preparation of packaged programs for a firm. If special programs are required, they may be costly because they necessitate additional research and development in conjunction with the client. F.C., a manager at Nairobi's Data Centre, described this situation.

8.20 F.C., Data Centre:

If we tailor a custom-designed package // they're paying for the development of that package // which is where my department comes in. We have sixteen programmers in the company // very versatile in what sort of systems they can write. As I said // we've got sixteen standard packages // which actually have been developed over the years. The rest of our systems apply to customer usage // single customer usage. So // if a client joins us and goes onto a standard system // he's really not paying for development which is quite costly. He just pays what we call an entry fee into the system. And he's ready to go, and he starts paying on a monthly basis whatever we run through for him. Whereas, if we have to write a system // that becomes much more costly. So // it really depends on the client.

Thus, according to F.C., the standard type 2 exchange is based upon services demanded by the client. The computer firm does not provide additional consultation without request. This means that the contracting company must have a clear idea of what its computerization needs are. This situation may lead to selective underutilization of the computer. In a discussion of financial planning, F.C. provided an example of the limitations of her firm's intervention.

8.21 F.C.:

Spreadsheets and financial plans // the clients basically take care of that themselves. I mean // we can only very generally project his turnaround or his costs // or something like that. Of course // we do quotations, but the client has got to take into consideration his budgeting and what his projected returns are going to be. But they do a lot of that manually // themselves // because they find it // I suppose // easier. But obviously on the microcomputer they do have spreadsheets and things like that. So // if people go to the micro and they require the spreadsheet facility // we show them how to use the system, and they apply it as they need it.

S.A., a computer consultant for a large charge card company in Nairobi, was more pessimistic about the role of computer vendors in firm-client relationships. He was particularly critical of certain vendors who, he believed, exploited their customers by advising them to purchase inappropriate systems at a high cost and by counseling them to do unnecessary work on systems that they already owned.[8]

8.22 S.A., computer consultant, N.C.C., Nairobi:
I had an experience with a computer company selling a computer to a firm that had one already and giving advice such as typing all of the material into the computer again. This type of activity is very evident at the lower end of microcomputers. Hardly is there any vendor willing to give comprehensive after-sales support.

Although S.A.'s example is not an instance of a broken contract, it suggests the possibility of an incomplete contractual exchange. According to the Ivory Coast's second national informatics plan, this case is evidenced by a means contract (a mixed type 1 – type 2 exchange) that could result in a negative sanction by the client, the government, or the economic market.

Under a fragile computer contract, type 3 exchanges designate the domain of "business as usual," but these exchanges may indicate problems with the computer system. The presence of a potentially inappropriate system and the transition from one system to another generate internal problems with the computer staff. P.M., the computer controller at the retail trading company in Nairobi, saw the old NCR system used by the company as a source of staffing difficulties and considered these problems as normal for an "upcoming department" (cf. excerpt 5.7). Initially, P.M. thought that his difficulties could be resolved by hiring more employees to handle data entry. Later he actually persuaded the company to change computer systems and to restructure the staff. Similarly, the public relations director at the small charge card company in Nairobi considered hiring a larger computer staff to be a solution to the problems that the company would inevitably face as it expanded. A.M., the assistant computer manager at L.C., an insurance firm in Nairobi, was satisfied with the company's IBM 34 system but also foresaw the need to employ more data-entry workers. He described the insurance company as a place where "data entry is going on all the time." Added A.M., "In a insurance company, data entry is a full-time job" (cf. excerpt 5.13). A.M. went on to say that worker retention was not a problem at L.C., but that there was some resistance to employing additional computer personnel.

8.23 A.M., assistant computer manager:
So far we have not lost any workers here from the end of 1982 until now. It is only the computer manager that we have lost to a parastatal organization.

B.J.-R.: That means that the work conditions here must be very good.

A.M.: Well // I am not in a position to comment on it.

A more positive picture emerged at the car dealerships such as T.D., where the manager reported full satisfaction with the computer system and personnel. He stated that sales records had been broken "three or four times" since the installation of the IBM 34 system (cf. excerpt 5.2). Thus, in certain cases, type 3 exchanges reflect successful computerization in the Kenyan setting. If the computer system meets the company's business requirements, type 3 exchanges reflect the smooth flow of work and increased productivity.

The areas in which the shortcomings of a fragile computer contract were most clearly in evidence emerged at the extreme ends in type 1 and type 4 contractual exchanges. In the first instance, most complaints centered around the government's policies of economic regulation affecting computer importations and sales. In the second case, type 4 exchanges exhibited broken contracts and gradual or unsuccessful computerization. These problems surfaced in complaints about lack of vendor support in computer training (a firm-client relationship) and inadequate data-entry staff within the firms (a firm-personnel exchange). When my Kenyan interviews are scrutinized as a whole, a pattern of sporadic, individual computer-related successes emerges against a backdrop of fragile type 1 exchanges that limit the access of companies to appropriate and adequate computer facilities. In spite of these problems, computer training institutions are growing, and the students and administrators at these facilities appear to be confident about the country's technological future.

Figure 8-4 presents a comparison of the progressive and fragile computer contracts in terms of the four types of contractual exchanges. The narrative program suggests that these exchanges may reappear at different stages in the process of computer adoption. Type 1 exchanges, however, determine the extent to which computers are available for sale to firms at reasonable prices. The economic circulation of computers affects the quality of computer-related services. The market, however, does not determine the availability of computer education, which seems to be on the rise in Kenya in spite of the high cost of computers. This situation emphasizes the importance of computers as a scarce technological resource and the role of education as a symbolic means of access to a scarce commodity. Since a fragile contract is precarious, violation of each type of exchange is threatened.

Although the progressive computer contract involves government regulation of computerization – a process that is often in conflict with economic competition – this type of contract is not under high risk of violation. Problems with the progressive computer contract emerge in the area of education where the aspirations toward new technological advances are most prominent. In Ivory Coast, there is a possibility that engineers and high-level computer experts will "glut" the market while computer operators and data-entry personnel may remain a scarce resource.

Educational resources *do not* reflect market conditions in the business world. Although Ivory Coast has built an educational program into its national informatics plans, detailed planning and scrutiny of educational institutions focuses on INSET and the national university. Consequently, private schools and institutes as well as the firms themselves assume the responsibility for training computer operators and data-entry personnel. Although the University of Nairobi's Institute of Computer Science did not emerge from a national informatics plan, its relationship to private institutions resembles the educational situation in Ivory Coast. The major difference is that lack of national focus on computer training means that there is little risk of overproducing engineers in Kenya. Instead, computer operators and data-entry workers may enter the job market in large numbers with credentials of mixed quality.

The progressive and fragile computer contracts frame the conditions for national advances in technology. These conditions form the background for contractual exchanges that constitute the daily business operations in firms marketing and using new technologies. From the perspective of planning, these contractual exchanges suggest that computerization will advance toward a positive attribution and a successful terminal contract in cases where regulation is well planned. If regulation remains unplanned, new technologies may be viewed as unleashed, hidden enemies. In the African context where new technologies originate from foreign sources, lack of official planning may have disastrous and costly results, such as those exhibited when type 1 and type 4 contractual exchanges fail. Although the progressive computer contract is not problem-free, it structures a series of narrative complications in which the government and private firms control technological resources. The fragile contract leaves this control up

Computer Contract	Cases	Type 1 Exchanges	Type 2 Exchanges	Type 3 Exchanges	Type 4 Exchanges
Ivory Coast: Progressive Computer Contract	1.	Government regulation of result and means contracts in national plans	SSII contracts with large companies	Upward mobility of engineers and senior programmers	Accessibility of new computer systems
	2.	Calls for bids and indirect government regulation of sales	Competition among computer vendors and SSII	Frozen internal mobility of computer operators	Focus on local software development
Kenya: Fragile Computer Contract	1.	Strict government regulation of import duties and taxes on computers	Data bureau contracts with large and small firms	Individual computer-related successes	Inappropriate computer systems and high computer costs
	2.	Absence of national computer plan	Problems with after-sales support by vendors	Problems with outmoded computer systems and shortage of computer personnel	Broken sales and maintenance contracts

Figure 8-4. Comparison of Computer Contracts In Ivory Coast and Kenya

to chance and allows individual computer-related successes to develop in the face of enormous bureaucratic and economic obstacles.

Text and Paratext in Computer Use

Contractual exchanges raise the question of the relationship of a text to the social activities that surround it. I have emphasized the ways in which oral language shapes learning about computers. Along these lines, Zuboff (1988:177) argues that the oral medium provides one of the most effective means of learning about new technologies in the workplace. Discourse, in a sense, "debugs" new technologies and provides resources for their application. Contractual exchanges formalize the mandate for the computer's distribution and regulate applications of new technologies.

The concept of contract is central to a sociosemiotic theory of computer adoption because contractual exchanges encompass social interactions, economic transactions, discourse, and formal documents. For example, a contract may be sealed by a simple gesture such as a handshake. Because contractual exchanges shape the environment (habitus) for computer use, these documents and verbal statements form the context for the computer and its texts. Insofar as the exchanges are manifest in actual social activities, they constitute a paratext for computer use. Literary theorists describe the paratext as the types of discourse, literary genres, and ways of speaking that influence the interpretation of a specific text (cf. Genette, 1982:7-8).[9] If we view the computer program and its applications as primary texts, it is possible to analyze contractual exchanges as regulators of the text and as the source of discourse genres, or ways of speaking about the computer. The four contractual exchanges outlined in this chapter situate how the computer is to be perceived and used. The government-firm exchange treats computers as valued objects that come under direct or indirect control of the state. The paratext invokes the obligation to follow specific regulations. The firm-client exchange is governed by both obligation and trust that are attributes of the marketplace in which computers and computer-related services are bought and sold. Firm management in the type 3 exchange develops work tasks and titles surrounding computers. Once again, administrative regulation limits the uses of technological power that emerge in the type 4 exchange. Nevertheless, the computer technology itself places constraints on the organization of the workplace through the requirement that certain technological tasks must be performed.

A parallel may be drawn between contractual exchanges and another type of verbal exchange, the syllogism. Based upon research with subjects from a Kpelle village in northern Liberia, Michael Cole and Sylvia Scribner (1974:160-170) found that village subjects answered syllogisms by using situational rather than formal logic. Instead of suspending disbelief in order to tease out the formal logic of the syllogism, subjects inserted their personal experiences and recalled situations related to the syllogism but not required to solve it. The syllogism served as a paratext that inspired the subjects to develop independent narratives. An interesting example is a syllogism about how a town chief handles disruptive behavior in his village.

Experimenter:	If Flumo or Yakpalo drinks cane juice, the Town Chief gets vexed. Flumo is not drinking cane juice. Yakpalo is drinking cane juice. Is the Town Chief vexed?
Subject:	People do not get vexed with two persons.
Experimenter:	(Repeats the problem.)
Subject:	The Town Chief was not vexed on that day.
Experimenter:	The Town Chief was not vexed? What is the reason?
Subject:	The reason is that he doesn't love Flumo.
Experimenter:	He doesn't love Flumo? Go on with the reason.
Subject:	The reason is that Flumo's drinking is a hard time. That is why when he drinks cane juice, the Town Chief gets vexed. But sometimes when Yakpalo drinks cane juice, he will not give a hard time to people. He goes to lie down to sleep. At that rate people do not get vexed with him. But people who drink and go about fighting – the Town Chief cannot love them in the town (Cole and Scribner, 1974:163).

In their analysis of this verbal experiment, Cole and Scribner (1974: 163-164) state that the subject substituted knowledge about his village and his personal acquaintances for the abstract characters in the syllogism. Therefore, he rejected the premise of the syllogism and replaced it with a social situation familiar to him. He responded to the syllogism on a social and factual basis rather than in abstract terms. This situational logic is not restricted to West African villagers, nor is it necessarily characteristic of them. Instead, it is a feature of everyday reasoning and oral discourse (cf. Fernandez, 1980:44-59). Although individuals working with computers must grasp the algorithmic logic of computer languages, their reasoning and discourse about the relationship of the computer text to the social text, or paratext, resembles the situational logic described by Cole and Scribner.

As with syllogisms, contractual exchanges exist on the borderline between text and action. Thus, it is not surprising to find references in the Ivorian informatics plans to firms that make contracts to provide equipment or services that they do not yet possess (cf. *Annexe II, Plan Informatique National, 1986-1990*:59). These contractual exchanges are not necessarily cases of violation or interdiction. Rather, they illustrate that contracts are prospective and proleptic. They are based on the expectation that equipment and services will be provided when the time comes. The prospective character of contractual exchanges is based upon their situational application. Although documents delineating these exchanges may follow formal legal logic, the exchanges are enacted in social situations. Moreover, these exchanges are paratexts for a complex series of social activities that arise in the marketplace and in work settings where computers are used. Although the progressive and fragile computer contracts are descriptions of patterns of discourse and social action, they are not models for action.

The Simulacrum and the Model

Sociosemiotics is the study of semantic universes, or the totality of significations that form the backdrop of meaning in social context.[10] As cultural objects, computers bring into focus the meaning of new technologies for their users. As I have already argued, computers are self-referential signs that both contain and generate their own texts. Simulation devices, such as computers and new visual media, operate by substituting signs and models for events and processes (cf. Baudrillard, 1981:11). When this substitution has been accomplished, the signs reference themselves rather than an external reality. Thus, for example, a photograph or videotape of an event acquires a reality of its own. No longer does the representation stand on behalf of another object: it *is* the focal object. This process leads to what Baudrillard (1976: 112) has termed the "fetishism of the lost object," or an unlimited referencing process in which one layer of signs invokes another when the original point of reference is lost.

In Chapter Seven, I explored how computers influence the world views of their users by examining the process by which students learn to work with computer programs and how computer operators apply programs in a commercial context. The computer user engages in metonymic reasoning, or the substitution of part for whole. When applied to the domain of sociosemiotics, this type of metonymic thinking involves a process that I shall refer to as "bypassing." For example, a single computer command refers to and sets in motion a number of operations. Several steps are bypassed in this process. When the computer user begins to think about social life in terms of bypassing, many steps may be combined and collapsed to reach a particular goal. This type of thinking resembles utopian thought in a religious context. In order to reach a religious or cognitive utopia, believers engage in acts of contrition, follow religious rules, and commit themselves to doctrinal tenents. These acts bypass stepwise rational procedures that may be used to attain similar goals. The wish for status advancement and a new world of prosperity underlying the discourse of computer trainees in Kenya and Ivory Coast involves the process of bypassing. The new world and its cognitive utopia can be attained through learning basic programming and moving into the opportunity structure offered by the computerized office place. Bypassing involves more than the superficial use of computer jargon. It entails commitment to a radically altered world view in which the elements of thinking used in manipulating computers are extended into the social realm. Through bypassing, the programmer trainee dreams of becoming a systems analyst or computer manager and pursues these goals, but not necessarily in a means-ends sequence. If several procedures and steps can be eliminated, the new cognitive and social utopia is within immediate reach. Although these steps exhibit an internal logic, they have been condensed and combined to make the social goals appear more accessible. This process is fundamental to the simulation of postmodernity by which the computer symbolically reproduces key elements in the world of its users.

Two semiotic concepts are critical for this analysis: the idea of modeling systems and the concept of indexical expressions. In Chapter Three, I outlined the modeling process surrounding computers in three parts: (1) the macrostructure provided by programming languages and computer software; (2) everyday discourse about computers; and (3) the full narrative program of computer adoption. The computer with its formal programming languages constitutes a secondary superstructure, and discourse about the computer in social context is the "natural language" infrastructure. The machine and its programs are applied to accomplish specific tasks and solve problems within an organization. Discourse about the machine and the social setting shapes the ways in which the computer is used as an organizational agent. This discourse, however, is multiform and often ambiguous. It may be viewed as part of what Sebeok (1987:78) has termed a tertiary modeling system in which "nonverbal and verbal assemblages blend together" as a model for cultural expression. In the workplace, this discourse is manipulated in such a way as to justify and rationalize bureaucratic practices. Although the computer may be intended to modify social practices, it does so in ways that must be justified bureaucratically. Thus, as the discourse excerpts suggest, the computer's self-referential capacity is distinct from what people make of the computer in the workplace. These interpretations are based on innuendo and indexical references to the attributes of the machine.

Indexical expressions are utterances that require further references and contextual information in order to be understood. These expressions include indefinite pronouns, anaphoric references, e.g., "this and that," and nonverbal signs that involve pointing to or referencing an object in context. Philosophers of language including Peirce, Russell, and Bar-Hillel have discussed the importance of indexical expressions and their essential link to the contexts in which they appear. Yehoshua Bar-Hillel (1954:359-379) analyzes indexical expressions in terms of the truth value of statements. For example, the truth value of a statement such as "it is raining" depends upon the context in which it appears. Peirce (1931) considers an "index" to be a sign directly connected to the object it references such as a thermometer reading or a medical symptom. More recently, indexical expressions have been characterized as aspects of naturally occurring communication that rely on common knowledge invoked or pointed to in an interaction but never specified (cf. Garfinkel and Sacks, 1970:348-350).

This broader characterization of indexical expressions is of particular interest when we examine the computer as a sign and a modeling system in the workplace. In one respect, the computer becomes what its users want it to be in these contexts: a simple tool, a work master, an organizer, and a reflection of the future. In another respect, however, the computer's coded texts – its programs and functions – must be reckoned with seriously and require experts for their interpretation.

Semioticians have linked indexical expressions to the referential functions of language through which an utterance, when it is expressed in the third person, indirectly refers back to the original subject of discourse (cf. Greimas and Courtés, 1979:311-313). This referential process creates an illusion of "reality" or "truth" by anchoring the utterance in space and time. Referential illusion is a technique commonly used in proleptic, or future-oriented, descriptions in which the computer's efficacy as a problem-solving device is invoked. A characteristic feature of accounts employing referential illusion is the absence of a means-ends, or processual, framework in the description. Take for example the assertion: "A DBMS [Data Base Management System], by assisting you to organize and systematize a precious resource, makes your business more efficient and ultimately more profitable" (Nordin, 1986:12). This claim makes referential illusion to the power and effectiveness of the computer and its data-base system, but does not demonstrate *how* the more efficient system actually increases profits. The missing step between system's installation and practical application creates the illusion of an orderly and efficient reality by mere reference to the presence of the system. Problems encountered in learning to use and apply the system are not discussed.

Different levels of knowledge of and familiarity with computers influence how they are described and referenced in social context. For example, Turkle (1984:17) refers to the proliferation of computer-related jargon in the United States and its diffusion into everyday speech. This jargon may be considered a sociolect, or set of linguistic practices, emerging from computer users. "Buffering" information, "debugging" personal relationships, and "reprogramming" one's life are among the metaphorical usages that Turkle found for computer terms. These metaphors influence managerial planning within organizations where computers are found and have been loosely applied to the realm of interpersonal relationships. Computers subtly model social reality through indexical and metaphorical references to them in everyday and technical talk. This process, however, may not be as straightforward as Turkle suggests. The new jargon first emerges in specific social contexts where computers are used. A process of diffusion resembling the creolization of languages occurs as computer terminology is popularized and spreads to a wider circle of speakers.[11]

A.I., the young senior programmer at Nairobi's Data Centre, revealed how difficult and uncertain it is to learn to talk about computers when he recalled his training and stated: "I remember the best thing I found about it is anytime anybody is talking about anything in computers anyway // you're not really lost." Here, A.I. is referring to his early courses in BASIC and COBOL and his initial understanding of computer programming. Although he is now a senior programmer, he speaks with hesitation of his journey toward knowledge and learning what the experts "are really talking about." He is not interested in borrowed terms but instead in understanding how computer applications relate to his job. A.I.'s response resembled that of a computer vendor whom I interviewed. The vendor explained that a knowledge of programming gave him a way of displaying competence with the product he had to sell, but this knowledge was not an essential part of his work (cf. excerpt 2.12).

Another form of specific referencing arises in the adoption process when computers are first installed in the workplace. For example, Pinckney, Cohen, and Leonard describe the process by which computers were installed in Kenya's Ministry of Agriculture.

8.24 *Microcomputers in Kenya's Ministry of Agriculture*

> Requests were broken down into those from old projects, new projects and parastatals since each group was treated differently. That financial data was referred to during the meeting whenever a new line item was discussed. Agreed upon figures, including the total allocations so far, the percentage of requests that had been granted, and the average percentage cut needed to keep MOA within the Treasury ceiling, were entered into the microcomputer and new forms distributed at the beginning of the next session (Pinckney, Cohen, and Leonard, 1984:18).

The consultants outlined a process for using the computer in budgetary reform and reorganization. The budget cuts were projected and made for each line item. The computer was used to complete projections and establish a budgetary plan. Discourse about the computer referenced the entire fiscal planning process and reorganization of the ministry. During the process of implementation, the ministry switched from packaged programs and spreadsheets to a more complex computer system that limited knowledge of implementation to the senior computer experts. What is most important in this explanation is that figures were negotiated, agreed upon, and planned *before* calculations were made on the computer. The computer, in this case, served as a symbolic device for managing organizational reform. It is likely that such reforms could have been envisioned without the computer. The new technology, however, offered an authoritative way of buttressing the reforms that became a social reality when "agreed upon figures" were entered into the computer as guidelines for planning. The transformation from the old budget to the new one was mediated by the symbolic force of the computer.

This example suggests that, in its capacity as a cultural modeling device, the computer is a simulacrum for organizational change. Modifications in bureaucratic goals, titles, and work tasks are both referenced to and symbolized by the computer. In turn, contractual exchanges specify how the computer is interpreted and introduce the roles of agencies and actors who treat the computer as a commodity and an object of value. These exchanges are part of the narrative program for computer adoption and their acceptance or violation leads to the final attribution of value, or definition of the computer's role in an individual enterprise and on the national scene. If S_1 represents the subject, without a computer, S_2 the subject with a computer, and 0 is the valued object (the computer with its programs, data, and files), we may make the following representation of the narrative program for computer adoption as a model of cultural transformation:

$$\text{Transformation: } (S_1 \cup 0) \longrightarrow (S_2 \cap 0).$$

The computer unites the subject and the valued object. Thereby, it models postmodernity, or the future, for those who are willing to make a technological leap.

Information and Ideology

Contractual exchanges may now be re-examined with reference to the unfolding of the narrative program for computer adoption and the discourse that sustains the program. The computer contract establishes and stabilizes the role of new technologies in business and government. At the confrontation stage, the potential of computers appears to be boundless both under the progressive and the fragile computer contracts. Lacking a means-end component, the discourse of computer confrontation links technology to utopian outcomes. Problems arise at the moment of the first qualifying test when the obstacles of purchasing and using computers are faced.

The shortcomings and daily struggles of the world of work with computers are strangely absent from accounts that glamorize the new technology as a positive ideological force and an instrument of social reform.[12] In contrast to these accounts are the type 1 and type 4 contractual exchanges that reflect the effects of overregulation, ambiguous government policy, and inappropriate computer technology. Computer consultant Clay Wescott (1985:2) describes the Kenyan government computer services in Nairobi where he states that "data processing capabilities are lagging far behind the government's needs." Three and four-year data backlogs and outmoded equipment were not unusual at the Kenyan government computer installation in 1984 and 1985. The ambivalence of the fragile computer contract reinforces conditions of scarcity in the computer market. In contrast, the progressive contract alters not only the availability of the computers but also the attitudes toward them by linking new technologies to an ideology of economic development and autonomy.

An analysis of contractual exchanges demonstrates divergent patterns in the progressive and fragile computer situations. These patterns range from formal statements of government policy to actual economic transactions and organizational changes. The computer's role as a symbol of change and the values attributed to the technology are framed by contractual exchanges. Direct regulation of access to technology appears to lead to a positive evaluation of computers while ambivalence and indirect regulation result in mixed messages, confusion, and frustration for computer users. This interpretation is by no means airtight and conclusive. Individual computer successes in Kenya have actually occurred because of the innovative business strategies necessitated by government economic regulation. The vitality of syncretism and technological adaptations made out of necessity should not be overlooked. These adaptations are inevitable even in the best of circumstances. Nevertheless, the importance of the computer contract to the existence and diffusion of new technologies emphasizes the crucial interplay among government, firms, and individuals in fitting the computer into an organizational structure.

A positive response to new technologies is not based solely on individual acceptance but is instead an artifact of the interaction of a number of forces that shape the technological paratext and influence how computers are acquired, regarded, and used in the narrative program of computer adoption. In the concluding chapter, these structural forces will be analyzed from the vantage point of terminal signs through which social and political meanings are attributed to new technologies.

Chapter Nine
Terminal Signs

> Ideology is dead; only simulacra remain
> (Jean Baudrillard, 1976:9)

Introduction

When I sat down to write this chapter, I had just received a letter from the son of one of my research assistants. It had been ten years since I conducted an employment survey of his urban village in Zambia. He was a member of a small band of children who followed me from door to door and watched the recording of interviews with a sense of awe. Now, he informed me, he had passed his secondary school examinations, and he was ready to travel to the United States to study computer science. Still active in the local religious group that I had studied a decade earlier, this young man felt that it was now time to deal with his future and the future of his country. He found the answer to his quest in computer technology. What, I asked myself, is the appeal of computers to the youth of Africa? Why had this young man turned to technology in his search for change and an improved way of life?

Computers are a symbolic source of empowerment in a situation where the hope for progress continually recedes. Yet, certain dominant discourses about development in Africa cast technology as the villain. According to Walter Rodney (1972:35):

> The capitalist countries are technologically more advanced and are therefore the sector of the imperialist system which determined the direction of change. A striking example to this effect is the fact that synthetic fabrics manufactured in the capitalist metropoles have begun to replace fabrics made from raw material grown in the colonies. In other words, (within certain limits) it is the technologically advanced metropoles who can decide when to end their dependence on the colonies in a particular sphere. When that happens, it is the colony or neo-colony which goes begging cap in hand for a reprieve and a new quota. It is for this reason that a formerly colonized nation has no hope of developing until it breaks effectively with the vicious circle of dependence and exploitation which characterizes imperialism.

This view places technological power in the hands of the West and offers no way out of the dilemma of domination for African nations. An equally pessimistic approach is presented by the French scholars who focus on "failed development" in Africa. A disciple of René Dumont's approach to failed development in Africa, Jacques Giri (1986:117-131) argues that the past two and a half decades of independence on the African continent have resulted in an unprecedented economic and political decline that is rapidly approaching a crisis.[1] He documents rising national debts, disasters in industrial development, and a fall in the standard of living in many African nations since their independence. According to Giri, the "El Dorado," Africa's utopia of political independence and modernity, has been irretrievably lost. In the place of this utopia are governments seeking the appearance of development based on the special interests of the ruling classes. Giri's solution is a sane political economy in African nations and a dynamic thrust toward development shaped by the people, with active participation of the rural sector (cf. Giri, 1986:183-184). Giri's narrative for African development is utopian and unrealistic because it presumes a massive mobilization of political and economic resources based upon national systems of communication that are only beginning to emerge. The case studies presented in this book suggest an alternative scenario of technological development.

Part I of this book has set the stage for the study by introducing the narrative program for computer adoption. This program is a schema that facilitates the analysis of spoken discourse data, official documents, and ethnographic descriptions of work and educational settings in which computers are used. In Part II, I have presented ethnographic studies of computer adoption in Kenyan and Ivorian settings. These settings have been compared with reference to the fragile and the progressive computer contracts that characterize different types of national policy and its local implementation in firms and schools. In Part III, I have explored the implications of these case studies through a microanalysis of human-machine interaction and a sociosemiotic study of the contractual exchanges essential to technological diffusion. My analysis concludes with a consideration of terminal signs, or the interpretive strategies and processes of signification that accompany Africa's entry into the postmodern era.

Signs of Postmodernity

The image of the modern, industrial era is one in which the presence of heavy, complex technologies transforms the order of society by increasing industrial productivity and creating a rational economy. According to Jean Baudrillard (1976:10), industrial machines correspond culturally with a conscious, rational, and referential process of signification. The machines of postindustrial era, in particular the computer, transform productivity into a cognitive and symbolic process. Following the narrative program for computer adoption, an attribution of value to this new machine, the computer, is the terminal utterance that brings the narrative of technological change to closure.[2] The terminal utterance signals the acquisition of the object of value by the performing subject. This process is reflexive because the very presence of the computer enhances and alters the position of the subject who makes use of it. As the narrative of computer adoption closes, terminal signs are appropriated by the computer user with the prospect of changing the individual and the social contexts of computer use. According to Greimas (1970:176), the terminal utterance, or sign, obliterates sender-receiver and subject-object distinctions by uniting the subject with the appropriated object in a final attribution of shared value.

The reflexive character of signs of postmodernity is at the heart of new cultural forms. We have seen how these signs are manipulated in the discourses of confrontation and domination that constitute the new mechanical solidarity and the social practices that link office interactions to public policy. Rather than addressing the critique of postmodern culture as a whole and the debates that this critique engenders, I posit an equivalency between terminal signs and the signs of postmodernity.[3] As instances of postmodern culture, terminal signs exhibit the transformation of the subject when a new technology has been mastered. In Chapter Seven, I argued that technological mastery is always incomplete because it is characterized by system opacity, which creates a fundamental knowledge gap between the computer user and the machine. Moving into the domain of social and organizational analysis, in Chapter Eight I examine the processes of acquisition and technological mastery in terms of contractual exchanges that delimit the uses of computers. These contractual exchanges are the vestiges of a modern society in which social relations are regulated by the terms of implicit and explicit contracts. The computer contract establishes the presence and legitimacy of computers as technological tools in Kenyan and Ivorian businesses. These contractual exchanges represent an effort to domesticate new technologies and to restrict the technological power of firm managers and workers.

The computer contract defines technology as essential to business and transforms modern enterprises, where bureaucracy was sustained by the manual processing of documents, into dynamic, postmodern firms dependent upon computers. Whether they are automobile parts or insurance claims, the commodities handled by these firms become the servants of technology. Locating an automobile part or an item in the stock inventory is secondary to the technological capacity to find the object. Technological competence creates bonds of special interest among firm directors and computer managers. Their allegiance to a specific firm is compromised by their deeper commitment to computers and the opportunities for horizontal and vertical mobility generated by technological mastery. This allegiance to technology challenges the contractual basis of the computer's presence in firms. When it is appropriated as a terminal sign, the computer itself, and not rational planning, holds the key to the future.

A dominant characteristic of terminal signs is their projection of the future by uniting the subject and the object of value in a newly empowered state. Four terminal signs have been identified for this study: (1) modeling devices, including the computer program and computer texts; (2) representational signs, encompassing representations of the future created by new technologies; (3) totalizing signs, represented by the computer terminal itself; and (4) analeptic signs, consisting of images that link the past to the future, as illustrated by the metaphor of the last train of the twentieth century. In these four types of signs, the process of signification turns back upon itself and creates an image of an image. These images are evanescent, infinitely receding, and difficult to capture. Since subject and object are united into a single signifier in terminal signs, the subject recedes from view as it is transformed. When computer managers in Ivory Coast and Kenya echo the refrain that they "do everything and nothing," they provide a glimpse of this receding human subject whose existence and self-definition are dependent upon the vicissitudes of technology.

At the moment of closure, the narrative program for computer adoption is reversed and may be read in terms of the final appropriation, or attribution.[4] The new schema of attribution ———→ domination

————→ confrontation permits a reinterpretation of the previous conflicts in the narrative and their significance. Valorization of new technologies under the progressive computer contract generates a positive reading of all prior efforts to use computers. In spite of obstacles and qualifying tests encountered during the course of the narrative, computers are the source of progress and success. The disjunctures implicit in the fragile computer contract, however, may result in a negative reading of prior events. Terminal signs are subject to contrasting interpretations, for they both close the narrative of computer adoption and make way for a new, future scenario and another potential contract.

Terminal signs contain the seeds for the review and deconstruction of the narrative program of computer adoption. The narrative program is not just a description of how computers are used in the specific cases of small businesses or government agencies. It does not provide a model for technological success. Instead, it offers a way of reflexively interpreting and evaluating the process of computer adoption. When the computer has been appropriated as an object of value, the program eliminates itself and opens the way for another narrative. Terminal signs link past knowledge and experience to the unknown and the future. In doing so, they transform knowledge into information. These signs point the direction to new ways of knowing, and they heighten our awareness of the problem of knowledge control (cf. MacCannell and MacCannell, 1982:126-128).

Computers provide a way of cataloguing serial information and expand our information resources. Data banks are at our fingertips. Although a vast array of information is available, questions arise about the sorting devices and relevant categories for assembling this information. The results obtained from a computerized bibliographic search provide an example of this information explosion. The trained eye must sort through a multitude of references in order to glean knowledge from information. Serialized information is not categorized knowledge. The reliability of this information and the questions of its authorship are also at issue. Old standards of authentication are no longer applicable (cf. Foucault, 1984:113-138).

When public discourse about new technologies and about the future enters into the picture, government policies frame attitudes about information and set formal parameters for the institutional control of knowledge. Certain descriptions of postmodernity reflect acute anxiety about the dual control of information and technology by the state (cf. Lyotard, 1979:78-88).[5] Jürgen Habermas (1971:112-113) and Alvin Gouldner (1979:39-43) argue that increased public discourse about the evolution and effects of technology will eventually lead to state regulation of technology in a broad participatory democracy. Lyotard's approach to postmodern metanarratives suggests that government control, while necessary, may produce irrational and contradictory results.

Building upon the implications of the 1978 Nora/Minc report in France, the Ivorian informatics plans attempt to affect state control by regulating access to computers, tightly monitoring progress in telecommunications, and designing a master plan for technological education. This master plan is sustained by the symbol of the contractual society, a society that is challenged and, in certain respects, superseded by the growing presence of new technologies and the models of the future that they imply.

Modeling Devices

In Chapters Three and Eight, I analyzed the computer's role as a modeling system. This analysis may be expanded by a re-examination of the modeling sign and its manipulation. The model projects a future state to be attained by the appropriation of the valued object. Learning to use the computer through a rite of appropriation makes symbolic reference to this future state.[6] The power of the modeling sign and the social practices that surround it are not diminished when the computer is abandoned. It is the anticipation of the benefits of the computer's use that constitute the model.

An analogy may be drawn for the modeling sign in some contemporary artworks from Ivory Coast. Aware of the power of the signs of postmodernity, Anoh Acou, a young Ivorian Carver in Grand Bassam, fashioned wooden telephones, radios, business suits, and jogging shoes. His next project is a wooden computer. The wooden apparel cannot be worn, and the telephones and radios do not transmit audible messages. Although they are amusing to passing tourists, these objects have a more fundamental significance for local consumers. Owning a silent wooden telephone or radio simulates possessing the functioning object. A simple referential interpretation of the modeling sign is not sufficient for an understanding of what is taking place. The referent for the wooden suit is only superficially a real suit cut from fine imported fabrics. The wooden suit is a model for the postmodern condition. Possessing the model suit promises the status and opportunities that accompany the real suit. The telephone and radio draw the consumer nearer to the networks of technological power. These carvings hold the promise of the benefits of an affluent, contemporary society.

Figure 9-1. Simulated Suits.
Carvings of modern Western apparel in the latest styles by Ivorian carver Anoh Acou, Grand Bassam, Ivory Coast model postmodernity.

Anoh Acou, whose artistic representations of technology are so intriguing, is neither the leader of an arcane cargo cult nor a grassroots follower of Andy Warhol. He is an instinctively astute reader of postmodern signs. Taking the process of symbolic representation that characterizes traditional African art to its logical, contemporary conclusion, Anoh has created a new reality by simulating it. The modeling device depicts the future through prospective and heuristic imagery. In addition, the sign-model presupposes an emotional response of longing and acquisition. New powers emerge from possessing the model without the requirement of mastering its technological apparatus. We do not have to worry about the breakdown of telephone lines or the failure of radio communications. As a simulation, the model

Figure 9-2. Carved Technology.
The technological gadgets and apparel carved by Anoh Acou of Grand Bassam, Ivory Coast bring the local consumer into closer contact with the riches of contemporary life.

does not stop at representing the real thing. The model contains a signifier that displaces its referent.

Representational Signs

These signs reproduce the postmodern experience. They are signs of the media operating through allusion, indexical reference, and visual display. The creators of postmodern plastic arts have tried to encode the repetitive images of visual media in static art objects. Computer simulations represent information in visual displays. The concept of representation implies that signs and utterances in a natural language stand on behalf of another system. A denotation, or core meaning, for a representation can ideally be located. Charles Sanders Peirce introduced the concept of an interpretant as fundamental to the structure of the sign. The interpretant is a second order representation that mediates between the signifier and the object.[7]

The series of representations may be unlimited. Representational signs of technology build upon this chain of unlimited signification. In contrast to the model, the system of representation holds the promise of translation into another series of signs and an alternative reality. Representations stand on behalf of this reality without modeling it.

The computer program, for example, furnishes a metalanguage that can be applied to a number of domains. It may be used as both a model and a representation. On the other hand, the computer print-out provides final results. It represents the steps that computer users take in applying the program. Yet, this printout is only one feature of a larger process such as inventory control or statistical reporting in the business world. The printout appears to have finality and an unalterable meaning (cf. Figure 9-3). What it represents is clear to the computer user. Nevertheless, the printout is meaningful only in a specific context, when accompanied by other printouts and a larger set of goals within an enterprise. In Chapter Five, I described the efforts of P.M., the computer controller at the Kenyan retail trading company, to generate printouts containing stock, inventory, and financial records. This information was already out of date when it was processed by the computer. The printout at the retail trading company became a sign for ideal business transactions and symbolized uses to which the computer had been relegated in the enterprise.

The printout of computer lessons in Figure 9-3 contains a record of the students' learning processes. It represents the results of applying a simple program to a question/answer exercise. The computer program discussed in Chapter Seven required students to fill in factual questions by identifying historical figures. Two steps were involved: (1) following the program's format and (2) producing a correct cultural identification. Students generally managed to follow the program with little difficulty, but they had trouble identifying historical figures. The correct answers relied on cultural knowledge not contained in the program. Lacking a cultural reference point for their answers, students substituted a technological referent devoid of explicit cultural content and moved on to the next stage of programming. The structure of students' responses resembles Baudrillard's description of the survey polling technique. Baudrillard (1976:96-97) states:

> We live in the mode of the poll (referendum), precisely because there is no referent. Every sign, every message . . . is presented to us as a question/answer. The entire system of communication is transformed from the complex syntatic structure of language into a binary signaling system of question/answer – of perpetual testing.

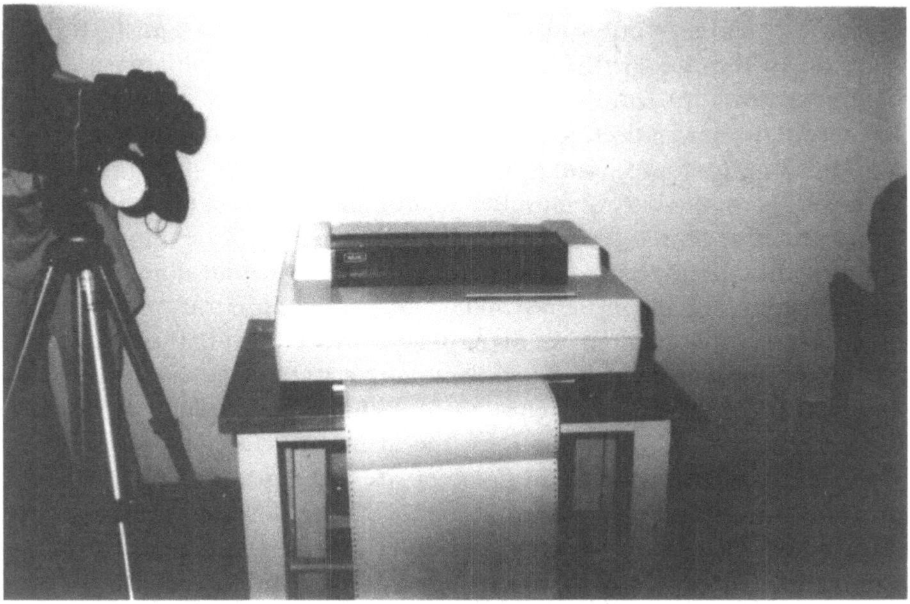

Figure 9-3. Computer Printout as a Representational Sign.
This computer printout of a completed lesson plan at a computer school in Nairobi may be analyzed as a representational sign.

This binary signaling system draws upon cultural knowledge but has technology as its ultimate referent. Therefore, a search for meaning is always elusive. The absence of a referent is actually an instance of a receding cultural referent in which technology becomes an interpretant for the referential sign.

Totalizing Signs

A totalizing sign molds unity from disparate referents. Victor Turner (1967:29-31) has described this process as the condensation of meaning and the unification of symbolic opposites into a single representation. Jean-Paul Sartre (1976:42-48) applies the concept of totalization to the production of tools and other cultural objects. He argues that a painting or a symphony may be seen as a totality within the cultural imaginary. This cultural whole, however, is born of a historical process set in motion by practical actions.[8] According to Sartre (1976:45), the whole is a "vestige of past actions." It is an illusion produced by the cognitive process of totalization. Garfinkel (1967:33) suggests that totalizing signs be viewed as accounts that organize social situations. However, in my opinion, the power of the totalizing sign lies in its figurativity. The totalizing sign or symbol combines cultural elements from different symbolic metasystems. Ilona Szombati-Fabian and Johannes Fabian (1976:7) characterize the totalizing symbol as "the kind of precious key every anthropologist hopes to find." I have argued that the totalizing sign is a vehicle for examining the anthropological concept of syncretism by providing a means for analyzing the collision of diverse symbolic forms and systems (cf. Jules-Rosette, 1984:224).

Totalizing signs unite past, present, and future in a continuous unity. The mami wata figure, or mermaid figure, combining the traditional image of an African water spirit with signs of modern wealth, totalizes symbols of old and new (cf. Figure 9-4). The mermaid is a foreign woman, Mediterranean or Indian. She wears Western jewelry as a sign of wealth and is accompanied by a snake. Promising boundless wealth to anyone who enters into an exclusive spiritual and sexual contract with her, the mermaid materializes into a venomous snake if the contract is broken through infidelity. The mermaid figure expresses the repressed fear of foreign wealth and neocolonial control. Similar totalizing signs appear in computer advertising, in which images of familiar objects and relationships are juxtaposed to high technology.

Although she is not explicitly linked to technology, the mermaid promises riches that are the products of Western technological advances. She evokes an ambivalence toward technological appropriation and a fear of domination by progress and Western wealth. As a terminal sign, the mermaid appears at the final stage of the narrative of modernity. She is a particularly apt representation of the fragile computer contract in which technology is accepted on the surface level but is rejected and feared on a deeper level. The logic of the sign relationship embodied by the mermaid is also evident in the progressive computer contract where the domination of technology through policy planning is the objective. An imaginary figure, the mermaid forecasts an uncertain future of precarious wealth at the price of personal enslavement to the bearer of riches.

As a totalizing sign, the computer terminal resembles the mermaid figure. Its mere presence in the workplace unites the hope for progress with the reality of daily office practices. In Chapter Two, I emphasized the power of computer success stories as forecasts of the boundless improvements that new technologies promise. Regardless of the uses

Figure 9-4. The Mermaid or Mami Wata.
The mermaid promises boundless wealth from a foreign source and threatens the destruction of the subject who breaks a contract with her.

of the computer, the terminal embodies and displays these hopes. It also evokes the fears of technological entrapment. In its unification of the wish for freedom and the fear of domination, the computer terminal is a totalizing sign. The terminal is the mami wata of a new age.

Totalizing signs transform history into a cultural representation. They imbue the inert tool/object with a life of its own and with the capacity to transfigure personal aspirations. The totalizing technological sign does not reference the actual functions of the computer as a tool. Rather, it thrusts the computer into a complex cultural imaginary in which control of work practices, vastly improved communications systems, and economic prosperity are promised by the tool as a sign. The computer terminal becomes a sacred object in the temple of postmodern worship. In Chapter Four, I noted that the metaphor of initiation is used to describe computer training in Ivory Coast. Computer initiation involves entry into a sacred contract with the totalizing tool/object. In advertisements for many of the Ivorian SSII, the color blue repeatedly appears as a signifier for new technolgoies.

Figure 9-5. The Computer Terminal as a Totalizing Sign.
As a totalizing sign, the computer terminal represents the hope for postmodern technological advances and the fear of domination by technological change.

The blue image of a computer terminal projects the machine into a fantasy world. This color creates a negative photographic print of external reality and a blueprint for a future that is reached by a leap from the nascent industrial society into African postmodernity.

Analeptic Signs

As with totalizing signs, analeptic signs link the past to the future, but they emphasize the importance of the past as a model for what is to come. The metaphor of the last train of the twentieth century, used as a slogan by the Ivorian government, employs the image of a heavy industrial technology as a means of promoting computerization. Like a train, the computer creates a sense of forward motion and improves networks of communication. As the last train of postmodernity, the computer connotes the end of an old era and a final and irrevocable opportunity for change.

Analeptic signs idealize the past by characterizing it as orderly and integrative. Yet, in the case of new technologies, the past is not adequate to meet future challenges. Education is a domain in which analeptic signs emerge with startling clarity. Ivory Coast has designed INSET as an experimental educational institute for engineers and computer technicians. Although this institute draws upon the hierarchy of technological qualifications contained in the old secondary school and university systems, it modifies criteria for achievement and advancement in terms of mastery of new technological systems and the requirements of businesses in the national economy. INSET, Ivory Coast's university of the future, may be viewed as both a model for postmodern education and an analeptic sign that links the pedagogical past to the technological future.

Alternatively, analeptic signs suggest a disjunction between the past and future by indicating that technological progress requires a radical break from the past. The fear of this hiatus between past and future is reflected in Kenya's fragile computer contract, which emphasizes limitations on the importation and accessibility of new technologies. By restricting the access of individuals and small businesses to computers, the fragile contract maintains the stability of the past. But this *status quo* is disrupted by individual efforts to technologize it and to challenge the barriers erected by the fragile contract. Those who perceive a radical disjunction between past and future see new technologies as dangerous and dominating unless they are in the right hands. The new information order is viewed as a harmful explosion rather than a source of empowerment for those who participate in it.

New technologies introduce the possibility of systems failure. When an agency or enterprise has become dependent upon computers, its entire operations are threatened by breakdowns and temporary malfunctioning. The broken machine brings into relief the salience of technology and the psychological bond established between technological managers and their automated support systems. At computer training institutions in Ivory Coast and Kenya, considerable attention is devoted to the smooth functioning of the machines. Blagué Valet of the SIEPAT school in Abidjan began his computer initiations by opening the machine for students. At the Starehe Boys Centre in Nairobi, students were incessantly reminded of the dangers of mechanical breakdowns and the misuse of computer hardware and software (cf. Figure 9-6). These reminders emphasize the fragility of the human-machine bond and the possibility that technological failure threatens a regression to the past.

Analeptic signs require constant renewal of the computer contract through transactions that maintain and upgrade technology. These signs emphasize the precarious status of new technologies in African nations where computers, spare parts, and repair facilities are scarce. The progress promised by modeling devices and totalizing signs may be disrupted easily, thereby imposing a new domination of the human user by the machine.

Postmodern Narrativity African-Style

Science embodies its criteria of proof in paradigms, systems of representation, and techniques of quantification. Sociologists who have tried to establish an equivalence between these systems of repre-

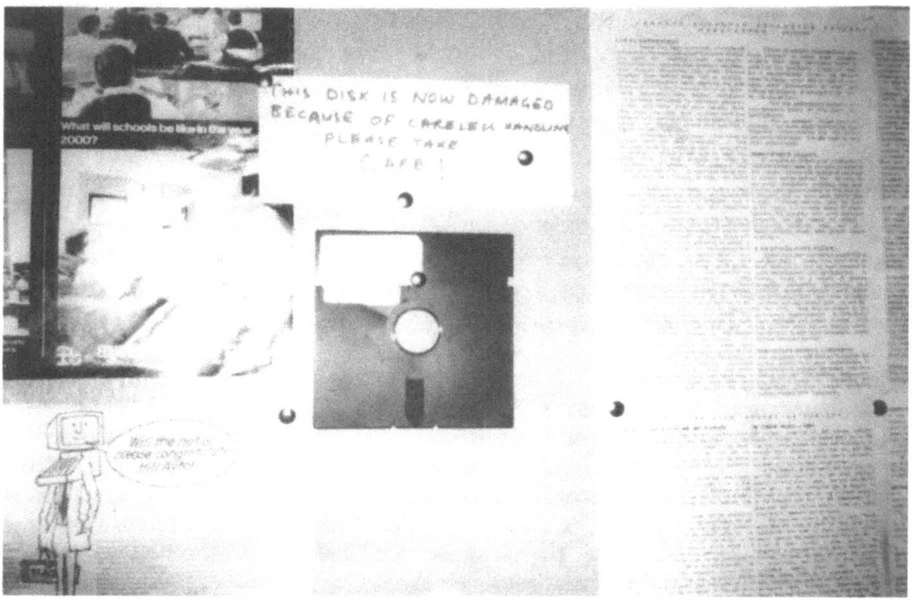

Figure 9-6. The Damaged Disk.
Students at the Starehe Boys Centre in Nairobi were warned against the dangers of damaging computer hardware and software. The broken machine is an analeptic sign that evokes a past without computers.

sentation and the unfolding of laboratory life as a narrative have met with incredulity and resistance from the scientific community.[9] Narrativity destabilizes scientific principles and results by questioning how they are constructed. Narratives, according to some theorists of postmodernity, are inherently in conflict with a type of science that does not question its underlying paradigms (cf. Polanyi, 1958:292-298; Lyotard, 1979:7-9).[10]

Yet, in this book, I have used the structure of narrative as a vehicle for analyzing technological change. The narrative program for computer adoption identifies emergent sequences in the application of technology. The program emphasizes the ways in which everyday language and social practices shape uses of technology in the workplace and lead to a questioning of technology's role in the state. Neither a metaphorical construction grafted onto daily activities in the workplace nor a heuristic device, the narrative program is an observational filter that reveals the extent to which narrativity is fundamental to human behavior. The narrative of technological innovation is based upon a scenario in which technology is both the valued object and the hero. Thus, in its dual narrative role, technology eliminates the lack suffered by the acting subject and, in turn, assumes the position of a subject to be acted upon and controlled. Viewed from the optic of the narrative program, the primacy of technology as a cultural form becomes understandable, and so too does the human subject's ambivalence toward it.

Sociologists who wish to place knowledge above technique in the scenario of social change are often puzzled by technology's primacy as a cultural form. For example, Talcott Parsons argues that knowledge, by which he means science, should guide technology in a rational society, for if the status of knowledge weakens, technological changes cannot be assimilated. Parsons (1970:620) states:

> In this context it is easier to appreciate the very qualified sense in which the modern world has become a "technological" society. There is, to be sure, a valid sense in which it is a handy characterizing slogan, a sense parallel to the slogans "capitalistic" and "socialistic," and indeed, "bureaucratic." Even at the level of characterization rather than analysis, however, it omits, or as interpreted "plays down," a variety of other features. First it seems clear that the institutionalization of the advancement and transmission of knowledge is at least as important, if not more so, as that of technology. It seems unlikely that, in the long run, the technological tail will be able to wag the scientific dog; if it does the phenomenon will require special explanation, it will not "obviously follow" from the nature of their relationship.

Although he appears to have sorted out the ideal relationship of technology to science, Parsons is caught in a semiotic paradox because technology is both the subject and object of discourse and action, the message sender and the receiver, the hero and the villain of the narrative. Drawing upon traditional African folktales, it is possible to consider another scenario in which this dilemma makes more sense.

The Kenyan computer-related success stories that I presented in Chapters One and Two may be compared to traditional African folktales. These folktales pit humans and their technologies against nature. Although technology often wins, there is always a price to pay. In these tales, the tools fashioned by people dominate nature, but only for a moment. Humans, nature, and spiritual forces are intertwined in constant and deadly combat. If people are victorious, it is due to their ability to transform themselves into superhuman beings. In these folktales, the human subject and the superhuman object are conjoined in a final test and attribution. A comparison of two Kamba folktales from Kenya with a Kenyan computer-related success story illustrates this point.

The Kamba folktale, *An Extraordinary Pumpkin*, depicts the battle of humans with their technology against nature and superhuman forces.[11] This narrative recounts the story of Wavata, a young boy (S_1), who shoots an extraordinarily large pumpkin (0) with his small bow and arrow. His father plucks the pumpkin and uses it as a gourd for brewing beer. Each person sent into the house to stoke the fire warming the beer disappears into the pumpkin. When Wavata sees the last person disappear, he runs away. As he does so, the pumpkin bursts and an ancestral spirit called an Iimu appears. John Mbiti (1966: 64) recounts the conclusion of the tale.

9.1 *An Extraordinary Pumpkin*

There came out also an Iimu, who immediately started chasing Wavata because he had shot him while he was on the vine of the pumpkin. He sang as he chased him.

"All those that foresook me,
I will eat till none is left,
Except the fearless ones."

The Iimu ate all the sheep, goats, and cattle, and every other person in that area except Wavata. Wavata continued to run

> until he was so tired that he could not run any more. He waited for the Iimu to approach him. When the Iimu stopped and came near, he shot at him with his bow and arrows until he died. Before he died, the Iimu told Wavata to cut off his little fingers, burn them, grind them into thin powder, and then throw it into the air, Wavata did so, and all the people who had been eaten by the Iimu, together with their cattle, came to life again, and there was great rejoicing that day (Mbiti, 1966:64).

In this tale, an initial complication is introduced when Wavata uses technology to dominate nature. A qualifying test occurs as the forces of nature overpower human beings, and the pumpkin swallows the individuals who heat the fire. The main narrative test takes place when Wavata, who realizes what is going on, attempts to escape the pumpkin but is caught by the Iimu. Combining courage and technology, Wavata shoots and destroys the superhuman force. At the attribution stage of the narrative, the Iimu tells Wavata to cut off his little fingers and grind them into a thin powder. When he obeys the supernatural force, Wavata's technological victory is recognized, and the Iimu's victims return to life. Technology permits Wavata to battle against seemingly insurmountable powers and to win. Yet, his victory involves a sense of loss. This narrative reflects an ambivalence toward technology, which disturbs the natural order of the universe. When this natural order has been disrupted, the technological challenge continues until the end in a deadly battle. In the right hands, technology wins, but there is a price for victory.

In another folktale, *A Man Who Could Transform Himself*, Mbiti (1966:133-134) recounts a Kamba narrative about the human desire to possess supernatural powers. In this narrative, the power to overcome human limitations is bestowed by a magical technology. An analogy may be drawn to telecommunications which create the capacity of human displacement over vast distances. Rapid, wireless communication resembles the possibility of invisible movement in which the individual flies through the night to a distant location. In the second tale, the young man Mbokothe and his brother are orphans who live together. Before they died, their parents bequeathed them two cows. Mbokothe decides to consult a famous medicine man about how to make the best use of his inheritance. The medicine man bestows upon Mbokothe the magical power to transform himself into a very large bull, and his brother drives the bull to the marketplace for sale. The bull is sold to an old man in exchange for two cows and five goats. When the man drives the bull home for slaughter, however, it escapes. The buyer chases the bull into the forest and is shocked to see that it has transformed into a frightening beast that is half bull and half lion. The buyer flees, and Mbokothe returns to find the two cows and five goats at his homestead.

Encouraged by the success of their first venture, Mbokothe and his brother try the scheme again. The brother sells the bull for ten goats and rejoices in his success. On this occasion, however, Mbokothe meets his match. Mbiti presents the conclusion of the tale.

9.2 *A Man Who Could Transform Himself*

> But this time, it was an unfortunate matter because the man who bought the bull had also been to the same medicine-man, and had received a similar treatment. When they came near the home of the man, the bull ran off, as on the previous occasion, and the owner chased it towards the forest. He came very near to catching it, and Mbokothe decided to transform himself into a lion, thinking that the man would be scared if he saw a lion. But the other man also changed into a lion, and continued to chase Mbokothe. When Mbokothe saw that he was about to be caught again, he transformed himself into a bird and flew away. But the other man changed into a kite, and both of them flew in the sky, chasing each other.
>
> Again Mbokothe saw that he was on the point of being caught, so he alighted on the ground, and transformed himself into an antelope, and continued to run. His pursuer changed into a wolf, and the two ran on, till Mbokothe yielded, and said to the other man, after they both changed back into human beings, "Now let's go to my home and I'll give you back your goats." They went to his home, and Mbokothe gave back the man his ten goats, but he and his brother retained the two cows and five goats which they obtained the first time. That is the end of the story about these two men who could transform themselves into animals (Mbiti, 1966:134-135).

This tale mirrors the competition among computer professionals and vendors. Empowered by the magical forces of technology, computer experts and vendors compete on the open market. They are armed with the tools of technology, and the battle becomes fierce, as it was between Mbokothe and his second victim. One party must yield, for they are both equally matched. In the end, the matter is resolved by a contractual agreement, and restitution is paid. This narrative concerns fair market practices in a traditional economy. It introduces the potential influence of magical powers on economic exchanges. These powers include the magic of technology, which can increase the human capacity to deceive, intimidate, and control. Mbokothe meets his match in another man who possesses identical powers. A comparison may be made between the powers possessed by these two heroes and the consequences of technological diffusion across many firms. One cannot be certain whether technology's magic resides in the hands of the government or of a single person.

The final attribution in this tale is a mutual recognition of magical powers possessed by both men and a call for resolution through agreement. This resolution may be compared to the contractual exchanges described in Chapter Eight. If the two men refuse to harness their magical powers and use them constructively, only a battle to the end resolved by a truce restores order.

These two narratives from Mbiti's collection of Kamba folktales bear a striking resemblance to a brief account of computerized farming techniques presented in a 1985 issue of *East African Computer News*. In Chapter One, this account was analyzed as a case of positive computer attribution that describes the relevance of computers to the daily lives of Kenyan farmers (cf. excerpt 1.10). Farmers are advised to install computers as an accurate means for monitoring the milk producing capacity of their cows. I shall repeat the conclusion of this tale.

9.3 *Getting Rid of a Particular Cow*

> A farmer, for example, may want to use the PC to keep records of cows on his farm. He may want reports telling him at a glance how much milk each cow has produced each month. Over a period of several years, the farmer may want reports on all cows showing which one has produced least. Such reports will assist him to determine when to get rid of a particular cow (*East African Computer News*, 1, 2, October 1985d:13).

Not only is the personal computer portrayed as a recordkeeping device, it is also an essential aid in decision making. As in the story of Wavata, technology arms the human subject with the ability to control and dominate nature. Life and death decisions about cows are made based upon computerized records of each cow's milk production. The assembly of consistent, hard records rather than impressionistic accounts of productivity gives the farmer an omniscient overview of this situation. Supported by the magical powers of technology, the farmer conquers nature.

Juxtaposition of the traditional tales and the contemporary account of the computer's uses emphasizes the ways in which technology's powers are appropriated in the battle of humans to control their environments in Kenyan narratives. In addition, this juxtaposition highlights the nonrational, narrative dimensions of modern discourse about computers. It is possible that the author of the news item in *East African Computer News* was merely searching for a familiar format in which to recount computer successes and, therefore, utilized a folktale style. A more convincing interpretation of the homologies between the traditional and the contemporary tales, however, may be based on their use of terminal signs. In each case, a final attribution depends on the acquisition of magical powers that are tied to both technology and the marketplace. In the traditional tales, human courage and social agreements are the source of resolution. In the contemporary tale, the computer empowers the human subject with the capacity to make a final decision that is technologically accurate and magically sustained by a superhuman force.

These stories raise the question of whether postmodern narrativity is actually new to Africa. Buried within African cultural traditions, in this case the Kamba folktale, are the kernels of a postmodern, utopian dream. The human subject recedes in confrontation with magical forces. Although fashioned by a village boy and a medicine man, these technological forces allow their users to transcend the human condition and to become all-powerful. The seeds of a postmodern dream of technological empowerment are alive in these traditional narratives.

Both the narrative structure of computer tales and the process of bypassing described in Chapter Eight suggest that Africa is a fertile cultural domain for the computer revolution. The promise of release from economic deprivation through technological skill and achievement is an old African narrative. The embodied and ritualistic character of work with computers resonates well with African ceremonial life, as the discussions of computer initiation have indicated. In its unraveling of the modern rationality of the industrial age, postmodernity and its machines force us to re-examine a more traditional cultural heritage. Beyond their status as short-term solutions to an economic crisis, computers are especially suited to African settings as potent cultural tools for interpreting and coping with the society of the future.

Africa and the Computer

Most African nations have not yet achieved widespread diffusion of computers and new information technologies. These countries, such as Mali and Sudan, rely upon other African nations and the West for the fragile implantation of new technologies on their soil. Although they differ in their orientations toward policies and goals for computerization, Kenya and Ivory Coast are still at the early stages of computerization. Kenya's approach, characterized by a fragile computer contract, has been conservative and restrictive. In certain respects, Kenya's gradualist approach to computerization resembles that of its neighbor, Tanzania. In 1974, Tanzania temporarily imposed a ban on computer importation and simultaneously formed a government-based Computer Advisory Committee (CAC) to regulate the diffusion and use of computer technology. Critics of this policy have referred to the temporary computer ban as "uniformed" but have been more supportive of the process of centralized computer control (cf. Lewis, 1988:287-289).[12] In terms of economic policy, Kenya's approach has paralleled Tanzania's early efforts at import control. However, on the level of policy planning, Kenya's strategy emerges as more ambivalent and less focused than that of Tanzania. Kenya's version of insider controlled capitalism partially explains the government's attitude toward computer policies. Lack of regulation creates an area of freedom for local and foreign investors who wish to champion new technologies. At the same time, lack of computer regulation inhibits the growth of technological monopolies and has caused international computer companies to withdraw from Kenya due to the restrictive market for their goods. The modernity of Nairobi's business world is characterized by an uneasy balance between opportunities for progressive expansion and suspicion of technological innovation along with foreign intervention.

Following the example of France, Ivory Coast has sought to place state controls on the computerization of society and to use computers as the basis for building a postindustrial economy. Having achieved a national awareness of computer technology in the domain of government planning, Ivory Coast considers itself to be in the second stage of computerization. An infrastructure, consisting of government ministries and national committees, is now in place to regulate computerization. Rather than limiting importation and hampering the economic conditions for computer diffusion, the country has emphasized greater access to computers and the technological expansion of education and private industry. This strategy has involved a combination of government and private sector efforts to distribute and upgrade technology. Rather than treating formal education in science and technology as an adjunct to this process, Ivory Coast has developed special schools and institutes to nationalize a cadre of engineers and computer professionals. Although this approach to indigenizing computers appears to start from the top and filter downward by emphasizing the technological modernization of a new elite, it involves a lateral coalition among government, industry, and small businesses. Although foreign consultants have been used in this computerization process, the government's primary goal has been the nationalization of technological expertise.

Computer consulting companies have spread at a rapid rate in Ivory Coast, and a new domain of economic growth has been projected, if not actualized. Ivory Coast is primed to take advantage of advances in computers, electronics, and engineering, even though the economic infrastructure for technological expansion is fragile. This planning strategy draws upon the modeling and representational signs of postmodernity by building a simulation of the future as a guideline for computer policy.

Development plans are actualizations of the terminal signs of postmodernity. They evidence not only the "future-orientation" described by Mazuri (1978:323) but also the maintenance of previous representations of development adopted by countries as national actors on the world scene. Kenya's ambivalence toward new technologies is part of its overall commitment to old forms of agricultural development. In terms of government policy, an effort to nationalize Kenyan businesses and industry has not been dependent upon new technological advances. In Ivory Coast, economic nationalization, which remains an unachieved goal, has been integrated into the informatics revolution. The government has taken the lead in determining how new technologies should be used and modified to benefit the country as it approaches the twenty-first century.

Africa in the Postmodern Era

A 1985 survey conducted by the General Secretariat of Informatics in Ivory Coast shows that 84% of the population has no contact with electronic technologies and is affected only indirectly by their presence in the country. In both Kenya and Ivory Coast, the question of whether these new technologies will affect the general population in a positive manner is a source of public debate and concern. One issue in this debate is the possibility for indigenizing new technologies and using them to reinforce the position of nationals as managers, engineers, and computer professionals. This issue, however, must be seen as secondary to the question of whether these technologies will actually become widespread enough to be the source of consistent employment for a new group of technological professionals.

Talcott Parsons (1967:151-154) speaks of a "value-science integrate" that he believes determines the importance of scientific knowledge as a source of advancement and a national resource in the United States. According to Parsons' view, the political and social value placed upon science links it to economic and educational advances. The cultural criteria for this value-science integrate are by no means well established in African nations. In contrast, concern with appropriate technologies indicates a need to harness technology to the service of pressing social and economic problems that characterize what Giri has termed the African crisis: the eradication of deadly diseases, removing the causes of life-threatening famines, providing basic services to remote areas, and increasing employment and economic productivity. Will the new information technologies be of immediate use in solving these problems or will they become political tools of a growing bureaucracy and symbols of administrative expansion? Will these technologies be the source of creating a new professional elite that merely feeds off the ephemeral fruits of urban change and the promise of economic progress?

One scenario, that of Ivory Coast, projects a future in which new technologies supplement agriculture and replace heavy industry as the keys to economic prosperity. By regulating these technologies, the country intends to strengthen its role as an actor on the world scene and to enter into full and rapid communication with the rest of the world. According to this scenario, domestic changes will occur, whether or not they are planned. Computer technologies will eventually influence agriculture, medicine, and the overall standard of living of the population in the postindustrial era. Postmodern signs, however, exist independently of these social and economic changes. They project the hope that technology, like Wavata's magical bow and arrow, will permit human actors to have superhuman powers. Terminal signs imbue the scenario of postmodernity with a utopian quality and project a positive resolution to the narrative of technological change.

Conclusions

Awareness of the rhetoric and representations of development and underdevelopment in Africa is a motivation for policies concerning new technologies. As Ivory Coast illustrates, national planners now situate themselves in the context of the world economy. Examples are taken from the West and other periphery nations. Administrators in Ivory Coast consider the possibility of adopting strategies initiated by China, Taiwan, Korea, India, and Brazil a means of making the leap into the postindustrial era. New technologies are the propelling force that gives national actors a chance to compete on the world scene. My study has linked national technological scenarios to the experiences of workers in enterprises where computers have been employed for the first time. Responses of new computer users range from "dusting the terminals" to making the technologies work for them as sources of empowerment and career advancement. There is surely a gap between these localized strategies and public discourse, which does not reflect the nuances of technological adaptation. Exploring this problem by analyzing the everyday practices of computer users in the workplace has been a major focus of this book. The implications of these everyday strategies draw this narrative to a conclusion. The results belie anything but mimicry of the West. New environments have been forged for the use of computers in African enterprises. But these environments are controlled by mixed government policies, contradictions between stated goals and economic capabilities, and multiple visions of the possible futures created by adopting new technologies.

Several lessons constitute the moral of my narrative. Some of these lessons are methodological. Others are substantive.

1. The uses and interpretations of new technologies vary greatly across African nations. A single case study of Kenya, for example, would have presented the impression of a tentative and ambivalent process of technology transfer in Africa. By dint of a comparative approach, a contrast has been established between the fragile and the progressive computer contracts under different cultural and social contexts of computer adoption.

2. Typically, policy assessments and studies of technological innovation examine the economic factors surrounding technological change. By using a sociosemiotic model for computer adoption, I have been able to study interpretations of computers as signs and symbols and to incorporate everyday behavior, verbal accounts, and official documents describing new technologies into my analysis. This approach has integrated microsociological insights about the narrative of computer adoption with the macronarrative of technology's role in national development.

3. As a consequence of this method, we have seen that African governments have been foresighted enough to establish national plans for the implementation of computer use. Through contractual exchanges and economic transactions, these regulations have influenced daily practices in the workplace. In the absence of such regulations, computer adoption has continued, as the Kenyan case illustrates, but major obstacles have arisen. Therefore, the final attribution of cultural value to new technologies has been mixed.

4. Through case studies, including a detailed analysis of audiovisual material, I have demonstrated the strong link between computer education and the recoding of titles and tasks in the workplace. With respect to a macroanalysis, this link indicates that the nationalization of computer education, which is underway in both Kenya and Ivory Coast, influences the definition and availability of computer-related jobs. On the microsociological level, this analysis suggests that mastering new technological procedures is independent of learning about the cultural content of encoded messages. This possibility emphasizes the dangers of partial technological mastery and inspires the hope for an indigenous cultural interpretation of new technologies in Africa.

Regardless of the attributions made about computers as this narrative closes, it is clear that widespread automation, with its benefits and shortcomings, has not yet reached the African continent. Computers remain tools of administrative organization and business management. Although they allow specialists, managers, and engineers to vie for new sources of prestige and income, the overall impact of new technologies can best be interpreted as the dream of a postmodern future of prosperity and the reclaiming of modernity's broken promise. Manipulated by the new technicians of the sacred, this utopian dream may create new representations of postmodern culture while only imperceptibly altering the life chances of many individuals on the African continent for a long time to come.

Notes

Introduction

1. The narrative program involves the action of one subject on another through a process that entails four stages and concludes with a contractual resolution. A.J. Greimas describes the narrative program as a method for textual analysis in several studies (Greimas, 1970:172-283; Greimas, 1987:93-105). The narrative program is also outlined by Joseph Courtés (1976:63-73) and Anne Hénault (1983:76-95).
2. Emanuel A. Schegloff (1987:225-229) argues for an exclusive focus on the structure of everyday talk as a methodological strategy for exhibiting the social context of discourse. According to this model, speech exchange systems cannot be correlated *a priori* with aspects of the social organization of settings in which particular speech forms emerge. In contrast, I have endeavored to show the links between spoken and written discourse in social settings where new technologies are introduced (cf. Ong, 1977:42-57; Tannen, 1982:1-16). Using the work of Kenneth Burke as a point of departure, Joseph Gusfield (1981:51-108) has examined the unfolding of public policies as dramatic narratives. Although my approach to narrative and discourse differs from his, I too attempt to link public policy to a larger cultural narrative that grounds and shapes it.
3. I studied Kenyan carving cooperatives in Nairobi, Wamunyu, Makindu, and Mombasa (cf. Jules-Rosette, 1984:105-141). The largest of these cooperatives, the Akamba Industry in the Changamwe section of Mombasa, has approximately 1,720 members. Although carvers completed pieces on an individual basis, apprentices did the work of rough cutting, sanding, and polishing for them. Thus, a factory-like atmosphere, although certainly not an assembly line, was approximated. The revenue of the cooperatives seldom reached more than $30,000 a year. Nevertheless, these enterprises were far larger in terms of both capital and number of workers than the cottage industries that I previously studied. They received state support and external aid and required export permits to market their goods. Hence, in many ways, the activities of these cooperatives brought them into the formal, modern business sector in spite of the informal economic activities of individual cooperative members.
4. Although Ilsa Schuster (1979:96) describes the collective interests of single, subelite women as "sex, drink, money, clothes, and status," she does not present any speculations on the percentage of such women who are actually in the clerical force. Her discussion gives an impression of the feminized workplace by virture of its focus on the problems of adaptation faced by subelite women in the wage labor force. In the Kenyan case, which bears

some historical resemblance to the Zambian situation, women constitute less than 10% of the white-collar employees in the finance, insurance, real estate and business sector. This employment pattern also reflects the potential number of reskilled female computer users in the business sector.

5. Suzanne Grant Lewis (1988:265-283) sees the new group of African computer experts primarily as a profession. She analyzes the influence wielded by the Computer Association of Tanzania (COMPAT) with reference to the establishment of professional standards by a coalition of experts from the public and private sectors. Although Lewis concedes that this group reinforces its status as a technocratic elite, she does not examine computer professionals in the context of the growing clerical subelite in East and southern Africa.
6. Michel Crozier (1971:10-11) compares the dramatic increase in the number of white-collar employees in the United States between 1940 and 1960 to the proportionately similar but numerically much smaller increase in France from 7.5% to 12.5% of the work force during the same period. He accounts for the differences by stating (Crozier, 1971:11): "Substantial modernization was possible, without many transfers into office jobs, to the extent that the percentage of nonmanual workers had until now been excessive in relation to the real development of the economy." Although the 1976-78 coffee boom in Kenya represented a period of administrative expansion, the proportion of white-collar employees in the banking and finance sectors has remained small in relationship to the total economy.
7. Clerical workers in manufacturing, transport, and administrative services are not listed separately in figures from Kenya's Central Bureau of Statistics. It is, therefore, difficult to estimate a total figure for white-collar workers in these areas. A picture emerges of a white-collar group consisting of approximately 38,547 workers as of 1981. This number is proportionately much smaller than it is in the United States, where office workers already numbered over nine million by 1960. The Kenyan profile probably bears more resemblance to the patterns described by Crozier for France during the 1960s. We may project this analysis to numbers of workers possibly dealing with major computer installations. As of 1981, Kenya had an estimated 127 mainframe and minicomputer installations as compared with 100,000 word processing machines listed in Western Europe according to OECD figures and 200,000 mainframe installations in the United States as of 1978 (cf. Scott, 1981:40; OCDE, 1982:25). Since that time, the numbers have increased dramatically in the U.S., Western Europe, and Kenya.
8. The relationship of new computer titles to office hierarchy and social organization will be discussed in more detail in Chapter Five. Job entitlement will be analyzed with reference to the administrative control exercised by technical experts and the access of these experts to specialized domains of knowledge within their organizations.
9. An article entitled "Computer Training in Kenya" appearing in *East African Computer News* (1, 1, May 1985b:7-12) briefly mentions the Computer Association of East Africa which was active during the early and mid-1970s. Since that time, the association has disbanded. Although COMPAT was established

in Tanzania in 1986, a comparable professional association did not survive the 1970s in Kenya.

10. The Association Professionelle des Informaticiens de Côte d'Ivoire was officially registered with the Ivorian government as of 1973 and published its full constitution on April 13, 1974. Based in Abidjan, the association represents the interests of individual computer experts, vendors, and computer consulting companies. Its board of directors consists of twenty-four members: eleven computer professionals, five vending companies, four computer consulting companies, two university representatives, and two representatives from the Ivorian government (cf. "Notre Association en Quelques Lignes," *L'Informatique en Côte d'Ivoire* 2, 3, 1980:4).
11. Recent studies demonstrate the feminization of the clerical labor force in the United States and a decrease in the salary differentials between white-collar and blue-collar employees (cf. Crompton and Jones, 1984:129-166). Trends in this direction have just begun in Kenya and Ivory Coast.
12. Resolution 4/19, adopted in 1980 by the UNESCO General Conference in Belgrade, contains eleven points oriented toward reducing the imbalance in information and communication technologies between developing countries and the rest of the world. The resolution urges economic, political, and social worldwide cooperation for the development and dissemination of new information technologies.
13. In 1981, a working group led by Elliot Berg submitted a report to the World Bank on the economic situation of the African continent. The report entitled *Accelerated Development in Sub-Saharan Africa: An Agenda for Action* (1981) argues in favor of Africa's more active participation in the world economy. The "Berg Report" has been sharply criticized by African leaders and scholars as contributing nothing new to the development agenda. The report places much of the blame for problems of development on the administrative organization of African governments. This report is often compared with the 1980 "Lagos Plan" developed by the Organization of African Unity as a strategy for the economic autonomy of Africa (cf. Giri, 1986:177-188).
14. The computer's role as a modeling device and a simulation of postmodern culture is discussed in more detail in Chapters Three, Seven, and Nine (cf. Sebeok, 1976:23; Zaliznjak, Ivanov, and Toporov, 1977:47-58; Baudrillard, 1981:9-17).

Chapter One

1. The concept of habitus developed by Pierre Bourdieu (1977:72-95) and the process of structuration described by Anthony Giddens (1984:16-27) are similar. Both authors are concerned with the ways in which social practices structure and reproduce institutional forms. These approaches may be used in conjunction with an analysis of cultural narratives in order to demonstrate the relationship between ideas about social institutions and the everyday practices that constitute these institutions.

2. Theodore Roszak (1986:68-70) discusses the computer's simulation capabilities as one of the major features of human interaction with computers. Simulation involves the manipulation of hypothetical models. In his discussion of simulation, Baudrillard (1981:41-42) contrasts the reduplication of images of events, such as telecasts of wars, violence, and atrocities, and the concept of image reproduction. Reduplication represents a mass mediated image of an event that actually took place, while reproduction simulates a reality through mechanical and aesthetic means. The computer's simulation of work tasks involves a process of image reproduction instead of the reduplication of an event.
3. In her study of U.S. clerical workers performing word processing and data-entry tasks, Baker (1986:11-12) describes the trance-like state that word processors enter while sitting in front of the terminal. Although this mechanization of human activity is evident in the habitual mastery of any technical skill, it is enhanced by the computer's functions. I have taken this idea a step further by analyzing the ability of workers to relate their experiences of computer use to larger organizational decisions in terms of a process of reflexive monitoring (cf. Giddens, 1984:5-6). This process of monitoring is particularly important as an aspect of the manipulation of information accrued from master files and central data banks. The social and political control of such information is a special concern for the development of national computer plans and policies.
4. At the fourth stage of the narrative program, according to Greimas (1970: 174-176), all of the relationships that have previously constituted the narrative are reviewed. The narrative program is reversed starting from the attribution stage and moving backward to the domination and confrontation stages. It is then possible to assess the relationships between the sender and receiver of communications at all stages. From a sociological perspective, the contract stage allows us to observe how the computer is integrated into the final resolution of technological change. At this point, it is possible to assess what relationships are established between humans and machines in various forms of discourse ranging from communications of public policy to everyday talk in the workplace. In this narrative resolution, terminal signs are used to attribute meaning to technology.
5. In Ivory Coast's *Plan Informatique, 1986-1990* (p. 27) a groundbreaking speech by the President of the Commission Nationale pour l'Informatique is quoted. This speech identifies computers as "the last train of the twentieth century" and discusses them as the keys to winning the "battle for the redistribution of productive forces in the world." This speech, a portion of which is quoted in my text, contains representations of development and the future.
6. Ali Mazuri quotes on two occasions from a revealing article appearing in the June 7, 1976 edition of *Weekly Review*, Kenya's weekly news commentary magazine (Mazuri, 1977:288-281; Mazuri, 1978:321-322). This article supports a restrictive use of new technologies and characterizes them as a force under foreign control, inhibiting the equitable distribution of the fruits of economic

progress and development. This article is provocative because it reveals the Kenyan government's restrictive and crisis-oriented approach to adopting computers in the public and private sectors.

7. Robert Scott's survey of computers in Kenya presents estimates of the number of mainframe installations in the country between 1961 and 1981. Although Scott qualifies that exact figures were not available to him, he estimates the number of mainframe installations at 127 as of 1981. A similar study was commissioned by the Ivorian government. This study estimated the number of mainframe installations to be 275 as of 1981. Suzanne Grant Lewis and Mohammed Sheya (1987:9-16) conducted a preliminary survey of microcomputer installations in Tanzania. They located 460 microcomputers, of which 419 were institutionally owned. Most of these microcomputers had been imported after 1983 (cf. Lewis, 1986:5). Over 60% were used in the public sector. Comparable microcomputer estimates are not available for Kenya. According to *Annexe I* of the *Plan Informatique National, 1986-1990* (p. 26), Ivory Coast had 221 microcomputers in the public sector as of 1985. The overall estimate for all microcomputer systems in Ivory Coast as of 1985 is approximately 1,221.
8. The Management Committee of the Nairobi Handicraft Cooperative consists of twelve elected members. All cooperative members must be carvers. Often, the members of the Management Committee are clan elders who are also experienced carvers. Staff members who do not belong to the cooperative handle daily accounting and bookeeping needs under the general supervision of the Management Committee (cf. Jules-Rosette, 1984:113-127).
9. I am indebted to Mr. Brian Wray of the Aga Khan Academy's Computer Project for bringing to my attention the relative definition of enterprise scale as it applies specifically to the process of computerization in Kenya. Indeed, "small" businesses in the international context may constitute important local firms in the Kenyan setting.
10. Government policies toward the informal sector are the subjects of public debate and discussion in Kenya. In *Part I* of the *Development Plan, 1974-1978* (pp. 278-280), the Kenyan government emphasizes its support of the nationalization and growth of local enterprises. Informal sector enterprises are essential to this effort. In an article in Kenya's *Daily Nation* (1984:9), G.K. Ikiara, a lecturer in the Department of Economics at the University of Nairobi, emphasizes the importance of the government's attempts to encourage investment in the informal sector through providing construction and marketing subsidies.

Chapter Two

1. In 1985, the government of Ivory Coast conducted a survey on the public response to computers and new technologies. The *Plan Informatique National, 1986-1990: Annexe II* (1986:7) reports that 84% of the Ivorian population

had little or no contact with computers when the survey was administered. Of the subjects questioned, 54% had heard of computers, and 65% believed that computers would contribute to a rise in unemployment. This survey was the first step in the government's efforts to educate the public about the economic benefits of computer use and automation.

2. In 1972, Michel Ginguay published his *Dictionnaire d'Informatique*, containing the English equivalents for a wide range of French computer terms. A more recent effort is the *Dictionnaire Informatique* published by Renée Fisher and Philippe Kruchten in 1983. This dictionary provides French equivalents for English computer terms. In actual French parlance, English loan words are often used. The second edition of *The Random House Dictionary of the English Language* includes 12,000 new terms related to computers and information processing. Many of these terms have now entered everyday, non-technical discourse.
3. Baudrillard (1970 and 1981), Lussato (1981), Turkle (1984), and Roszak (1986) propose that the computer is a simulacrum of human activites. As emphasized in the Introduction, scholars analyzing the role of the computer in the Third World contend that this capcity of the computer is not necessarily universal (Mazuri, 1978:320-342; Stover, 1984:59-73). Because the computer is an imported technology, which these authors argue may be ill-suited to the tasks and orientations of non-Western workers, they view it as a source of sociocultural disruption that increases economic dependence and introduces ways of thinking alien to the work settings in which it is used.
4. Discourse set 1.7 in Chapter One illustrates the symbiosis that develops between the human operator and the computer. In this case, the operator describes nearly falling asleep in front of the terminal as he enters a trance-like state. In an editorial from the *East African Computer News* (1, 4, December 1985/January 1986:5), Maxwell Kareithi describes "symptoms including tiredness, headaches, giddiness, nausea, irritability, anxiety, and depression" resulting from prolonged work in front of a terminal (see discourse set 2.8). The mechanization and automatic quality of the computer operators' work entail a surrender of self that creates reactive states ranging from total absorption to nausea as byproducts of the efficiency and routinization of daily work tasks.
5. In concluding his comparative study of computerization in Nigeria, Tanzania, and Uganda, John Wallace (1977:13-41) advocates more broadly based training for systems analysts and the preparation of clerical staff and other office personnel to perform the duties of office managers and systems analysts within their specific domains of work.
6. In organizations where computers are essential to daily work tasks, the control to which James Coleman (1980:145-151) refers is transferred not only to superordinate technical and managerial experts, such as systems analysts and computer managers, but also the machines themselves. Consequently, the computer becomes an anthropomorphized object that actually takes on a position of symbolic authority within an organization.
7. Pamela De Lagy described this example of the "sacred" and isolated mainframe computer at a Sudanese research institute in a 1987 personal communication.

8. Stephen Tyler (1985:93) refers to the attempts at impersonal and objective description in ethnographic research as strategies for establishing a cognitive utopia in which the social world described is a product of the ethnographer's vision. Accounts of organizations that idealize the role of the computer establish similar utopian visions of the power of technology.

Chapter Three

1. Educational control of the computer profession is evident in Ivory Coast's national informatics plans and in the criteria set by the APICI and GIMI, the professional computer associations. In Kenya, development of the computer profession based upon technical competence involves a tension between support of popular access to knowledge and the strict control of technical information and competence.
2. Computers, as the technical locus and apparatus of information, have become the valued objects of the new technicians of the sacred. These new technicians engage in a "technological colonization" of knowledge in their attempts to regulate access to the machines and the skills necessary to operate them (cf. de Certeau, 1980:28). The network of competent computer users, while loosely articulated, has sectarian features. Even those individuals with some computer skills may be excluded from the inner circle of experts who are known to and call upon each other when certain positions need to be filled.
3. The text of this article appears in discourse set 2.8. In this article, Maxwell Kareithi argues that computer operators are exploited, overworked, and underpaid. He proposes that the solution to this problem is the integration of computer operators into the technological decision-making processes of firms.
4. An editorial in the August 12, 1984 edition of Kenya's *Daily Nation* requests that the government stop treating computers as luxury items and lower the import duties. A 1985 article appearing in *Signature*, the official magazine of Kenya's *Diners Club*, complains of the high duties and overheads faced by the computer industry in Kenya (Eldon, 1985:7). These editorials imply that computer purchases by individuals and small enterprises in Kenya are limited by official government policies concerning computer importation.
5. A development discussion paper prepared by the Harvard Institute for International Development (Pinckney, Cohen, and Leonard, 1984:2-4) explains that pressures from donor agencies such as the Integrated Agricultural Development Project of the World Bank pushed the Ministry of Agriculture to reform its budgetary management practices. Administrative pressure in the form of ministerial allocations also influenced the agency's decision to review its process of budgetary planning. A revised version of this paper is published as "Kenya's Introduction of Microcomputers to Improve Budgeting and Financial Management in the Ministry of Agriculture" (Pinckney, Cohen, and Leonard, 1987: 67-93).

6. Clay G. Wescott, a computer consultant for the Management Section of Kenya's Ministry of Finance and Planning states that Kenya represents "the only case we know of worldwide" where a national budget has been prepared on microcomputers (Wescott, 1985:1). Preparation began on Kaypro and Osborne computers and was later transferred to IBM personal computers. In an article entitled "Microcomputers for Improved Government Budgeting: An African Experience," Wescott (1987:95-116) summarizes his experience as a computer consultant with Kenya's Ministry of Finance and Planning during the mid-1980s.
7. These restrictions on the growth of new religious movements were, in part, legacies from the colonial era during which religious movements were viewed as seedbeds of social and political protest (cf. MacGaffey, 1983:25-50; Jules-Rosette, 1985:15-35). As sources of cultural innovation, these movements may be compared with the new insulated cultures surrounding high technology. External acceptance of these cultural innovations implies control of their organizational leadership.
8. In *Symbols of Change* (1981:171-196), I analyzed the responses of two indigenous African churches to externally funded urban relocation schemes. The members of these groups drew upon their religious ideals of community to adapt to and modify externally generated development plans (cf. also Jules-Rosette, 1979:225-238).
9. Ali Mazuri (1978:337) states: "But even if an African country has to deal primarily with the West when it comes to computers, it makes sense to exploit competitive tendencies between western monopolies. . . . Just as international business competition once facilitated western imperialism, so international business competition could facilitate decolonization – if the victims of imperialism can learn how to exploit the opportunities presented to them." Mazuri, however, does not explore the consequences of such diversification and reverse exploitation in terms of developing viable and well integrated computer systems for government and industry.
10. The concept of modeling systems was elaborated at a 1966 conference of Soviet scholars held near Tartu. This concept has much potential for developing a communicative frame of reference for the study of computers as cultural objects (cf. Sebeok, 1976:22-23; A.A. Zaliznjak, V.V. Ivanov, and V.N. Toporov, 1977:45-58).
11. Graeme Salaman (1979:142) emphasizes the ways in which technology is used within work organizations to control productivity and reinforce hierarchy when he states "[M]echanization, automation, and computerization are forms of control. Technology is a mechanism of control. These operate to constrain and direct employees' behavior, limit or reduce their choices, determine the quality and quantity of their work activity, and all without the apparent involvement of management. These controls just become part of the everyday management of work, and as such are hard to resist or oppose."
12. This application of the narrative program to the computerization of business and government is intended to underscore the ways in which the discourse and interactions surrounding computers may be used to identify and shape the course of technological change in Africa.

Chapter Four

1. A report on the Conference for Integration of African Informatics appears in *L'Informatique en Côte d'Ivoire* 2, 3, 1980:25-29. This publication is the official journal of the Association Professionelle des Informaticiens de Côte d'Ivoire. The results of the conference include the five thematic resolutions outlined in my discussion. In addition, two subcommittees functioned as computer commissions during the conference. The first commission made a series of recommendations intended to increase African access to international telecommunications resources and data banks. The second commission explored the politics surrounding new technologies and recommended strategies for using computerization to implement national development goals in African nations.
2. This organizational chart is presented in the *Plan Informatique, 1981-1985*: 20 bis. It outlines the ideal working relationship of the various subcommittees and research groups within the Commission Nationale pour l'Informatique (CNI). In 1983, some of these responsibilities were assumed by the Secrétariat Général à l'Informatique (SGI), which is the government body charged with regulating and implementing national computer policies.
3. The *Plan Informatique National, 1986-1990* is far more sophisticated than the first plan and twice as long. The word "national" has been added to the title of the second plan. Perhaps this fact indicates even more confidence in the plan's success as a blueprint for the national development of new technologies. The second plan outlines strategies for implementing the goals set by the first plan. It also contains a more detailed statement of these goals with respect to education, employment, and private sector enterprises. Regulation of computerization and telecommunications is a key objective of the second plan.
4. Portions of Ivory Coast's second informatics plan are influenced by the report that Simon Nora and Alain Minc submitted to the French government in 1978. In particular, the French TRANSPAC system, a data transmission system providing increased access to administrative information for taxation and public services, is considered as a model for SYTRANSPAC in Ivory Coast. TRANSPAC has also been used as a model by other African countries, such as Cameroon, to implement telecommunications systems and computerize customs clearance operations (cf. *J.A.-Bis*, 10, September 1985:20-21).
5. In most cases, I have used Chamber of Commerce publications as a source for the capital assets of private computer schools and computer consulting firms in Ivory Coast. Assets are listed in CFA francs, the currency shared by the Communauté Française Africaine (French African Community). At the time of my study, the exchange rate was 310 CFA francs per U.S. dollar. The value of the CFA franc is based on the French franc. Countries sharing this currency in the French African community include: Benin, Burkina Faso, Cameroon, Central African Republic, Chad, Congo, Gabon, Ivory Coast, Mali, Niger, and Sénégal (cf. Ungar, 1985:331-332).

6. In addition to using 1987 Chamber of Commerce figures for the capital assets of the SSII that were studied, I used figures provided by the companies themselves to cross-check the published listings. Several of the companies gave me brochures outlining their annual turnover and financial standing. When these materials were available, they were used to update and supplement the Chamber of Commerce data and other published information. In most cases, Chamber of Commerce listings provided a breakdown of the number of employees in the SSII with respect to nationality. These figures were cross-checked against interview data.
7. *Annexe II* of the second national informatics plan contains a survey of fifty SSII and a breakdown of their consulting activities and software distribution services (cf. Appendix A). These companies are of particular interest to the government because they serve as a means for buffering and controlling the growth of multinational computer vendors. The government also views the SSII as a resource for educating private sector enterprises about the most effective uses of computerization for their firms.
8. In "La Bataille des Géants" appearing in *Jeune Afrique Économie* 98, June 1987:79-82, Patricio Tupper presents a ranking of the top twenty international computer vendors and analyzes the effects of their competition for the computer market in African countries. It is interesting that four of the top five companies, IBM, DEC, UNISYS, and NCR are expanding their markets in Ivory Coast while two of these four – IBM and UNISYS – had withdrawn their direct distributorships from Kenya by 1986.
9. A brief commentary within a larger article entitled "L'Afrique à l'Heure de la Micro," describes the increase in microcomputer sales throughout Africa (*Jeune Afrique Économie* 98, June 1987:83-84). DEC is presented as a company on the rise with plans for expanding its sales operations in Ivory Coast, Cameroon, Sudan, and Kenya. In a position paper entitled "L'Informatique en Côte d'Ivoire," D. Desbois and G. Vidal (1984:379) describe IBM, BULL, and UNISYS as the three largest vendors in Ivory Coast. They argue that IBM's strength is in the sale of large systems and minicomputers while BULL and UNISYS have more success with smaller systems, multi-user systems, and microcomputers. The article by Desbois and Vidal predates the official presence of DEC in Ivory Coast. Inforafric, the Ivorian distributors for DEC, officially began its operations in April of 1987 and is regarded by the other vendors as a progressive company to be watched.
10. Ivorian state regulation of international computer vendors contradicts patterns of investment that political scientists describe for Ivory Coast. For example, in an analysis of the state and capitalist development in Ivory Coast, Bonnie Campbell (1987:281-303) presents an argument that has become commonly accepted among political scientists. She claims that Ivory Coast permits the development of local capital only under conditions that subordinate it to foreign capital. This pattern typifies neocolonial development strategies. Campbell (1987:287) states:

    In contrast to the Kenyan interpretation and to summarize the argument that will be developed below, the Ivorian state appears to have played

> the role of an intermediary, creating quite specific places to be ceded in time and only under certain circumstances to local interests. This permits the emergence of local capital but only at such a time and in such a manner as to be in subordinate association and, therefore, compatible with foreign capital. If the exercise of state power may be seen as the means by which those who achieved power set about reproducing the conditions of accumulation, then what is most striking in the case of the Ivory Coast is the continuity of the manner in which this has been done and the success, defined solely in terms of the reproduction of these conditions, with which it has met so far.

The attempt to establish increased local control of the SSII could be a point of departure for a new development strategy emphasizing the intervention of indigenous capital in Ivory Coast's computer industry.

Chapter Five

1. In Kenya's third national development plan, extending from 1974 to 1978, agricultural development is outlined as a target area for national economic growth, with the goal of a 6.7% growth rate in agricultural production. ***Part I*** of the ***Development Plan***, 1974-1978 (p. 197) states: "The most important feature of agricultural development strategy will be to increase the rate of public expenditures on programmes aimed at helping large numbers of farmers to intensify production." Regional agricultural growth has remained a key objective of the fourth development plan covering 1979 through 1983 (cf. Ungar, 1985:180). These goals have required increasing and upgrading systems of recordkeeping and communication in order to affect changes in agriculture and industry.
2. The projections for computerization in Ivory Coast are outlined in the nation's first and second national informatics plans discussed in Chapter Four. The plans present use of computers in the public and private sectors as a key feature of national economic development in Ivory Coast.
3. The title data processing manager (DPM) has different connotations across firms. In some cases, the DPM is a general manager of computer personnel. The role of titles and their relationship to the organization of enterprises that have adopted computers is a productive area for comparative research. Work titles provide a point of departure for analyzing how computers are interpreted as agents of organizational change.
4. The Industrial and Commercial Development Corporation (ICDC) is a parastatal agency that supports economic development in private sector industrial, business, and agricultural concerns. According to Steven Langdon (1981: 36), government officials from ICDC have served on the board of directors of many foreign subsidiaries in which the Kenyan government has acquired investments. ICDC has been instrumental in the purchase of equipment and machinery for companies that it has assisted.

5. Using the Data Centre, one of Nairobi's largest data processing firms, and similar data bureaus to set up a batch computing system is a common practice in private businesses. When the areas most in need of computerization have been established by the computer bureau, companies often purchase their own systems. Some data bureaus specialize in accounting, financial planning, and surveys of the Kenyan computer market (cf. Price Waterhouse Associates, 1986).
6. Records kept by an organization may be used for contractual and expedient purposes. Garfinkel (1967:200) emphasizes that the "assiduousness" with which records are maintained by an organization has less to do with the quality of the recordkeeping, which may be uneven, than it does with the objectives for which the records are stored. Computer experts are the entitled organizers and keepers of records in the Kenyan firms profiled in this chapter. Their titles reflect their competence to manipulate computer texts and their organizational interests in assembling these texts.
7. During my field research on the Nairobi Handicraft Cooperative in 1981 and 1982, I was informed that all cooperative members receive a 15% commission on the carvings that they turn in to the cooperative for sale (cf. Jules-Rosette, 1984:120-124). When the computer was installed, the bookkeeper confirmed to me that 15% was still the standard commission. The entry of members' payments into the computer revealed that carvers often received payments made in Kenyan shillings based on a sliding scale. Although this arrangement was not made public, apparently it was acceptable to the cooperative management and to the carvers. Note that the exchange rate for Kenyan shillings at the time of this study was 15.88 shillings per U.S. dollar.
8. The cooperative setting was not rigidly organized as an experimental environment. The introduction of the computer allowed me to structure the information and work procedures presented to the bookkeeper and his computer assistant and to monitor the use of the computer over a year's time with the assistance of a volunteer. Thus, I have chosen to refer to this setting as the computer experiment, even though controlled experimental conditions were not maintained.
9. Based on case studies in the areas of British banking and insurance, Crompton and Jones (1984:49-57) have found that the presence of computers delimited the tasks of clerks by confining their work to data entry and computer functions related to specific segments of the company's operations. They conclude that work segmentation, specialization, and repetitive data-entry tasks characterize batch computing operations and that these activities are not significantly altered by the presence of on-line computer systems.
10. According to Jean Baudrillard (1981:16-17), on the level of symbolism and representation, capitalism is characterized by the simulation of valued objects. The simulated objects and structures are reproduced as further simulations. Using this model, recoding is the social and cultural simulation of the mechanical operations engendered by new technologies. Recoding of titles and tasks is evident in the six Kenyan case studies. This process sheds light on the changing relationship between titles and work tasks in firms that have recently installed computers.

Chapter Six

1. An article in *East African Computer News* (1, 1, May 1985b:7-12) lists fifteen public and private institutions at the secondary school and university levels as providing computer education. The University of Nairobi and Kenya Polytechnic are considered to provide excellent training. The article, however, contains a criticism of the mixed quality of computer education in Kenya and urges that industry take the lead in upgrading computer education throughout the nation.
2. This assessment of training preferences is based upon my interviews with a cross-section of enterprise heads and computer managers in Nairboi. Interviewees questioned about the availability of local training referred to the University of Nairobi and Kenya Polytechnic as the best public institutions with programs in computer studies. This evaluation is corroborated by responses published by the *East African Computer News* (1, 1, 1985b:11-12).
3. As we saw in Chapter Five, Kenya is now in the process of decentralizing many governmental functions and placing more responsibility for regional planning in the hands of district officials. Decentralization involves the upgrading of provincial schools. In the future, computers may be used more extensively at the provincial level both in administrative planning and education. In 1986, some computer vendors informed me that they were gearing up for increased computer sales after 1987 as a result of the government planning initiatives. The government began to take steps toward appointing a national committee on computer policy in late 1986. These measures mark a slight reversal of the restrictive and gradualist official attitudes toward public support of computerization evident in Kenya during the early 1980s.
4. One research engineer was enthusiastic about the quality of training provided for INSET engineers. He emphasized that before 1984, training of this caliber had to be sought abroad. His views were confirmed by other enterprise heads in Abidjan (cf. excerpt 3.8 in Chapter Three).
5. In their analysis of the impact of new technologies on the job market in the United States, Rumberger and Levin (1984) argue that the need for computer specialists and technicians has been overestimated by educational institutions. They contend that the skill requirements for automated offices and industry will level off as computerization becomes more widespread in the future. This situation may eventually affect the labor market in African countries where computer use and automation are currently increasing.
6. GIMI is structured as a "horizontal" syndicate that unites and represents the interests of a cross-section of business organizations. This type of union structure is common in France (cf. Branciard, 1978:482-486).
7. The problem implied by the interview in this article is the matching of credentials, skill levels, and expectations among new computer trainees entering the labor force. Each enterprise has its own computer hardware, software, and applications of computerization. The problem of standardization of equipment is particularly acute in Kenya where old and new machines, which are often incompatible, may be found in the same workplace. The

cost of retraining credentialed workers from university settings to use the equipment in a particular enterprise or industry may simply be much higher than that of training beginners with basic computer literacy to take on the same responsibilities (cf. *East African Computer News* 1, 1, May 1985b:12).

8. In an article entitled "Learning by Computer," Mary Friedland (1985:21-23) summarizes her interviews with members of the faculty at the Starehe Boys Centre. She analyzes David Wilson's innovative role as the initiator of the computer curriculum at the school. In addition, she indicates that Starehe's Computer Project may well become one model for computer literacy in Kenya. The results of my interviews with the administration and instructors at the Starehe Boys Centre correspond with Friedland's positive conclusions.
9. The Aga Khan Academy's Pilot Project was reviewed along with the Starehe Boys Centre's Computer Project in the article by Mary Friedland cited above. Brian Wray, who had worked with David Wilson on the Starehe project, consciously decided to take a different approach to computers in the Aga Khan experiment. The goal of using computers as a supplement to academic training became the focus of the Aga Khan Academy's Pilot Project. This approach contrasts with the prevocational orientation at the Starehe Boys Centre.
10. The "daylight robbery" reaction is a commentary on the relationship among expectations for computer training, job availability, and the pool of potential trainees. A more detailed study of the extent to which school leavers resort to computer training from the commercial colleges in order to enter the job market is necessary to cross-check these impressionistic responses to the commercial colleges as a training alternative.
11. The profiles of young computer professionals presented here are based on interviews conducted in August 1986 at the Data Centre in Nairobi. These interviews bear out the overall impression that the type of training received by Starehe graduates is effective and particularly well suited to entry-level positions in the computer industry.
12. Robert Scott (1984:19-20) addresses the problem of training from the perspective of coordination and suggests the implementation of a nationalized computer policy with standardized local examinations. Employers appear to be less concerned about the standardization of examinations and the credentialing crisis than they are with competent, cost-effective job performance. In fact, the credentialing crisis in the computer world occasionally operates to the employers' benefit by allowing them to set internal standards for competence and avoid higher pay scales for employees with certificates and diplomas from various public and private institutions.

Chapter Seven

1. A preliminary analysis of the data for this chapter was included in a paper entitled "'Who Was Mozart?': Computers and the Semiotics of Invention" presented at the annual meetings of the Semiotic Circle of California in

Berkeley, California on January 30, 1987. I am indebted to Pierce Julius Flynn who collaborated on the transcription and data analysis for the 1987 paper. His collaboration and insights on the coding of audiovisual data have been invaluable to me in working on this chapter.

2. The narrative program for computer adoption has been used as a framework for semiotic analysis throughout this volume (cf. Greimas, 1970:172-174). In this chapter, the narrative program has been applied to actual occasions of working with the computer at school and on the job. In particular, I have focused on the domination or mastery stage of the program, which has been analyzed in terms of the relationship of control emerging between the computer operator (S_1) and the machine (S_2). This application of the narrative program contrasts with a more general use of the narrative program to describe computer policy decisions at the national level.
3. Inner *dureé* is part of the subject's lived experience and is a feature of the temporal ordering of working with computers (cf. Schutz, 1962:85-86). This lived time contrasts with time as measured by the clock and with linear sequences of text as they appear on the computer screen.
4. When the computer user brushes away the other student's hand and focuses on the computer operations in image 1.4, he employs a gestural signifier that is part of a mixed code. This gestural signifier is both a mimetic, or ritualized gesture, and a practical motion (cf. Greimas, 1970:75-78).
5. Historically, the tool-user interaction has raised interesting theoretical questions for sociological analysis. Max Weber, for example, believed that human action oriented exclusively toward inanimate objects was not social action. Weber (1947:112) stated:

   > Not every kind of action, even overt action, is "social" in the sense of the present discussion. Overt action is non-social if it is oriented solely to the behavior of inanimate objects.

   Thus, in Weber's view, much of human-machine interaction would fall in the non-social category. My analysis, however, suggests that the computer's capacity to respond generates a wide range of social action patterns surrounding its use and its interpretation in specific work settings.
6. In his analysis of the cognitive schemata used in story reading and comprehension, David Rumelhart (1984:12-16) times reading in milliseconds. The reading of an eight line story takes less than a second. When difficult programming problems were absent from the human-computer exchanges, I found that short computer line entries took between three and six seconds in contrast to brief human instructional sequences that took between four and nine seconds. This overlap in timing, along with the routinization of the human instructional exchanges, suggests a strong similarity between brief human-human monitoring sequences and human-machine interchanges. One of the reasons for the brevity of instructional monitoring sequences is that the teacher already knows what the student should be learning. In Charles Goodwin's terms, the teacher is a knowing recipient. According to Goodwin (1981:156), the knowing recipient possesses the information that the com-

municator is expressing and "might merely monitor the adequacy of the speaker's presentation." Such monitoring takes the form of brief, indexical, and truncated communications.

7. Ray Birdwhistell (1972:399) emphasizes that gestures should be analyzed as meaningful signs on equal footing with the messages communicated in verbal exchanges when he states:

 > There is a temptation to say that when one channel carries a distinction which is not made by the other, the fuller channel carries the "real" meaning. This implies that a given performance has a particular meaning. . . . From the examination of extensive sound-filmed interactional sequences, I have every reason to posit the proposition that in human experience there are at all observational times many streams of meaning in process.

8. The T.D. automobile dealership in Nairobi was described in Chapter Five as a business in which a smooth transition was made to the use of a new technology. The IBM 3400 computer system was well suited to the company's needs, and computer operators were reskilled from the available sales and clerical staff. Automobile mechanics also employed the computer system to log in their requests for spare parts. The operator's demonstration of the uses of the computer system for inventory and stock control in this chapter displays his confidence in the mastery of the new technology.
9. At the attribution stage of the narrative program, computer professionals establish the meaning of the computer as a valued object in order to legitimize their work procedures and professional standing. They become the new technicians of the sacred. In turn, they transmit the discourse of confrontation by explaining the benefits of new technologies to neophyte computer users.
10. Shoshona Zuboff (1988:174-178) argues that action-centered skills that rely on the intangible dimensions of a worker's experience draw upon oral culture, or talk about how the new technologies operate. I have argued that the oral dimension of working with computers affects both the computer classroom and the computerized work environment by challenging old hierarchies and opening up new lines of communication. Moreover, the new cybernetic orality is well suited to African settings in which oral traditions are still a living reality that can be integrated into new communicative codes.

Chapter Eight

1. According to this early definition, contract is treated as the set of rational understandings and agreements that hold the civil state together (cf. Hume, 1748; Rousseau, 1762; Gildin, 1983:29-66).
2. A.J. Greimas (1970:181-182) describes the resolution of the narrative program in terms of the establishment of a positive, or conjunctive, relationship between two subjects and the attribution of a new set of values to the narrative. This conclusion leads to the initiation of a new contract between the participating subjects.

3. In his discussion of Émile Durkheim's *Division of Labor in Society*, Talcott Parsons (1937:333-334) argues that Durkheim's concept of *conscience collective* modifies the interpretation of a "contractual" society to a society in which acceptance and consensus are based upon normative rules and shared values as opposed to contractual agreements. Parsons (1937:334) states: "No longer is the common-value element tied to a state where there is similarity of individuals and lack of differentiation. Above all, the freedom itself which is the basis of 'contractual' society is seen to be capable of being related positively to a *conscience collective*."
4. In Chapter Four, this firm director's complaints about the closed computer bidding system in Ivory Coast were outlined in greater detail. He expressed the view that limitations on the acceptance of bids from companies without French capital might severely restrict economic competition and the availability of computer hardware and software in Ivory Coast (cf. excerpt 4.31).
5. The importance of contractual exchanges as a basis for the distribution and use of computers is emphasized by the explicit discussion of result and means contracts in *Annexe II* of the *Plan Informatique National, 1986-1990*. This discussion underscores the government's concern with regulating computer sales, services, and in-house training in the private sector.
6. *Annexe II* of the *Plan Informatique National, 1986-1990*, p. 59 describes means contracts as difficult for the government to enforce. The plan states that these contractual exchanges are controlled in the private sector by the pressures of competition in the economic market rather than through direct government regulations.
7. In this case, the computer's use was governed by a written agreement between the handicraft cooperative and the researcher. The service contract would have been a written document concerning the computer's maintenance. Thus, in this case, type 2 contractual exchanges involved both implicit agreements and formal written texts.
8. This computer consultant's experiences at a large charge card company in Nairobi were discussed at length in Chapter Five (cf. excerpts 5.20 and 5.21). He was concerned with the implementation of a computer system adequate to meet the volume of work of his company, and he advocated working closely with vendors on problems of computer installation and the in-house training of computer operators.
9. Gérard Genette (1982:9-10) describes a literary genre as part of the paratextual contract between the author and the readers. He argues that the choice of a genre creates an informal narrative contract between the author and the audience. The author is bound by the constraints of the genre, and the readers employ the identification of the genre as a means of deciphering the text. The genre frames the text and reinforces the unstated agreement of communication and interpretation between the reader and the audience.
10. In their definition of sociosemiotics, Greimas and Courtés (1979:355-357) designate the study of a shared semantic universe of cultural representations as an important focus for research. They conceive of the analysis of a semantic universe as the intersection between sociological accounts of shared representations and a semiotic analysis of interpretive processes. With regard to the

study of computers in social context, sociosemiotics involves the analysis of how new categories emerge for describing and interpreting new technologies. A sociosemiotic study utilizing the narrative program requires an examination of the contractual understandings and exchanges that develop during the course of computer adoption.

11. Creole languages emerge from the intersection of a dominant foreign language and a local dialect (cf. Decamp, 1971:19-22). In spite of the controversies over the etiology of these languages, it is clear that both lexical and grammatical borrowing occurs from all languages and dialects involved. The diffusion of computer terminology may be compared to a creolization process. The lexicon is developed from a specific set of technical terms. Grammatical changes occur when these terms in their nominal form are transformed into verbs, such as "debugging" and are used to describe the effects of the new technology.
12. Rob Kling and Suzanne Iancono (1988:235-236) discuss the ideological basis of pro-computer movements in the United States. They claim that the proponents of new technologies see computers as a "central medium for creating the world they prefer" (Kling and Iancono, 1988:235). Accounts of computer activists resemble the discourse of computer confrontation described in Chapter Two. In computer movements, the benefits of new technologies are described without reference to the processes by which these technologies are adopted and mastered. This rhetoric of computer adoption may be compared to utopian religious discourse in which an idealized outcome is described without reference to the specific means by which the outcome is to be achieved.

Chapter Nine

1. In *False Start in Africa*, René Dumont (1969:152-171) argues that by neglecting agricultural development and failing to devise economic plans that integrate the rural sector into the national economy, African nations will eventually face famine and major economic crises. Jacques Giri (1986: 117-131) extends Dumont's thesis further by documenting cases of economic decline and faulty government planning across the continent. He calls for more balanced economic planning that shifts the focus from the emerging urban elite to the rural sector.
2. A.J. Greimas (1987:77) describes the terminal narrative utterance as follows:

> Up to now, we have considered the final narrative utterance of performance (NU_3) – which is, at the surface level, the equivalent of logical assertion at the level of deep grammar – to be an attributive utterance (AU). One might wonder, however, if such a formulation is satisfactory. Such an attribution or acquisition of an object by the subject – seems very much to be a reflexive doing: The performing subject attributes to himself, seeing himself as the subject of the descriptive utterance, an object-value. If this is so, reflexive attribution is nothing more than a special case of a much more general attribution structure, one that is well known in linguistics as the *communication schema* or, even more generally, as the

structure of exchange. It is, as we know, represented in its canonical form with three actants – sender, object of communication, and receiver:

$$\text{TU} = \text{F: transfer } (\text{S} \longrightarrow \text{O} \longrightarrow \text{R})$$
[Terminal utterance]

The terminal narrative utterance is a reflexive act that triggers the review of the entire narrative program. At this point, the subject makes a final attribution of meaning to the valued object.

3. Jean Baudrillard defines postmodernity as an attribute of the interpretation of meaning. At this point, signs no longer reference an external reality and turn back upon themselves. Baudrillard (1976:18) states: "[T]ous les signes s'échangent désormais entre eux sans s'échanger du tout contre du réel. . . ." Jean-François Lyotard (1979:7) defines postmodernity as the "incredulity of metanarratives" and a suspension of belief in explanatory discourses that ground meaning in science and history. Although I have drawn upon these ideas in my discussion, I prefer to focus on an examination of terminal signs as the keys to understanding postmodern narratives in relation to the acquisition and use of new technologies.
4. At the moment of the terminal narrative utterance, the narrative program is reversed. Its outcome, the final attribution, presumes that the processes of confrontation and domination have unfolded in a particular way (cf. Greimas and Courtés, 1979:131). Reversal of the narrative program implies that the final attribution accounts for, or judges, the outcomes of the qualifying test and the main test in the narrative (cf. Figure 8-1). In addition, reversal of the narrative program at the point of outcome necessitates a rereading of the content of the entire narrative (cf. Greimas, 1976:9-11; Hénault, 1983: 80-82). Consequently, with reference to computer adoption, successful completion and a positive attribution of value to the computer initiate a rereading of all previous struggles to adopt new technologies. Under the progressive computer contract, these problems are seen as adjustments and learning experiences rather than as insurmountable obstacles to computerization.
5. Jürgen Habermas (1971:112-113) and Alvin Gouldner (1979:37-39) argue that a commitment to rational public discourse about technology leads to a form of enlightened social control. Habermas' ideal speech situation is one in which technocrats exercise informed control of technology following the guidelines of practical reason and rational morality. For Gouldner, these moral commitments lead to control of legislation and public discourse about technology through democratic participation of the populace. Lyotard's approach challenges the possibility for rational discourse by introducing the metanarratives that frame scientific and technological advances. In spite of the questions posed by these tacit metanarratives, state regulation of technology is the conclusion of Lyotard's critique of the relationship between postmodern narrativity and the evolution of the state. I have suggested that the progressive computer contract, which includes state regulation of informatics and telecommunications, is an effective approach to technological change in African nations.

6. In his study of Ndembu rituals in northern Zambia, Victor Turner (1969: 19) argues that therapeutic rites in Zambian villages make symbolic reference to the supernatural entities believed to have caused an illness. Through this symbolic reference, the patient comes to understand the sources of illness and realizes the possibilities for cure. Similar symbolic reference is made to the computer as a superhuman force in rites of technological initiation.
7. According to Umberto Eco (1976:69) "the very definition of 'sign' implies a process of ***unlimited semiosis***." One interpretant is replaced by another in an infinite series. This process of substitution is the final interpretant in lieu of an "absolute object." The final interpretant is the last resort of meaning and the terminus of the process of semiosis.
8. In developing his theory of totalization, Jean-Paul Sartre describes the role of tools. Sartre (1976:45) states: "In the case of practical objects – machines, tools, consumer goods, etc. – our present action makes them seem like totalities by resuscitating, in some way, the ***praxis*** which attempted to totalize their inertia." In this case, praxis refers to the reasons for the invention of the tool and the history of its use.
9. Ethnographic studies of science in which the narratives describing laboratory work are analyzed as part of the process of scientific construction include Latour and Fabbri (1977:82-95); Latour and Woolgar (1979); Garfinkel, Lynch, and Livingston (1981:131-158); Knorr-Cetina (1983:115-140); and Lynch (1985). In these studies, analysis of the construction of scientific findings through practical activities is a primary concern.
10. Michael Polanyi (1958:292-294) analyzes the ways in which scientific paradigms limit the phenomena under study and the status and truth-value of scientific findings. Jean-François Lyotard (1979:88-97) takes this argument a step further by examining the relationship of scientific narratives and meta-narratives to the discovery process. Both authors question the moral basis and outcomes of scientific research.
11. In his anthology of folktales collected among the Kamba people of Kenya, John Mbiti (1966:29-30) groups the narratives into stories about adventure, marriage, ancestral spirits, nature, and intergroup relationships. He argues that the effects of Christianity, Western technology, and government intervention are reflected in these stories. In particular, he argues that stories about the ***Aimu*** (ancestral spirits), such as ***An Extraordinary Pumpkin*** quoted in this chapter, are influenced by Christianity, which he claims that the Kamba associate holistically with a foreign technological culture. The story of the transforming man reflects similar influences. These narratives provide a cultural framework for the interpretation of new technologies. According to Mbiti, through the vehicle of the folktales, new cultural forms are "interpolated into a traditional background" (cf. Mbiti, 1966:33).
12. In her doctoral dissertation on the adoption of microcomputers in Tanzania, Suzanne Grant Lewis (1988:287) argues that Tanzania's temporary ban on computer imports in 1974 "grew out of a limited understanding of the problems with computer systems at the time and a mistaken perception of the foreign exchange savings possible from restricting computer imports." Kenya's

import duties and taxes on computers may have arisen under similar circumstances. In both cases, foreign technology has been viewed as both a threat to and a potential source of income for the local government.

Glossary

Actant: An agent or subject who acts or is acted upon in the narrative program for computer adoption. Both the computer user and the computer are actants.

Analyste-système: Systems analyst in charge of organizing computer systems, analyzing the flow of computer data and tasks, and implementing computer programs in Ivorian firms. (See *systems analyst*.)

APICI: The Association Professionelle des Informaticiens de Côte d'Ivoire is a professional association established as of 1973 in Abidjan, Ivory Coast. This association promotes the interests of computer experts, firms, and vendors. It sets standards for the computer profession in Ivory Coast.

Appel d'offres: A call for bids of computer hardware, software, or services. Government bids are a major source of revenue for computer firms in Ivory Coast.

Attribution: Final stage of the narrative program for computer adoption. The computer user attributes meaning to the new technology. (See *narrative program for computer adoption*.)

Batch computing: Grouping of computer programs and data together for serial data processing, also known as noninteractive processing, is a technique used by many African firms. Materials are prepared by a firm for batch computing at a data bureau.

Break down [machines]: Popular Kenyan English term describing the relationship of computers to office workers by designating the functions of the machines for organizational tasks.

Bypassing: Using the steps involved in computer logic to organize and interpret social life. (See *simulation*.)

Cash register: A microcomputer in the parlance of Kamba carvers at the Nairobi Handicraft Cooperative.

CCD: The culture of critical discourse encompasses the language and forms cultural expression of the new technocratic class. (See *postmodernity*.)

CFA franc: Currency shared by the Communauté Française Africaine (French African Community). In 1987, the exchange rate was approximately 310 CFA francs per U.S. dollar. Countries using the CFA franc include Benin, Burkina Faso, Central African Republic, Chad, Congo, Gabon, Ivory Coast, Mali, Niger, and Sénégal.

CNI: The Commission Nationale pour l'Informatique was established by presidential decree on January 26, 1983 as a government commission to plan and implement national computer policies in Ivory Coast. The commission was preceded by a planning committee known as Commission Nationale pour le Traitement de l'Information established on January 14, 1980. (See *OCM; SGI*.)

COMPAT: The Computer Association of Tanzania was established in 1986 as an alliance of computer professionals, educators, and vendors. The association sets standards for the computer profession and lobbies to promote the interests of computer experts.

Comptable: Accountant in Ivorian computer firms. The accountant reports to the firm's data processing staff.

Computer Association of East Africa: Association established in Kenya during the early 1970s to promote the interests of computer professionals. The association was short-lived and had limited socioeconomic impact.

Computer controller: Individual in charge of computer staff and operations in Kenyan firms. This title is often used interchangeably with systems analyst.

Confrontation: First stage of the narrative program of computer adoption. The computer user confronts the new technology. (See *narrative program for computer adoption*.)

Contract: The formal and informal understandings of obligations shared by subjects in an interaction. Contract is the final stage of the narrative program for computer adoption. (See *narrative program for computer adoption; terminal signs*.)

Contrat avec obligation de moyens: Ivory Coast's second national informatics plan defines a contract providing means of support as a contract between a computer vendor or consulting firm and its clients to provide material and human means of support for computer installations. Contracts providing means of support cover computer hardware and software, instruction, and support personnel.

Contrat avec obligation de résultats: Ivory Coast's second national informatics plan defines a contract with obligation of results as a contract between a computer vendor or consulting firm to provide computer services for clients. These services include feasibility studies, financial plans, and payroll and accounting packages.

Data bureau: This type of firm handles the data processing needs of other companies in Kenya. Its activities include feasibility reports, financial planning, inventory programs, stocks, and word processing. The SSII in Ivory Coast perform similar data processing functions for client companies. (See *SSII*.)

Data-entry staff: Workers charged with inputting computer data either on the keyboard or through keypunch operations. These workers are also known as computer operators. (See *pupitreur*.)

Data Processing Manager (DPM): Manager in charge of all computer staff and operations in a firm. The DPM's tasks include overseeing systems analysts, programmers, and computer operators. The DPM functions in both a general managerial and a technical capacity. This individual structures organizational tasks and titles and writes operations manuals with the assistance of programmers.

Deskilling: Transformation in the skill level required for work resulting from the recoding of job tasks and titles. (See *recoding*.)

Domination: Second stage of the narrative program. The computer user masters the new technology. (See *narrative program for computer adoption*.)

Entitled reading: Interpretation of computer texts by experts within an organization. (See *narrative program for computer adoption*.)

ESII: The École Supérieure d'Ingénieurs Informaticiens is one of the nine schools of the Institut National Supérieur de l'Enseignement Technique (INSET) in Yamoussoukro, Ivory Coast. ESII trains software researchers and specialists in three-year cycles and offers a postgraduate degree in computer science.

Evolués: Members of the rising elite of clerks, office workers, and entrepreneurs in French-speaking Africa.

Fragile computer contract: Configuration of contractual exchanges in a situation where new technologies are accepted with reluctance. Mixed attitudes toward computers characterize the fragile contract. (See *contract; narrative program for computer adoption*.)

GCSC: The Government Computer Services Centre coordinates computer operations in the public sector throughout Kenya. Mainframe computers are used to compile and store government statistics.

Génie logiciel: Popular term used to describe a software whiz, or expert in software research and development, in Ivory Coast's computer profession.

GIMI: The Groupement Interprofessionnelle pour le Management et l'Informatique is a syndicate representing the interests of computer firms, vendors, consultants, and professionals in Ivory Coast. GIMI focuses on setting professional standards for private sector businesses in Ivory Coast's computer industry.

Habitus: Material and social conditions that constitute a particular environment. This term is used to describe the workplace as an environment for new technologies.

Hypercompetent: Popular term used to describe a highly qualified and especially competent computer expert or technician in Ivory Coast.

IAI: The Institut Africain d'Informatique located in Libreville, Gabon offers advanced degrees in computer science and engineering. It is recognized as a major West African center for computer education.

ICDC: The Industrial and Commercial Development Corporation is a Kenyan parastatal agency through which the government provides loans and subsidies to private business and industry.

Informaticien: Computer expert in Ivory Coast. An *informaticien* may be a specialist in computer hardware or software.

Informatique: Information processing, computer science, and computer-related communications. The English equivalent is informatics.

Informatique de gestion: Business computer systems.

Ingénieur-concepteur: Research engineer in Ivory Coast.

Ingénieur d'étude: Student engineer in Ivory Coast.

Ingénieur principal: Head engineer in charge of research and systems design in Ivory Coast.

Iimu (plural aimu): Kamba language term for ancestral spirits who are believed to influence the destinies of their living descendants. In Kamba folktales, the force of human technology is pitted against the supernatural powers of the ancestors.

INSET: The Institut National Supérieur de l'Enseignement Technique consists of nine schools that offer advanced degrees in engineering and computer science. Established in 1980 by the mandate of Ivory Coast's first national informatics plan, INSET is located in the rural town of Yamoussoukro and has become Ivory Coast's university of the future.

INSTAT: Institut National pour la Sous-Traitance et l'Assistance Technique is a French national institute training administrators in scientific management. Many of Ivory Coast's computer professionals are graduates of this institute.

Institute of Computer Science, University of Nairobi: Located on the Chiromo campus of the University of Nairobi, the Institute was established in May of 1979. An outgrowth of the university's Faculty of Engineering, the Institute is the only educational institution in Kenya offering a postgraduate degree in computer science. The Institute monitors the computer needs of the university.

ISI: The Institut Supérieur d'Informatique is one of the nine schools of the Institut National Supérieur de l'Enseignement Technique (INSET) in Yamoussoukro, Ivory Coast. ISI trains students in three-year cycles and offers a master's degree in management of computer systems for business. (See *MIAG*.)

IUT: Institut Universitaire de Technologie is a French institute training engineers and computer specialists.

Keep in memory: Popular Kenyan English expression for storing and retrieving data from the computer. The technical term computer memory is used. Variations of this expression have entered current, nontechnical language in Kenya.

Kenya Polytechnic: A polytechnic institute in Nairobi offering a degree in computer studies. Graduates of Kenya Polytechnic's computer studies program are highly regarded in Kenyan business and industry.

Kenyan shilling: The local currency of Kenya, valued at an exchange rate of approximately 15.88 Kenyan shillings to the U.S. dollar in 1985 when data for this study were collected.

KIE: The Kenya Institute of Education is the government body charged with curriculum research, planning, and development for Kenyan public educational institutions.

KNEC: The Kenya National Examinations Council is the government body charged with establishing school curricula and national examinations in Kenya. All computer curricula must be reviewed by the KNEC.

Last train of the twentieth century: This expression was coined in 1983 in a speech delivered by the President of the Commission Nationale pour l'Informatique in Ivory Coast. The expression "last train" is now popularly used as a synonym for computers.

Main test: Major challenges faced in the mastery of new technologies. (See *domination; narrative program for computer adoption*.)

MALD: The Ministry of Agriculture and the Ministry of Livestock Development were combined in Kenya to form the Ministry of Agriculture and Livestock Development in September, 1983. The ministry uses microcomputers for budget preparation and fiscal planning.

Mami wata: African water spirit assuming the form of a mermaid figure with Mediterranean or Indian features. The mermaid spirit is believed to confer modern wealth, riches, and success in exchange for bondage to her.

Marker for conversational pause: The symbol (//) has been used to mark conversational pauses in the discourse sets transcribed and recorded throughout this book.

Mastery: Competent control of new technologies at the domination stage of computer adoption. (See *domination; narrative program for computer adoption*.)

MFP: The Ministry of Finance and Planning handles national budgetary planning in Kenya. In 1983, this ministry was among the first to employ mircrocomputers for budget preparation and review.

MIAG: The Maîtrise d'Informatique Appliquée à la Gestion is a master's program in business computer systems developed by INSET. Normally, students take examinations for this degree after a three-year course. (See *INSET; ISI*.)

MOA: The Ministry of Agriculture in Kenya was one of the first ministries to use microcomputers in fiscal and budgetary planning in 1981.

Narrative program for computer adoption: The sequence of events involved in computer adoption including initial confrontation with computers, mastery of technology, and contractual exchanges. The narrative program encompasses discourse and action. (See *attribution; confrontation; contract; domination; terminal signs*.)

Nationalization: Increasing the number of national citizens at all levels of the computer industry in African countries.

New mechanical solidarity: Changes in the division of labor and work organization resulting from computer technology.

New technicians of the sacred: Computer experts in Kenya and Ivory Coast.

OCM: The Office Central de la Mécanographie was established by the government of Ivory Coast on November 28, 1967. The OCM is a government agency that oversees the coordination of technological projects involving data processing in commerce and industry. Some of its earlier functions have now been assumed by the CNI. (See *CNI; SGI*.)

Onduleur: Surge suppressor used to regulate the electrical current for computer installations in Ivory Coast. Surge suppressor units are installed by several companies in Ivory Coast.

Ordinateur: French term for computer or data processing system. (See *système d'exploitation*.)

PDG: President Director General who serves as the head of a firm in Ivory Coast.

Postmodernity: Forms of cultural expression accompanying new technologies in Africa. (See *CCD; terminal signs*.)

Programmer: This individual writes and executes computer programs developed by the systems analyst. (See *senior programmer*.)

Programmeur-analyste: Person in charge of writing, implementing, and documenting computer programs in Ivory Coast. This individual is the equivalent of a senior programmer in Kenya. (See *senior programmer*.)

Progiciel de gestion: Business computer software covering accounting packages, payroll, inventory, and financial planning in Ivory Coast.

Progressive computer contract: Configuration of contractual exchanges in a situation where new technologies are promoted. Technological advances characterize the progressive contract. (See *contract; narrative program for computer adoption*.)

Pupitreur: Computer operator and data-entry worker in Ivory Coast. The *pupitreur* enters data by keypunch or keyboard methods. (See *data-entry staff*.)

Qualifying test: First challenge faced during the confrontation stage of the narrative program for computer adoption. The qualifying test encompasses the initial adaptations to new technologies. (See *confrontation; narrative program for computer adoption*.)

Reocoding: Restructuring the meaning and value of computer titles and tasks in the workplace. Tasks generally change before titles.

Sapeurs: Young urban professionals in Brazzaville, Congo.

Senior programmer: This person is also known as a chief programmer in Kenya. The senior programmer writes, implements, and documents computer programs in conjunction with the systems analyst. In some Kenyan firms, the senior programmer also functions as a systems analyst or a data processing manager. (See *DPM; systems analyst*.)

SGI: The Secrétariat Général à l'Informatique, established on January 26, 1983, is the government agency charged with implementing computer policies and projects in Ivory Coast. The SGI is the official agency that implements the policies developed by the Commission Nationale pour l'Informatique. (See *CNI; OCM*.)

Simulation: A computer program that models a system or process; modeling of social reality by computer-like means. (See *bypassing*.)

Social drama: Narrative of social conflict and resolution, containing the four stages of breach, crisis, redressive action, and reintegration. (See *narrative program of computer adoption*.)

SSII: The Sociétés de Services et d'Ingénerie en Informatique are engineering and computer consulting firms in Ivory Coast. Some of these companies are affiliates of French banks and computer consulting firms. The SSII are increasing in Ivory Coast and are in competition with vendors for the marketing of their computer support services. (See *data bureau*.)

Système d'exploitation: This French term refers to the operating system of the computer. It may be a tape operating (TOD), card operating (COS), or disk operating (DOS) system. (See *ordinateur*.)

Tactile grounding: Touching the computer screen ritualistically in order to orient to the machine. This practice is used in Kenyan computer training sessions and in the workplace. (See *confrontation; mastery*.)

Terminal signs: Signs attributing meaning to computers. Terminal signs occur at the closure of the narrative program of computer adoption. They include modeling devices, representational signs, totalizing signs, and analeptic signs. (See *contract; narrative program for computer adoption; postmodernity*.)

Travailleurs nonproductifs: The French term for nonproductive workers is used to describe the noncomputer administrative staff, clerical workers, and maintenance personnel of a firm in Ivory Coast.

Travailleurs productifs: The French expression for productive workers is employed to describe the engineers and data processing staff of firms in Ivory Coast.

References

Abercrombie, Nicholas and John Urry

1983 *Capital, Labour, and the Middle Classes*. London: Allen Unwin.

Association Professionnelle des Informaticiens de Côte d'Ivoire

1980 "Notre Association en Quelques Lignes," *L'Informatique en Côte d'Ivoire* 2, 3:4.

Attewell, Paul and James Rule

1984 "Computing and Organizations: What We Don't Know," *Communications of the ACM* 27, 12:1184-1192.

Baker, Phyllis

1986 "Being Bored and Being Busy: The Clerical Worker's Paradox." Paper presented at the Society for the Study of Symbolic Interaction, New York: August 30.

Balandier, Georges

1985 *Le Détour: Pouvoir et Modernité*. Paris: Librairie Arthème Fayard.

Bar-Hillel, Yehoshua

1954 "Indexical Expressions," *Mind* 63:359-379.

Barker, Jane and Hazel Downing

1985 "Word Processing and the Transformation of Patriarchal Relations of Control in the Office," in *The Social Shaping of Technology*, Donald MacKenzie and Judy Wajcman (eds.), 149-163. Philadelphia: Open University Press.

Barthes, Roland

1974 *S/Z: An Essay*. Translated by Richard Miller. New York: Hill and Wang. (Originally published as *S/Z*. Paris: Éditions du Seuil, 1970.)

1975 "An Introduction to the Structural Analysis of Narrative," *New Literary History* 6, 2:237-272.

Baudrillard, Jean

1968 *Le Système des Objets*. Paris: Librairie Gallimard.

1970 *La Société de Consommation*. Paris: Le Point de Question.

1976 *L'Échange Symbolique et la Mort*. Paris: Librairie Gallimard.

1981 *Simulacres et Simulation*. Paris: Éditions Galilée.

Bay, Christian

1978 "From Contract to Community: Thoughts on Liberalism and Post-industrial Society," in *From Contract to Community: Political Theory at the Crossroads*, Fred R. Dallmayr (ed.), 29-45. New York: Marcel Dekker Inc.

Bell, Daniel

1973 *The Coming of Post-Industrial Society: A Venture in Social Forecasting*. New York: Basic Books.

1981 "Introduction," in *The Computerization of Society*, Simon Nora and Alain Minc (eds.) vii-xvi. Cambridge, Massachusetts: MIT Press. (Originally published in French as *L'Informatisation de la Société*. Paris: Documentation Française, 1978.)

Birdwhistell, Ray L.

1972 "A Kinesic-Linguistic Exercise: The Cigarette Scene," in *Directions in the Ethnography of Communication*, John J. Gumperz and Dell Hymes (eds.), 381-404. New York: Holt, Rinehart, and Winston.

Bittner, Egon

1963 "Radicalism and the Organization of Radical Movements," *American Sociological Review* 28:928-940.

Boltanski, Luc

1982 *Les Cadres: La Formation d'un Groupe Social*. Paris: Éditions de Minuit.

Bourdieu, Pierre

1977 *Outline of a Theory of Practice*. Translated by Richard Nice. New York: Cambridge University Press. (Originally published as *L'Esquisse d'une Théorie de la Pratique*. Geneva, Switzerland: Librairie Droz, 1972.)

Bourdieu, Pierre and Jean-Claude Passeron

1977 *Reproduction in Education, Society, and Culture*. Translated by Richard Nice. Beverly Hills: Sage Publications. (Originally published as *La Reproduction*. Paris: Éditions de Minuit, 1970.)

Bowles, Samuel and Herbert Gintis

1976 *Schooling in Capitalist America: Educational Reform and the Contradictions of Economic Life*. New York: Basic Books.

Branciard, Michel

1978 *Dictionnaire Économique et Social*. Paris: Éditions Ouvrières.

Braverman, Harry

1974 *Labor and Monopoly Capital: The Degradation of Work in the Twentieth Century*. New York: Monthly Review Press.

Brodman, Janice Z.

1985 *Microcomputer Adoption in Developing Countries: Old Management Styles and New Information Systems*. College Park, Maryland: International Development Center, University of Maryland.

1987 "Key Management Factors Determining the Impact of Microcomputers on Decision-Making in the Governments of Developing Countries," in *Microcomputers in Public Policy*, Stephen R. Ruth and Charles K. Mann (eds.), 95-116. Boulder, Colorado: Westview Press.

Bujra, Janet

1979 "Proletarianization and the 'Informal Economy': A Case Study from Nairobi," *African Urban Studies* 3:47-66.

Bunn, James

1981 *The Dimensionality of Signs, Tools, and Models: An Introduction*. Bloomington, Indiana: Indiana University Press.

Bureau Intergouvernemental pour l'Informatique

1980 "Rapport Final des Travaux: Conférence d'Intégration sur l'Informatique Africaine," *L'Informatique en Côte d'Ivoire* 2, 3:25-29.

Burke, Kenneth
1962 *A Grammar of Motives*. Berkeley: University of California Press.
Campbell, Bonnie
1987 "The State and Capitalist Development in the Ivory Coast," in *The African Bourgeoisie: Capitalist Development in Nigeria, Kenya, and the Ivory Coast*, Paul M. Lubek (ed.) 281-303. Boulder, Colorado: Lynne Rienner Publishers.
Carim, D. Enver
1983 "Coming to Grips with the New Technology," *African Business* 58 (June):9.
Central Bureau of Statistics, Kenya
1981 *Employment and Earnings in the Modern Sector, 1979*. Nairobi, Kenya: Ministry of Economic Planning and Development.
Cicourel, Aaron
1958 "The Front and Back of Organizational Leadership," *Pacific Sociological Review* 1:54-58.
Clifford, James
1986 "Introduction: Partial Truths," in *Writing Culture: The Poetics and Politics of Ethnography*, James Clifford and George E. Marcus (eds.), 1-26. Berkeley: University of California Press.
Cohen, Abner
1981 *The Politics of Elite Culture: Explorations in the Dramaturgy of Power in a Modern African Society*. Berkeley: University of California Press.
Cole, Michael and Sylvia Scribner
1974 *Culture and Thought: A Psychological Introduction*. New York: John Wiley and Sons.
Coleman, James S.
1980 "Authority Structures," *Public Opinion Quarterly* 44, 2:143-163.
Collins, Randall
1974 "Where are Educational Requirements for Employment Highest?," *Sociology of Education* 47 (Fall):419-442.
1986 *Weberian Sociological Theory*. New York: Cambridge University Press.
1987 "Interaction Ritual Chains, Power and Property: The Micro-Macro Connection as an Empirically Based Theoretical Problem," in *The Micro-Macro Link*, Jeffrey C. Alexander, Bernhard Giesen, Richard Münch, and Neil J. Smelser (eds.), 193-206, Berkeley: University of California Press.
Computer Training Centre
1986 Unpublished brochure. Nairobi, Kenya: Computer Training Centre.
Cromptom, Rosemary and Gareth Jones
1984 *White-Collar Proletariat: Deskilling and Gender in Clerical Work*. London: MacMillan Press.
Crozier, Michel
1971 *The World of the Office Worker*. Translated by David Landau. Chicago: University of Chicago Press, (Originally published as *Le Monde des Employès du Bureau*. Paris: Éditions du Seuil, 1965.)

Courtés, Joseph
1976 *Introduction à la Sémiotique Narrative et Discursive*. Paris: Librairie Hachette.

Daily Nation
1984 "Computers Are No Longer Luxury Items," *Sunday Nation*. Nairobi, Kenya: August 12.

Danow, David K.
1986 "Temporal Strategies and Constraints in Narrative," *Semiotica* 54, 3-4:248-268.

Decamp, David
1971 "Introduction: The Study of Pidgin and Creole Languages," in *Pidginization and Creolization of Languages*, Dell Hymes (ed.), 13-39. London: Cambridge University Press.

de Certeau, Michel
1980 "On Oppositional Practices of Everyday Life," *Social Text* 3 (Fall): 3-43.

De Lagy, Pamela and Craig Calhoun
1987 "Computerization and Development Planning: A Sudanese Case Study." Paper presented at the Annual Meetings of the African Studies Association, Denver, Colorado, November 20.

Desbois, D. and G. Vidal
1984 "L'Informatique en Côte d'Ivoire," *Annales de l'Université Nationale de Côte d'Ivoire*, Série K (Sciences Économiques), VII, Abidjan: Centre Universitaire de Traitement de l'Information (CUTI).

Dreyfus, Hubert L. and Stuart E. Dreyfus
1986 *Mind Over Machine: The Power of Intuition and Expertise in the Era of the Computer*. New York: The Free Press.

Dumont, René
1969 *False Start in Africa*. Translated by Phyllis Nauts Ott. New York: Praeger Publishers. (Originally published as *L'Afrique Noire Est Mal Partie*. Paris: Éditions du Seuil, 1962.)

Duncan, Starkey
1972 "Some Signals and Rules for Taking Speaking Turns in Conversation," *Journal of Personality and Social Psychology* 23, 2:283-292.

Durkheim, Émile
1949 *The Division of Labor in Society*. Translated by George Simpson. New York: The Free Press. (Originally published as *De la Division du Travail Social*. Paris: F. Alcan, 1893.)

East African Computer News
1985a "Computers Assist Coffee and Tea Farmers," Nairobi, Kenya. 1, 1 (May):13, 15, and 27.
1985b "Computer Training in Kenya," Nairobi, Kenya. 1, 1 (May):7-12.
1985c "Punch Operators: How Secure Are Their Jobs?," Nairobi, Kenya. 1, 1 (May):6.
1985d "Introducing the Personal Computer," Nairobi, Kenya. 1, 2 (October): 5 and 13.
1985e "NCR Computers for the Diners Club," Nairobi, Kenya. 1, 2 (October):13.

1986 *East African Computer News*, 1986.
"Training Top-flight Professionals," Nairobi, Kenya. 1, 4 (December 1985/January):20-21.

Eco, Umberto
1976 *A Theory of Semiotics*. Bloomington, Indiana: Indiana University Press.

Ekman, Paul and W.V. Friesen
1969 "The Repertoire of Nonverbal Behavior: Categories, Origins, Usage, and Coding," *Semiotica* 1, 1:49-98.

Eldon, Michael
1985 "Shed a Micro-Tear. . . . ," *Signature* (April):7.

Ellul, Jacques
1964 *The Technological Society*. Translated by John Wilkinson. New York: Vintage Books. (Originally published as *La Technique ou L'Enjeu du Siècle*. Paris: Librairie Armand Colin, 1954.)

Eyerman, Ron, Lennart G. Svensson, and Thomas Söderqvist
1987 "Introduction," in *Intellectuals, Universities, and the State in Modern Western Societies*, Ron Eyerman, Lennart G. Svensson, and Thomas Söderqvist (eds.), 1-15. Berkeley: University of California Press.

Fernandez, James
1980 "Edification by Puzzlement," in *Explorations in African systems of Thought*, Ivan Karp and Charles S. Bird (eds.), 44-59. Bloomington, Indiana: Indiana University Press.

Fisher, Renée and Philippe Kruchten
1983 *Dictionnaire Informatique*. Paris: Éditions Eyrolles.

Flexner, Stuart Berg, ed.
1987 *The Random House Dictionary of the English Language: Second Unabridged Edition*. New York: Random House.

Foster, Philip J.
1965 *Education and Social Change in Ghana*. London: Routledge and Kegan Paul.
1977 "The Vocational School Fallacy in Development Planning," in *Power and Ideology in Education*. Jerome Karabel and A.H. Halsey (eds.), 356-365. New York: Oxford University Press.

Foucault, Michel
1984 "What Is an Author?," in *The Foucault Reader*, Paul Rabinow (ed.), 113-138. New York: Pantheon Books.

Friedland, Mary
1985 "Learning by Computer," in *East African Computer News*, Nairobi, Kenya. 1, 2 (October):21-23.

Gandoulou, Justin-Daniel
1984 *Entre Paris et Bacongo*. Paris: Centre Georges Pompidou.

Garfinkel, Harold
1967 *Studies in Ethnomethodology*. Englewood Cliffs, New Jersey: Prentice-Hall.

Garfinkel, Harold and Harvey Sacks
1970 "On Formal Structures of Practical Actions," in *Theoretical Sociology: Perspectives and Developments*, John C. McKinney and Edward A. Tiryakian (eds.), 338-366. New York: Appleton-Century-Crofts.

Garfinkel, Harold, Michael Lynch, and Eric Livingston
1981 "The Work of a Discovering Science Construed with Materials from the Optically Discovered Pulsar," *Philosophy of the Social Sciences* 5:131-158.
Genette, Gérard
1982 *Palimpsestes: La Littérature au Second Degré*. Paris: Éditions du Seuil.
Giddens, Anthony
1984 *The Constitution of Society: Outline of the Theory of Structuration*. Berkeley: University of California Press.
Gildin, Hilail
1983 *Rousseau's Social Contract: The Design of the Argument*. Chicago: University of Chicago Press.
Gill, Colin
1985 *Work, Unemployment, and the New Technology*. Cambridge, England: Polity Press.
Ginguay, Michel
1972 *Dictionnaire d'Informatique*. Paris: Éditions Masson.
Giri, Jacques
1986 *L'Afrique en Panne: Vingt-cinq Ans de "Développement*." Paris: Éditions Karthala.
Goffman, Erving
1959 *The Presentation of Self in Everyday Life*. Garden City, New York: Doubleday.
1981 *Forms of Talk*. Philadelphia: University of Pennsylvania Press.
Goldthorpe, John
1982 "On the Service Class, Its Formation, and Future," in *Social Class and the Division of Labor*, A. Giddens and G. MacKenzie (eds.), 162-185. Cambridge, England: Cambridge University Press.
Goodwin, Charles
1981 *Conversational Organization: Interaction Between Speakers and Hearers*. New York: Academic Press.
Gouldner, Alvin
1979 *The Future of Intellectuals and the Rise of the New Class*. New York: Seabury Press.
Graffins College
1986 "Graffins College: Computer Studies." Unpublished brochure. Nairobi, Kenya: Graffins College.
Greimas, A.J.
1970 *Du Sens: Essais Sémiotiques*. Paris: Éditions du Seuil.
1976 "Les Acquis et les Projets: Préface de A.J. Greimas," in *Introduction à la Sémiotique Narrative et Discursive*, Joseph Courtés (ed.), 5-25. Paris: Librairie Hachette.
1983 *Structural Semantics: An Attempt at a Method*. Translated by Daniele McDowell, Ronald Schleifer, and Alan Velie. Lincoln, Nebraska: University of Nebraska Press. (Originally published as *Sémantique Structurale: Recherche de Méthode*. Paris: Librairie Larousse, 1966.)

1987 *On Meaning: Selected Writings in Semiotic Theory*. Translated by Paul J. Perron and Frank Collins. Minneapolis, Minnesota: University of Minnesota Press.

Greimas, A.J. and Joseph Courtés

1979 *Sémiotique: Dictionnaire Raisonné de la Théorie du Langage*. Paris: Librairie Hachette.

Gusfield, Joseph R.

1981 *The Culture of Public Problems: Drinking-Driving and the Symbolic Order*. Chicago: University of Chicago Press.

Habermas, Jürgen

1971 *Toward a Rational Society: Student Protest, Science, and Politics*. Translated by Jeremy Shapiro. Boston: Beacon Press.

1975 *Legitimation Crisis*. Translated by Thomas McCarthy. Boston: Beacon Press.

Haferkamp, Hans

1987 "Complexity and Behavior Structure, Planned Associations and Creation of Structure," in *The Micro-Macro Link*, Jeffrey C. Alexander, Bernhard Giesen, Richard Münch, and Neil J. Smelser (eds.), 177-192. Berkeley: University of California Press.

Halbherr, Peter

1987 *IBM: Mythe et Réalité*. Lausanne, Switzerland: Pierre-Marcel Favre.

Harper, Malcolm

1984 *Small Business in the Third World*. Chichester, England: John Wiley and Sons.

Heim, Michael

1987 *Electric Language: A Philosophical Study of Word Processing*. New Haven: Yale University Press.

Hénault, Anne

1983 *Narratologie: Sémiotique Générale*. Paris: Presses Universitaires de France.

Hume, David

1962 "Of the Original Contract," in *Social Contract: Essays by Locke, Hume, and Rousseau*, Ernest Barker (ed.), 147-165. New York: Oxford University Press. (Originally published in *Moral and Political Essays*, 1748.)

Ikiara, G.K.

1984 "How Can We Solve the Informal Sector Problems?," *Daily Nation*. Nairobi, Kenya: Thursday, August 9:9.

Ingle, Marcus D., Noel Berge, and Marcia Hamilton

1983 *Microcomputers in Development: A Manager's Guide*. West Hartford, Connecticut: Kumarian Press.

Institute of Computer Science

1986 "Institute of Computer Science: Postgraduate Diploma in Computer Science." Unpublished brochure. Nairobi, Kenya: University of Nairobi, Institute of Computer Science.

International Bank for Reconstruction and Development

1981 *Accelerated Development in Sub-Saharan Africa: An Agenda for Action* (the "Berg Report"). Washington, D.C.: The World Bank.

Ivanov, Vjaceslav
1969 "L'Évolution des Signes-Symboles," *Semiotica* 1, 1-2:218-221.
J.-A. Bis
1985 "Au Cameroun – L'Ordinateur Douanier: Un Bon Encaisseur," *J.-A. Bis* 10 (September):20-21.
1987 *JAI 1987: Troisième Journées Africaines d'Informatique*, Abidjan: SGI.
Jones, Michael Owen
1985 "Is Ethics the Issue?," in *Organizational Culture*, Peter J. Frost, Larry F. Moore, Meryl Reis Louis, Craig C. Lundberg, and Joanne Martin (eds.), 235-252. Beverly Hills, California: Sage Publications.
Joshi, Heather, Harold Lubell, and Jean Mouly
1976 *Abidjan: Urban Development and Employment in the Ivory Coast.* Geneva, Switzerland: International Labour Office.
Jules-Rosette, Bennetta
1979 "Alternative Urban Adaptations: Zambian Cottage Industries as Sources of Social and Economic Innovation," *Human Organization*, 38, 3 (Fall):225-238.
1981 *Symbols of Change: Urban Transition in a Zambian Community.* Norwood, New Jersey: Ablex Publishing Corporation.
1984 *The Messages of Tourist Art: An African Semiotic System in Comparative Perspective.* New York: Plenum Publishing Corporation.
1985 "New Religious Consciousness and the State in Africa: Selected Case Studies," *Archives de Sciences Sociales des Religions* 64, 1:15-35.
1988 "Women and New Technologies in Comparative Perspective: Selected African Case Studies," Paper presented at the Conference on Gender, Science, and Technology. Davis, California: University of California, Davis, April 9.
Jules-Rosette, Bennetta and Pierce Julius Flynn
1988 "'Who Was Mozart?': Computers and the Semiotics of Invention." Paper presented at the Annual Meetings of the Semiotic Circle of California, Berkeley, California, January 30.
Kanter, Rosabeth Moss
1977 *Men and Women of the Corporation.* New York: Basic Books.
Kareithi, Maxwell
1986 "The Trials and Tribulations of Computer Operations Staff," *East African Computer News* 1, 4 (December 1985/January):5.
Kling, Rob and Suzanne Iancono
1988 "The Mobilization of Support for Computerization: The Role of Computerization Movements." *Social Problems* 35, 3 (June):226-243.
Knorr-Cetina, Karin D.
1983 "The Ethnographic Study of Scientific Work: Towards a Constructivist Interpretation of Science," in *Science Observed: Perspectives on the Social Study of Science*, Karin D. Knorr-Cetina and Michael Mulkay (eds.), 115-140. Beverly Hills, California: Sage Publications.
Kumar, Krishna
1980 "Social and Cultural Impact of Transnational Enterprises: An Overview," in *Transnational Enterprises: Their Impact on Third World*

Societies and Cultures, Krishna Kumar (ed.), 1-43. Boulder, Colorado: Westview Press.

Lall, Sanjaya
1983 *The New Multinationals: The Spread of Third World Enterprises*. Chichester, England: John Wiley and Sons.

Landowski, Eric
1986 "Pour une Approche Sémiotique et Narrative du Droit," *Actes Sémiotiques* 8, 71:1-37.

Langdon, Steven W.
1981 *Multinational Corporations in the Political Economy in Kenya*. New York: St. Martin's Press.

Latour, Bruno and Paolo Fabbri
1977 "La Rhétorique de la Science: Pouvoir et Devoir dans un Article de Science Exacte," *Actes de la Recherche en Sciences Sociales* 18:82-95.

Latour, Bruno and Steve Woolgar
1979 *Laboratory Life: The Social Construction of Scientific Facts*. Beverly Hills, California: Sage Publications.

Lecallo, Denis
1982 *Études et Documents, No. 49: Les Entreprises Publics en Côte d'Ivoire*. Abidjan: Ministère des Relations Extérieures.

Lemarchand, René
1964 *Political Awakening in the Belgian Congo*. Berkeley: University of California Press.

Lewis, Suzanne Grant
1986 "Training for Public Sector Use of Microcomputers in the Third World: The Experience of Tanzania." Paper presented at the Comparative and International Education Society's Western Regional Conference, University of Hawaii, Honolulu, Hawaii, December.
1987 "Computer Diffusion in Tanzania and the Rise of a Professional Elite." Paper presented at the 30th Annual Meetings of the African Studies Association, Denver, Colorado, November 20.
1988 "Microcomputers in Tanzania: A Study of Control and Influence in the Adoption Process." Unpublished Ph.D. dissertation. Stanford, Calfironia: Stanford University.

Lewis, Suzanne Grant and Mohammed S. Sheya
1987 *Survey of Computers in Tanzania – 1986*. Dar es Salaam, Tanzania: University of Dar es Salaam.

Little, Kenneth
1973 *African Women in Towns: An Aspect of Africa's Social Revolution*. London: Cambridge University Press.
1974 *Urbanization as a Social Process: An Essay on Contemporary Africa*. London: Routledge and Kegan Paul.

Luke, Timothy and Stephen K. White
1985 "Critical Theory, the Information Revolution, and an Ecological Path to Modernity," in *Critical Theory and Public Life*, John Forester (ed.), 22-53. Cambridge, Massachusetts: MIT Press.

Lussato, Bruno
1981 *Le Défi Informatique*. Paris: Librairie Arthème Fayard.

Lynch, Michael
1985 *Art and Artifact in Laboratory Science.* London: Routledge and Kegan Paul.

Lynch, Michael, Eric Livingston, and Harold Garfinkel
1983 "Temporal Order in Laboratory Work," in *Science Observed: Perspectives on the Social Study of Science*, Karin D. Knorr-Cetina and Michael Mulkay (eds.), 205-238. Beverly Hills, California: Sage Publications.

Lyotard, Jean François
1979 *La Condition Postmoderne: Rapport sur le Savoir.* Paris: Éditions de Minuit.

MacCannell, Dean and Juliet Flower MacCannell
1982 *Time of the Sign: A Semiotic Interpretation of Modern Culture.* Bloomington, Indiana: Indiana University Press.

MacGaffey, Wyatt
1983 *Modern Kongo Prophets: Religion in a Plural Society.* Bloomington, Indiana: Indiana University Press.

Maffesoli, Michel
1979 *La Conquête du Présent: Pour une Sociologie de la Vie Quotidienne.* Paris: Presses Universitaires de France.

Mannheim, Karl
1936 *Ideology and Utopia: An Introduction to the Sociology of Knowledge.* Translated by Louis Wirth and Edward Shils. New York: Harcourt, Brace, and World. (Originally published as *Ideologie und Utopie.* Bonn: F. Cohen, 1929.)
1950 *Freedom, Power, and Democratic Planning.* Hans Gerth and Ernest K. Bramstedt, eds. London: Routledge and Kegan Paul.
1952 *Essays on the Sociology of Knowledge.* Paul Kecskemeti, ed. London: Routledge and Kegan Paul.

Markus, M.L.
1984 *Systems in Organizations.* Marshfield, Massachusetts: Pitman Publishing Company.

Masini, Jean, Moises Ikonicoff, Claudio Jedlicki, and Mario Lanzarotti
1979 *Multinationals and Development in Black Africa: A Case Study in the Ivory Coast.* Northamptonshire, England: Saxon House.

Mauss, Marcel
1967 *The Gift: Forms and Functions of Exchange in Archaic Societies.* Translated by Ian Cunnison. New York: W.W. Norton and Company. (Originally published as *Essai sur le Don, Forme Archaïque de l'Échange. Année Sociologique*, Paris: Nouvelle Série, T.I., 1925.)

Mazuri, Ali A.
1977 "Development Equals Modernization Minus Dependency: A Computer Equation," in *The Computer and Africa: Applications, Problems, and Potential*, D.R.F. Taylor and R.A. Obudho (eds.), 279-304. New York: Praeger Publishers.
1978 "The African Computer as an International Agent," in *Political Values and the Educated Class in Africa*, Ali A. Mazuri, 320-342. London: Heinemann.

Mbiti, John
1966 *Akamba Stories*. Oxford: Clarendon Press.
Mehan, Hugh B.
1985 "Computer Literacy and Social Stratification," *Report #9*. La Jolla, California: Center for Human Information Processing, University of California, San Diego, November.
Menzies, Heather
1981 *Women and the Chip: Case Studies of the Effects of Informatics on Employment in Canada*. Montréal, Canada: The Institute for Research on Public Policy.
Mukerji, Chandra
1989 *A Fragile Power: The Role of Science in American Society*. Princeton, New Jersey: Princeton University Press.
Nora, Simon and Alain Minc
1981 *The Computerization of Society*. Cambridge, Massachusetts: MIT Press. (Originally published as *L'Informatisation de la Société*. Paris: La Documentation Française, 1978.)
Nordin, John P.
1986 "Data-Use It or Lose It," *East African Computer News* 2, 2 (March): 12-13.
Norman, Colin
1980 "Microelectronics at Work: Productivity and Jobs in the World Economy," *Worldwatch Paper 39*. Washington, D.C.: Worldwatch Institute.
Nzioka, Gideon
1986 "Computerized Banking in East Africa," *East African Computer News* 2, 1 (February):16, 17, and 23.
OCED: Groupe de Travail sur la Politique de l'Information, de l'Informatique, et de la Communication.
1982 *La Micro-éléctronique, la Robotique et l'Emploi, 7*. Paris: OCED.
Ong, Walter
1982 *Orality and Literacy: The Technologizing of the Word*. New York: Methuen.
Papagiannis, George J.
1985 "The Nairobi Aga Khan Academy Computers in Education Project: A Review." Geneva, Switzerland: Unpublished report, the Aga Khan Foundation.
Papagiannis, George J., C. Douglas, N. Williamson, and R. Le Mon
1987 *Information Technology and Education*. Ottowa, Canada: International Development Research Center, September.
Papert, Seymour
1980 *Mindstorms: Children, Computers, and Powerful Ideas*. New York: Basic Books.
Parsons, Talcott
1937 *The Structure of Social Action*. New York: The Free Press.
1967 "An Approach to the Sociology of Knowledge," in *Sociological Theory and Modern Society*, Talcott Parsons, 139-165. New York: The Free Press.

1970 "The Impact of Technology on Culture and Emerging New Modes of Behaviour," *International Social Science Journal* 22, 4:607-627.

Peirce, Charles Sanders
1931 *Collected Papers*. Cambridge: Harvard University Press.

Perrolle, Judith A.
1987 *Computers and Social Change: Information, Property and Power*. Belmont, California: Wadsworth Publishing Company.

Pichette, Serge
1979 *Problèmes Juridiques de Transferts de Technologie*. Montréal, Canada: L'École des Hautes Études en Administration Internationale.

Pinckney, Thomas C., John M. Cohen, and David K. Leonard
1984 "Kenya's Use of Microcomputers to Improve Budgeting and Financial Management in an Operating Ministry: An Updated Report," *Development Discussion Paper*, No. 169. Cambridge, Massachusetts: Harvard Institute for International Development.
1987 "Kenya's Introduction of Microcomputers to Improve Budgeting and Financial Management in the Ministry of Agriculture," in *Microcomputers in Public Policy*, Stephen R. Ruth and Charles K. Mann (eds.), 67-93. Boulder, Colorado: Westview Press.

Polanyi, Michael
1958 *Personal Knowledge: Towards a Post-Critical Philosophy*. New York: Harper and Row.

Price Waterhouse Associates
1986 *Kenya Computer Buyer's Guide*. Nairobi, Kenya: Price Waterhouse Associates, March.

Prince, Gerald
1987 *A Dictionary of Narratology*. Lincoln, Nebraska: University of Nebraska Press.

Republic of Kenya
1974a *Development Plan, 1974-1978, Part I*. Nairobi, Kenya: Government Printer.
1974b *Development Plan, 1974-1978, Part II*. Nairobi, Kenya: Government Printer.
1979 *Development Plan, 1979-1983, Part I*. Nairobi, Kenya: Government Printer.

République de Côte d'Ivoire
1981 *Plan Informatique, 1981-1985*. Abidjan: Commission Nationale pour le Traitement de l'Information.
1986a *Plan Informatique National, 1986-1990*. Abidjan: Commission Nationale pour l'Informatique.
1986b *Plan Informatique National, 1986-1990: Annexe I*. Abidjan: Commission Nationale pour l'Informatique.
1986c *Plan Informatique National, 1986-1990: Annexe II*. Abidjan: Commission Nationale pour l'Informatique.

Rodney, Walter
1972 *How Europe Underdeveloped Africa*. London: Bogle-L'Ouverture.

Rogers, Everett M. and Judith K. Larsen
1984 *Silicon Valley Fever: Growth of High-Technology Culture.* New York: Basic Books.
Roszak, Theodore
1986 *The Cult of Information: The Folklore of Computers and the True Art of Thinking.* New York: Pantheon Books.
Rothenberg, Jerome
1968 *Technicians of the Sacred: A Range of Poetries from Africa, America, Asia, and Oceania.* Garden City, New York: Doubleday.
Rousseau, Jean-Jacques
1918 *Du Contrat Social.* C.E. Vaughan (ed.), Manchester, England: Manchester University Press. (Originally published in 1762.)
Rumberger, Russell W. and Henry M. Levin
1984 *Forecasting the Impact of New Technologies on the Future Job Market. Project Report #84-A4.* Stanford, California: Institute for Research on Educational Finance and Governance, Stanford University, February.
Rumelhart, David E.
1984 "Understanding Understanding," in *Understanding Reading Comprehension,* J. Flood (ed.), 1-20. Newark, Delaware: International Reading Association.
Salaman, Graeme
1979 *Work Organizations: Resistance and Control.* London: Longman Group Limited.
Sartre, Jean-Paul
1976 *Critique of Dialectical Reason.* Translated by Alan Sheridan Smith. London: Humanities Press. (Originally published as *Critique de la Raison Dialectique.* Paris: Éditions Gallimard, 1960.)
Savané, Ibrahim Sy
1987 "L'Afrique à l'Heure de la Micro," *Jeune Afrique Économie* 98 (June): 83-84.
Schechner, Richard
1985 *Beyond Theater and Anthropology.* Philadelphia: University of Pennsylvania Press.
Schegloff, Emanuel A.
1980 "Preliminaries to Preliminaries: 'Can I Ask You a Question?'" *Sociological Inquiry* 50, 3-4:104-152.
1987 "Between Micro and Macro: Contexts and Other Connections," in *The Micro-Macro Link,* Jeffrey C. Alexander, Bernhard Giesen, Richard Münch, and Neil J. Smelser (eds.), 207-234. Berkeley: University of California Press.
Schuster, Ilsa
1979 *New Women of Lusaka.* Palo Alto, California: Mayfield Publishing Company.
Schutz, Alfred
1962 *Collected Papers, Vol. I: The Problem of Social Reality.* The Hague: Martinus Nijhoff.

Schwartz, Howard and Jerry Jacobs
1979 *Qualitative Sociology: A Method to the Madness.* New York: The Free Press.

Scott, Robert J.P.
1971 *Computers for Developing Countries.* Nairobi, Kenya: East African Publishing House.
1981 *Computers Kenya: Report on I.D.R.C. Sponsored Project.* Nairobi, Kenya: Institute of Computer Science, University of Nairobi.
1984 *Development of Computer Sciences and Application of Informatics in Education: A Survey of East Africa.* Nairobi, Kenya: University of Nairobi, Institute of Computer Science.

Sebeok, Thomas A.
1976 "Semiotics: A Survey of the State of the Art," in *Contributions to the Doctrine of Signs*, Thomas A. Sebeok, 1-45. Bloomington, Indiana: Indiana University Press.
1987 "In What Sense Is Language a 'Primary Modeling System?'" in *Proceedings of the 25th Symposium of the Tartu-Moscow School of Semiotics*, Henri Broms and Rebecca Kaufmann (eds.), 67-90. Helsinki, Finland: Arator Incorporated.

Secrétariat Général à l'Informatique
1987 *Troisième Journées Africaines d'Informatique.* Abidjan: JAI.

Simmel, Georg
1950 *The Sociology of Georg Simmel.* Translated by Kurt H. Wolff. Glencoe, Illinois: The Free Press.

Smelser, Neil
1987 "Depth Psychology and the Social Order," in *The Micro-Macro Link*, Jeffrey C. Alexander, Bernhard Giesen, Richard Münch, and Neil G. Smelser (eds.), 267-286. Berkeley: University of California Press.

Snow, David and Richard Machalek
1984 "The Sociology of Conversion," *Annual Review of Sociology* 10: 167-190.

Special Secretariat of Informatics of Brazil in Cooperation with the Ministry of Communications of Brazil
1983 *Transborder Data Flows and Brazil: Brazilian Case Study.* New York: United Nations Center on Transnational Corporations.

Sperber, Dan
1982 *Le Savoir des Anthropologues: Trois Essais.* Paris: Collection Savoir Hermann.

Stover, William James
1984 *Information Technology in the Third World: Can I.T. Lead to Humane National Development?* Boulder, Colorado: Westview Press.

Strom, Sharon Hartman
1987 "'Machines Instead of Clerks': Technology and the Feminization of Bookkeeping, 1910-1950," in *Computer Chips and Paper Clips, Vol. II: Technology and Women's Employment*, Heidi I. Hartmann (ed.), 63-97. Washington, D.C.: National Academy Press.

Suchman, Lucy A.
1984 "Human-Machine Communication." Palo Alto, California: Xerox Palo Alto Research Center.
Sudnow, David
1983 *Pilgrim in the Microworld.* New York: Warner Books.
Szombati-Fabian, Ilona and Johannes Fabian
1976 "Art, History, and Society: Popular Painting in Shaba, Zaire," *Studies in the Anthropology of Visual Communication* 3, 1 (Spring):1-21.
Tannen, Deborah
1982 "The Oral/Literate Continuum in Discourse," in *Spoken and Written Language*, Deborah Tannen (ed.), 1-16. Norwood, New Jersey: Ablex Publishing Corporation.
Toffler, Alvin
1980 *The Third Wave.* New York: William Morrow and Company.
Tupper, Patricio
1987 "La Bataille des Géants," *Jeune Afrique Economie* 98 (June):79-82.
Turkle, Sherry
1984 *The Second Self: Computers and the Human Spirit.* New York: Simon and Schuster.
Turner, Victor
1957 *Schism and Continuity in African Society: A Study of Ndembu Village Life.* Manchester, England: Manchester University Press.
1967 *The Forest of Symbols: Aspects of Ndembu Ritual. Ithaca:* Cornell University Press.
1969 *The Ritual Process: Structure and Anti-Structure.* Chicago: Aldine Publishing Company.
1974 *Dramas, Fields, and Metaphors: Symbolic Action in Human Society.* Ithaca: Cornell University Press.
1985 *On the Edge of the Bush: Anthropology as Experience.* Tuscon, Arizona: University of Arizona Press.
Tyler, Stephen A.
1985 "Ethnography, Intertextuality, and the End of Description," *American Journal of Semiotics* 3, 4:83-98.
UNESCO
1980 "The New World Information and Communication Order," *Resolution 4/19 of the Twenty-First Session of the UNESCO General Conference, Belgrade.* Reprinted in *NACLA Report on the Americas* 16, 4 (July/August):32.
Ungar, Sanford J.
1985 *Africa: The People and Politics of an Emerging Continent.* New York: Simon and Schuster.
Vansina, Jan
1985 *Oral Tradition as History.* Madison, Wisconsin: University of Wisconsin Press.
Wad, Atul and Michael Radnor
1984 *Technology Assessment: Review and Implications for Developing Countries.* Paris: UNESCO.

Wallace, John B.
1977 "Computer Use in Independent Africa: Problem and Solution Statements," in *The Computer and Africa: Applications, Problems, and Potential*, D.R.F. Taylor and R.A. Obudho (eds.), 13-41. New York: Praeger Publishers.

Warren, Bill
1983 "The Postwar Experience of the Third World," in *The Political Economy of Development and Underdevelopment*, Charles K. Wilber (ed.), 109-133. New York: Random House.

Weber, Max
1946 "The 'Rationalization' of Education and Training," in *From Max Weber: Essays in Sociology*, translated by Hans H. Gerth and C. Wright Mills (eds.), 240-244. New York: Oxford University Press.
1947 *The Theory of Social and Economic Organization,* translated by A. M. Henderson and Talcott Parsons. New York: The Free Press.

Weekes-Vagliani, W. with S. Pralong
1983 "In Service (On-the-job) Training as a Means of Developing Human Resources." Paris: OECD Development Centre.

Weekly Review
1976 "Computers: Benefit or Detriment?" Nairobi, Kenya (June 7):25.

Wells, Louis T.
1983 *Third World Multinationals: The Rise of Foreign Investment from Developing Countries*. Cambridge, Massachusetts: MIT Press.

Wescott, Clay G.
1985 "Microcomputers for Improved Budgeting in the Kenya Government." Nairobi, Kenya: Unpublished Report, August 6.
1987 "Microcomputers for Improved Government Budgeting: An African Experience," in *Microcomputers in Public Policy*, Stephen R. Ruth and Charles K. Mann (eds.), 95-116. Boulder, Colorado: Westview Press.

Wilson, Ernest J.
1988 "Privatization in Africa: Domestic Origins, Current Status, and Future Scenarios," *Issue* 16, 2:24-29.

Wynn, Eleanor Herasimchuk
1979 "Office Conversation as an Information Medium." Unpublished Ph.D. dissertation. Berkeley: University of California, Berkeley.

Zaliznjak, A.A., V.V. Ivanov, and V.N. Toporov
1977 "Structural-Typological Study of Semiotic Modeling Systems," in *Soviet Semiotics: An Anthology*, translated by Daniel P. Lucid (ed.), 47-58. Baltimore: John Hopkins University Press.

Zuboff, Shoshana
1988 *In the Age of the Smart Machine: The Future of Work and Power.* New York: Basic Books.

Index

A

Abercrombie, Nicholas, 14, 379
Abidjan (Ivory Coast), 107-109, 122-123, 127, 128, 130-138, 149, 151, 291, 334, 351n.10, 361n.4, 371, 386
  commercial center of, 107
  government center of, 107, 130
Accounting. *See* Accounts, financial
Acou, Anoh (carver), 326-328
A.C.P. (Association of Computer Professionals), 207, 231-232, 237
Actant, 31, 48, 258, 366-367n.2, 371
Actors, 28, 344
  as agents of social change, 10, 28, 48, 118, 215-216, 314, 317, 343-345, 359n.3, 371, 388
  in the narrative program of computer adoption, 93, 284, 289, 317
Accounts
  financial, 32, 36, 45, 47, 62, 69, 72-73, 87, 138-139, 141, 147, 270, 271, 297, 305, 316, 353n.8, 360n.5, 372, 376
  media, 30, 60, 63, 250, 251, 312, 328, 352n.2, 364n.7
  subjunctive, 53, 65, 72, 249, 250, 251
  verbal, 56, 278, 286-287, 315, 318, 341, 346, 355n.8, 365-366n.10, 366n.12
Africa
  colonial, 3, 14, 19, 23, 24, 26, 41, 96, 190, 321
  and computers, 5, 26, 27, 52, 96, 107, 111, 116, 142, 190, 205, 243, 301, 308, 342-345, 351n.13, 356n.9, 367n.5, 388, 392, 393-394
  definition of, 3
  neocolonial, 110, 321, 365n.9
  postcolonial, 12, 25, 80, 34, 344-345, 347, 351n.13, 393
Aga Khan Academy (Nairobi), ix, 203, 228-229, 353n.9, 362n.9, 389
Agriculture, 39, 41, 42, 49, 86-87, 117, 153, 157, 322, 340, 343-344, 359n.1, 359n.4, 366n.1
Analepsis
  of information, 19
  and prolepsis, 312, 315
  and signs, 8, 19, 324, 333-335
    See also Signs
APICI (Association Professionnelle des Informaticiens de Côte d'Ivoire), x, 21-22, 83, 94, 129, 131, 215-216, 35ln.10, 355n.1, 357n.1, 371, 379
Appel d'offres. See Bids, calls for
Apple Computers, 69-70, 124, 158, 203, 208, 251, 256, 259
  computer installations of, 86, 167, 226, 256
  ranking of, 148
Artificial intelligence, 116, 205
Attewell, Paul, 18, 379
Attribution, 37-38, 39, 51, 58, 66, 112, 367n.3
    mixed, 159, 286, 346

in the narrative program for computer adoption, 4, 37-39, 159, 200, 251, 276, 285, 317, 324-325, 346, 352n.4, 364n.9, 364n.2, 366-367n.2, 367n.4, 371, 375
positive, 37, 39, 51, 65, 66, 112, 276, 286, 297, 308, 318, 323, 340, 343
reflexive, 366-367n.2
terminal, 8, 112, 200, 277-278, 285, 308, 318-319, 322, 323, 324, 337, 340, 346
Authorship, 325, 341, 383
Automation, 11, 16, 17-18, 27, 32, 48, 55, 108, 112, 200, 210, 346
of agriculture, 112, 123
industrial, 49, 81, 112, 115, 116, 205, 206, 212, 214-215, 230, 245
office, 16, 19, 27, 36, 205, 334, 361n.5
and robotics, 48, 114-116, 389
Automobile dealerships, 34-36, 46, 60, 68-69, 73, 81, 91, 101, 161-170 173, 183, 186-187, 190, 199, 203, 219, 238, 244, 267-274, 324, 364n.8

B

Baker, Phyllis, 36, 352n.3, 379
Balandier, Georges, 76, 379
Bar-Hillel, Yehoshua, 314, 379
Barker, Jane, 201, 379
Barthes, Roland, 286, 379
Bata Shoes, 161
Batch computing, 19, 60, 141, 163, 187, 269, 271, 304, 360n.5, 360n.9, 371
Baudrillard, Jean, 16, 53, 55, 201, 258, 312, 321, 323, 329, 351n.14, 352n.2, 354n.3, 360n.10, 367n.3, 379
Bay, Christian, 282, 379
BBC Computers, 222, 226, 256
Belgium, 135
Bell, Daniel, 4, 12, 110, 379-380
Benin, 357n.5, 371
Berg, Elliot, 35ln.13, 385
Berg Report (*Accelerated Development in Sub-Saharan Africa*), 25, 35ln.13, 385
Berge, Noel, 219, 385
Bids
calls for, 149-151, 294-296, 309; 365n.4, 371
economic effects of, 149-151, 295, 365n.4
government, 113, 133, 149-151, 294-295, 371
Birdwhistell, Ray L., 263, 364n.7, 380
Bittner, Egon, 79, 380
Bolamba, Antoine-Roger, 14
Boltanski, Luc, 93, 380
Bookkeeping, 22, 69, 72, 171, 353n.8, 360n.7, 360n.8, 392
and computers, 22, 69, 72, 130, 171, 191-195, 219, 360n.7, 360n.8
Bordeaux, Henri, 14
Bourdieu, Pierre, 27, 220, 35ln.1, 380
Bowles, Samuel, 17, 20, 220, 380
Branciard, Michel, 36ln.6, 380
Braverman, Harry, 20, 185, 380
Brazil, 97, 116, 345, 392
Brazzaville (Republic of Congo), 15
Brodman, Janice Z., 33, 86, 88, 89, 380
BULL, 10, 133, 138, 142, 146, 148, 155, 294
ranking of, 148, 358n.9
Burja, Janet, 23, 380
Bunn, James, 258, 261, 380
Bureau Intergouvernemental pour l'Informatique (Ivory Coast), 111, 380
Bureaucracy, 14, 41
and government, 78, 301, 344
and lines of authority, 14, 20, 58, 71-73, 76

and organization of the workplace, 30, 53, 68, 71-73, 76, 79, 88, 98-99, 101, 158, 170, 184, 200-202, 275, 277
and technology, 53, 71, 76, 78-79, 88, 95-96, 98-99, 103, 173, 184, 301, 310, 314, 316-317, 324, 344
Burke, Kenneth, 54, 349n.2, 381
Burkina Faso, 109, 357n.5, 371
Burroughs, 10, 45, 139, 146, 154
See also UNISYS
Business Computer Systems (Kenya), 203
Bypassing, 31, 313, 371, 376
See also Cognition

C

CAI (Computer Assisted Instruction), 212
Calhoun, Craig, 51, 382
Cameroon, 92, 139, 145, 357n.4, 357n.5, 358n.9, 386
Campbell, Bonnie, 358n.10, 381
Capital
and assets, 42, 43, 44, 45, 67, 108, 124, 130-131, 133-134, 140-145, 197, 295, 349n.3, 357n.5
cultural, 85, 158, 215, 220
economic, 67, 116, 220, 378
educational, 215, 218, 220
foreign, 12, 42-43, 45, 49, 117, 121, 131, 133-134, 140-145, 146-154, 170, 295, 331, 358-359n.10, 365n.4
informational, 26
local 40, 43, 131, 133-134, 358-359n.10
reproduction of, 201, 358-359n.10
Capitalism, 201, 358-359n.10, 360n. 10, 381
Carim, D. Enver, 110, 381
Cash register, 193-194, 371
National (*see* NCR)
CCCE (Caisse Central de Coopération Économique), 149, 295
CCD (Culture of Critical Discourse). *See* Discourse
Central African Republic, 357n.5, 371
Central Bureau of Statistics, Kenya, 350n.7, 381
CETI (Centre d'Enseignement Technique en Informatique, Abidjan), 123
CFA franc, 124, 131, 133, 135, 140, 142, 357n.5, 371
Chad, 357n.5, 371
Chamber of Commerce
data from Ivory Coast, 124, 130-131, 133, 142, 145, 357n.5, 358n.6
data from Kenya, 130, 160
Charge card companies
International, 137, 178
in Ivory Coast, 137
in Kenya, 44, 73, 159, 178-183, 190, 198, 204, 305-306, 355n.4, 365n.8, 382
Chauhan, Nishit, ix, 235-237
China, 345
Cicourel, Aaron, 55, 381
CIERIA, 155
Class
computer experts as emerging, 23, 37, 57, 75, 95, 122-123, 152
economic, 48
elite, 14, 23, 42, 43, 78, 93, 214, 217-218, 243, 322, 343, 350n.5, 366n.1, 373, 381
postcolonial, 12, 80, 97
proletariat as, 13, 14, 15, 20, 23, 26, 96, 380, 381
reproduction of structures, 220
ruling, 322
social, 14, 23, 379, 384, 385
subelite, 14, 15, 23, 25, 26, 349-350n.4, 350n.5

technocratic, 16, 20, 25-26, 30, 37, 58, 78, 93, 96-97, 190, 218, 220, 225, 243, 350n.5, 367n.5, 371, 384
See also Computers and elite; Stratification
Clifford, James, 9, 381
CMIRH (Centre Mondial d'Informatique et Ressources Humaines), 205
CNI (Commission Nationale pour l'Informatique), 40, 52, 107, 113, 352n.5, 357n.2, 371, 374, 375, 376
Codes
communicative, 364n.10
of empowerment, 8
and models, 100-103, 315
and recoding, 19, 21, 58, 73-74, 89, 99, 101, 201, 346, 360n.10, 372, 376
semiotic, 100-103, 362-363n.1, 363n.4, 383
Cognition, 62, 75-76, 102, 253, 277, 312, 323, 363-364n.6, 391
See also Bypassing; Logic
Cohen, Abner, 14, 23, 381
Cohen, John M., 18, 19, 86, 87, 88, 316, 355n.5, 390
Cole, Michael, 311-312, 381
Coleman, James S., 71, 354n.6, 381
Collins, Randall, 102, 220, 253, 381
Commodore Computers, 232-233
Commercial colleges, 24, 79, 83, 95, 123, 218
in Ivory Coast, 123-128, 206, 214
in Kenya, 52, 123, 153, 207-210, 229-237, 362n.10
Communication
between students and teachers, 259, 260, 261, 262, 264, 265, 266-267
by computer, 257, 261, 344, 351n.12
human, 261, 283, 286, 315, 364n.10
networks of, 332
official, 352n.4
senders and receivers of, 28, 284, 285, 288, 289, 294, 323, 352n.4, 363-364n.6, 365n.9, 366-367n.2
systems of, 322, 330-331
verbal, 263, 315
See also Discourse; Interaction; Systems
COMPAT (Computer Association of Tanzania), 80, 83, 350n.5, 350-351n.9, 371-372
See also Tanzania, computer associations in
Comp-Rite (Apple Computers, Kenya), 203, 208, 226, 240
Computer Association of East Africa, 211, 229, 350n.9, 372
Computer controller, 73-74, 171-174, 188, 372
Computer professionals, xi, 14, 18, 19, 24, 30, 52, 53, 68-71, 80, 81, 83, 87, 93-94, 95, 97, 198
and employment, 213-215, 217, 344
exclusivity of, 71, 79, 85, 142, 216, 220, 344
networks, 14, 22-23, 80, 85, 87, 97, 103, 111, 202, 209, 217, 218, 343
as priesthood, 72, 81, 94, 97-98, 100
shortage of, 18, 209, 220, 309
training of, 18, 69, 94, 97, 115-123, 126-140, 209, 217, 218, 233-234, 255-256, 343
turnover of, 297
See also Professionals
Computer science, 120-121, 125, 132, 198, 210, 221, 373, 374
Computer Training Centre (Nairobi), ix, 203, 230-236, 237, 381
Computer users, 26, 30, 97, 112, 116, 118-123, 125, 371, 372
attitudes toward computers, 28, 31, 61, 62, 67, 71, 74, 76, 94, 97, 120-

121, 130, 159, 242, 244, 253, 272, 313, 318, 323, 334, 345, 352n.3, 354n.4, 355n.2, 355n.4, 363n.2, 363n.4, 364n.9, 373
authority of, 103, 199, 220
hackers as, 219, 242, 245, 252
health hazards to, 67, 354n.4, 355n.3
hierarchy of, 20, 58-60, 71, 188-191, 216, 272, 275-277, 309, 359n.3, 362n.12
as operators, 8, 36-37, 60, 66, 69, 83, 95, 99, 111, 119-120, 134, 147, 202, 213-217, 218, 245, 256, 267, 269, 270, 272, 275-276, 292, 308, 354n.4, 355n.3, 363n.2, 364n.8, 365n.8, 372
and work organization, 52, 67-68, 86-91, 94-95, 97, 103, 117-118, 160, 175-176, 185, 201, 220, 237-244, 275-276, 355n.3, 359n.3
Computerization, 11, 13, 18, 41, 57, 108-113, 116
of agriculture, 64-65, 66, 117, 340, 359n.1
attributes of, 11, 184, 198, 275, 313, 318
benefits of, 35, 48, 117, 168, 184, 210, 215, 273, 340, 343, 361-362n.7, 366n.12, 394
conditions for, 47, 112, 153, 159, 198, 201, 219, 289, 303, 305, 354n.5, 357n.3, 367n.4, 382, 386
and employment, 110, 217, 350n.7, 357n.3, 361n.5
fear of, 110, 158, 304
of government, 11, 13, 18, 20, 41, 47, 85-93, 98, 113-115, 131-132, 143, 149-151, 157, 158, 206, 216-217, 238, 308, 316-318, 342-343, 356n.12, 357n.1, 359n.2
of private enterprises, 16, 42, 44-48, 66, 81, 98, 101-102, 159-191, 195-196, 200-201, 217, 297, 305, 307, 313, 356n.12, 357n.1, 358n.7, 359n.2, 350n.3, 360n.5, 361-362n.7
selective, 47, 68, 187, 305
of society, 4, 6, 41, 46, 110-111, 116-117, 153-154, 206, 213-214, 298-299, 303-304, 342-343, 357n.1, 359n.2, 361n.3, 380, 389
Computers
activists for, 366n.12, 386
adoption of, 3, 5-7, 11, 26, 29, 41-42, 43-45, 49-50, 51-57, 66, 88, 92, 98-99, 100, 103, 109, 112, 115, 157, 158, 159-161, 165-166, 173, 183, 189, 193-198, 202, 249, 255-256, 285, 286-310, 313-314, 316-317, 318, 322, 324-325, 331, 342-345, 346, 355n.8, 359n.3, 360n.10, 361n.3, 361-362n.7, 364n.10, 365n.5, 365n.8, 365-366n.10, 366n.12, 368-369n.11, 375, 394
advertising, 107-109, 124, 331, 332
assembly of, 117, 127-128, 224, 303
and associations, 21-23, 52, 80, 83, 99, 215-216, 218, 355n.1
and banking, 38-39, 46, 47, 63-64, 78, 107-108, 130, 133-134, 135-136, 138, 179, 219, 291, 299, 350n.6, 360n.9, 389
and children, 252, 255, 389
and economy, 98, 111, 113, 210, 214, 308, 316, 342, 350n.7, 368-369n.11
and education (*see* Education)
and elite, 52, 72, 74, 78, 89, 93-94, 96-97, 122, 212, 220, 225, 243, 344
and employment, 17-21, 71, 97, 110, 118, 209, 213-214, 220, 222, 228, 234, 293, 308, 357n.3, 389

experiments with, 91, 97, 191-198, 228, 271, 304, 360n.8, 362n.8, 362n.9
experts with, 5, 18, 21, 30, 54, 56-57, 67, 74-75, 82, 83, 89, 92, 93, 99, 102, 107-123, 125-128, 132, 142, 160, 170, 176, 188, 202, 207-208, 210-217, 222, 225, 316, 343
and fantasies, 100-102, 313, 331-333
hardware for, 52, 141, 142, 145, 149, 151, 173, 219, 224, 288, 298, 334, 361-362n.7, 365n.4, 372, 373
importation of (*see* Regulation)
integration of, 40, 55, 69, 111, 116, 166, 174, 180, 185, 255, 267, 274, 288, 323, 341, 350n.7, 352n.4, 355n.3
as luxury items, 159, 303, 382
mainframe, 10, 19, 41, 46, 60, 62, 68, 72, 78, 143, 173, 350n.7, 353n.7, 373
misuse of, 160, 184-185, 300, 318, 334
personal, 38-39, 42, 52, 68, 91, 109, 237, 340, 356n.6, 382
printouts from 61, 164, 251, 253, 329
programs for, 16, 70-72, 87, 90, 99, 107, 124-125, 137, 141-142, 176, 178, 184, 189, 198, 205, 217, 220, 221-222, 223, 226, 228, 231-232, 250, 253-256, 257, 259, 260, 264, 267, 269-270, 271, 274-276, 293, 304, 310, 313, 316, 329, 371, 372, 375, 376
repair of, 26, 56, 58, 109, 120, 125, 149, 167, 293, 304, 334-335
sales of, 38, 44, 51, 91, 109, 129, 138, 146-151, 292, 302, 317, 365n.5, 390
as second subject, 249, 253, 393
and security systems, 163, 168
and social change, 55, 75, 100, 111, 116, 159, 318, 359n.3, 390
software for, 52, 92, 109-110, 129, 134, 137, 139, 145, 149, 151, 173, 177, 183, 219, 231, 243, 288, 299, 309, 313, 334, 358n.7, 361-362n.7, 365n.4, 372, 373, 376
training on, 52, 61, 66-71, 79, 81, 84, 92, 94, 98-99, 109, 118-128, 139, 153, 161, 170, 177, 198, 200, 205-206, 209-212, 213-237, 242, 244, 246, 249-250, 251, 253-256, 259, 260, 261, 262, 263, 264, 265, 275-277, 278, 308, 313, 321, 334, 361-362n.7, 362n.8, 362n.9, 362n.10, 362n.11, 362n.12, 363n.2, 365n.5, 374, 382
as tutor, 228, 249-253
underutilization of, 47, 55, 183, 191, 300, 305, 344
See also Education, Technology
Conference for Integration of African Informatics, 6, 41, 111, 113, 357n.1
Confrontation, 29, 31-32, 66-71, 341
and computer use, 52-53, 57, 60-63, 66, 68, 69, 71, 72, 88-89, 92, 111, 207, 220
in the narrative program for computer adoption, 4, 31-32, 43, 45, 51-57, 60, 63, 66, 71, 75-76, 78, 111, 183, 249, 251, 254, 263, 271-273, 285, 317, 352n.4, 357n.5, 367n.4, 371, 372, 375, 376, 383, 388
Congo, 14, 15, 143
Consultants, 22, 33, 60, 73, 128-151, 161, 343
and consulting firms, 16-17, 42, 44, 47, 99, 107-108, 115, 121-123, 128-151, 211, 216, 289, 291, 292, 298-300, 305, 343, 351n.10,

357n.5, 358n.6, 358n.7, 372, 376;
foreign, 7, 18, 20, 25, 33, 44, 49, 85-93, 108, 123, 133-135, 137-138, 152-154, 161, 205, 219, 238, 243, 291, 316, 343, 355n.5, 356n.6
local, 17, 91-92, 96, 121, 133, 139, 178, 180-183, 206, 306, 365n.8
Contract, 8, 38-40, 43, 284, 287-291, 299, 324, 325, 372, 376
acceptance of, 39, 284, 285, 317, 331, 340, 364n.1, 364n.2, 365n.3, 365n.7, 365n.9
broken, 39, 284, 290, 292, 294, 300, 304, 307, 309, 312, 331
and business transactions, 8, 14, 288-300, 301, 304, 308, 309, 323-324, 346
and computer sales, 109, 138-139, 141-142, 145-154, 287-288, 290-294
definition of, 284, 287, 364n.1
fiduciary, 283, 294
fragile computer, 4, 7, 19, 39, 43, 79, 89, 112, 159-160, 283-284, 286, 292, 294, 300-310, 312, 317, 322, 325, 331, 334, 342, 346, 373
government, 109, 112-113, 138, 142-143, 146-154, 288-289, 295-299, 309, 318
international, 135, 136
interpretation of contractual exchanges, 8, 44, 287, 288-310, 312, 317, 323, 340, 346, 365n.5, 365n.6, 365n.7, 365-366n.10, 373, 376
maintenance, 109, 125, 140-143, 145, 290, 294, 304, 309, 365n.5, 365n.7, 372
mandate in, 284, 285, 297
means, 288, 290-291, 295-296, 365n.5, 365n.6, 372
in narrative, 38-40, 43, 284, 286, 300, 325, 341, 365n.9, 372, 375
and obligations, 283, 287, 290, 294
progressive computer, 4, 6, 7, 40, 43, 79, 112, 284, 286, 289-300, 308, 309, 312, 317, 322, 325, 331, 346, 367n.4, 367n.5, 376
resolution of, 28-29, 38, 40, 68, 340-341, 349n.1, 352n.4, 364n.2
results, 288, 291-292, 295-296, 365n.5, 372
semiotic concept of (*see* Semiotics)
sexual, 331
social, 283, 287, 294, 323, 326, 352n.4, 361n.1, 365n.3, 379, 384
spiritual, 331
structures of, 283-288
terminal, 308, 325
theory of, 287
See also Computers; Exchange; Information; Semiotics
Control, 49, 56-57, 60, 71, 73
acceptance of, 340
budgetary, 18, 72, 85-92, 97, 200, 316, 355n.5, 356n.6, 374, 375, 390, 394
bureaucratic, 16, 20, 71, 73, 77, 92, 101, 103, 350n.8
economic, 42, 97, 148, 157, 298, 329, 340
government, 11, 49, 109-113, 288, 295-298, 300, 310, 318, 325-326, 342-343, 356n.9, 358n.7, 358-359n.10, 365n.5, 365n.6, 367n.5
ideological, 72, 97, 158, 201, 246, 325
of information, 21, 355n.2, 367n.5
organizational, 57, 60, 73, 76, 103, 111, 181, 185, 190-192, 201-202, 246, 254, 353n.3, 356n.7, 356n.11, 391
social, 72, 277-278, 325

technological, 11, 49, 56, 60, 66, 73-74, 76, 80, 92, 101-102, 220, 244-246, 275-277, 288, 298, 310, 325, 350n.8, 352n.6, 354n.6, 355n.1, 356n.10, 356n.11, 363n.2, 367n.5
See also Information
Conversion experience, 392
and computer education, 244-246
and technology, 244, 245-246
Cooper Motors, 46
Cooperatives
farming, 64-65
handicraft, 33, 44, 64-65, 74-75, 160, 191-198, 304, 349n.3, 360n.7, 360n.8, 365n.7
See also Nairobi Handicraft Cooperative
Coopers and Lybrand, 155
Côte d'Ivoire. *See* Ivory Coast
Cottage industries, 349n.3. *See also* Enterprises
Corporations
competition of, 47, 215, 342
local, 47, 49, 215
multinational, 41-42, 44, 47-49, 80, 98, 109, 152, 161, 170, 342, 356n.9, 358n.7, 386
Third World, 47, 64, 387, 388, 394
Courtés, Joseph, 9, 245, 254, 283, 287, 315, 349n.1, 365n.10, 367n.4, 382, 385
Credentials, 7, 15, 18, 22, 26, 77, 83, 97, 111, 115, 120, 128, 160, 189, 198, 207-209, 210, 217, 219-223, 225, 231, 234-235, 237, 243, 308, 334, 361-362n.7, 373, 374
crisis of, 79, 80, 115, 209, 221, 223, 229, 362n.12
and examinations, 115, 119, 120-121, 206, 217, 222, 224-225, 231-234, 237, 243, 321, 362n.12
See also Education; Schools
Crédit Lyonnais, 47, 136
Creoles (Sierra Leone), 15
Crompton, Rosemary, 20, 24, 82, 187, 198, 275, 351n.11, 360n.9, 381
Crozier, Michel, 17, 350n.7, 381
Culture, 4, 101
alien, 266, 368n.11
change in, 9, 23, 100, 318, 344, 356n.7
computer, 7, 99, 245, 336, 344-345, 360n.10, 390
and images, 214
interpretation of, 9, 23, 28, 346, 381
media, 206
office, 190
oral, 364n.10
postmodern, 8, 9, 12, 200, 323, 347, 351n.14, 375
and representations, 102, 261, 329-330, 332, 365-366n.10
reproduction of, 31, 220, 351n.1, 380
status, 78, 85, 99, 220
Western, 278, 345
See also Postmodernity; Symbols

D

Daily Nation, 303, 355n.4
Dakar (Sénégal), 134, 143, 145
Danow, David K., 102, 382
Data
archival, 111
backlog of, 19-20, 86, 173, 318
banks for, 117, 150, 324, 352n.3, 357n.1
collection of, 19, 111, 160-161, 191-193, 249, 256, 279, 321, 355n.8, 358n.8, 362-363n.1
computer programming languages

and, 70, 87-89, 100-101, 124, 164-165, 178, 205, 208, 219, 221, 223, 226-227, 231, 233-243, 250, 251, 253, 256-257, 259, 264, 267, 271, 274, 276, 293, 301, 304-305, 310, 313, 316, 329, 371, 372, 375, 376
dead, 31
inputting of, 69, 137, 252-253, 259, 260, 262, 264, 265, 266-271, 278, 306, 360n.9, 364n.8, 372
management of, 66, 70, 120, 135, 137, 208, 220, 222, 242, 315, 346
overload of, 186
processing of, 19, 27, 61-62, 67, 73, 110, 129-130, 135, 139, 166, 175-176, 186, 189, 208, 224, 240, 249, 256, 316-317, 360n.5, 371, 372, 389
transborder flow of, 97, 392
Data Bureaus, 16, 19, 24, 32, 34-35, 46-47, 55, 62, 68, 70, 71, 99, 130, 153, 175, 190, 238, 309, 360n.5, 371, 372
Data Centre (Nairobi), 46-47, 61, 70, 163, 175, 181, 187, 238-243, 304-305, 315, 360n.5, 362n.11
Data General Computers, 46, 148
Data Processing Manager (DPM), 20, 24, 34-35, 36, 58, 60, 63, 68-69, 73, 78, 99, 147, 161, 169, 175, 184, 186, 188, 199, 213, 244, 359n.3, 372, 376
Datsun Dealership, 167-170, 172, 192, 199, 204. *See also* Automobile dealerships
DEC (Digital Equipment Corporation), 148, 151
ranking of, 148, 358n.8, 358n.9
Decamp, David, 366n.11, 382
de Certeau, Michel, 355n.2, 382
Decision making, 54, 76, 340
democratization of, 26, 97
in firms, 19, 28, 35, 56, 83, 138, 145, 183, 186, 286, 352n.3, 355n.3
in government, 6, 89, 97, 286, 303, 380
See also Vendors
De Lagy, Pamela, 51, 354n.7, 382
Deontology, 22. *See also* Doctrinal purity
Desbois, Dominique, 358n.9, 382
Deskilling, 19-20, 24, 185, 187, 198, 201, 220, 372
and reskilling of workers, 18, 19-29, 112, 162, 169, 177, 185, 187, 194, 200, 209, 217, 223, 226, 246, 255, 295, 350n.4, 364n.8
Development
of computer industry, 11, 206, 305, 322, 373, 375, 385, 392
of computers, 39, 43, 47, 48, 66, 128-129, 139, 150-154, 358-359n.10, 392
economic, 4, 25, 49, 96-97, 215, 318, 322, 350n.6, 351n.13, 354n.3, 358n.10, 359n.2, 359n.4, 366n.1, 381, 384, 388, 390, 394
and employment, 97
failures of, 322
government projects for, 85-86, 89, 96-97, 102, 143, 322, 359n.1
industrial, 322
national, 30, 43, 110, 115-116, 157, 215, 217, 322, 346, 352-353n.6, 357n.3, 359n.1, 366n.1
representations of, 42, 48-49, 321, 345, 352n.5
strategies of, 12-13, 48, 49, 322, 351n.13, 356n.8, 356n.9, 357n.3 358-359n.10
theories of, 13
Discourse, 5, 9-10, 52, 55, 77, 393

of confrontation, 43, 45, 51, 54, 58, 68, 71, 75-76, 183, 249-263, 273, 322, 364n.9, 366n.12
conversation as, 52, 54, 249-250, 289, 300, 375, 382, 384, 394
culture of critical, 30, 57-58, 77, 367n.5, 371
of domination, 43, 45, 244, 249, 271-274
and images, 250, 252, 256-271, 324, 363n.4
in narrative, 51, 249, 283-284, 287, 365-366n.12, 382
neocolonial, 321
political, 30
postmodern, 12, 58, 72, 367n.3
public, 6, 12, 28, 29, 30, 50, 325, 345, 354n.2, 367n.5
and rhetorical strategies, 34, 38, 55-58, 67, 70, 75, 85, 92, 116, 158, 184, 345
and sociolects, 315, 322
and speech exchange systems, 9, 30, 54, 250, 289, 349n.2, 367n.5
strategies of, 29, 49, 52, 57, 58, 68-69, 71, 78, 284, 287, 310, 312, 345, 349n.2, 352n.4, 363-364n.6, 382
structure of, 50, 283-284, 349n.2
technical, 29, 30, 35, 37, 57, 69, 70, 79, 244, 274, 300, 310-312, 313, 315, 322, 337, 354n.2, 356n.12, 364n.10, 366n.11, 367n.5, 371, 374
transcription of, 250, 252, 288, 322, 375
universe of, 30, 253, 312, 365-366n.10
utopian, 5, 51, 75, 78, 366n.12
Divine kingship, 102
Doctrinal purity, 71, 77-78, 79, 84, 97, 103, 116, 313, 324
Domination, 15, 34-37, 51, 66, 69, 89, 111, 128, 154, 158
and computer use, 43, 45, 66, 69, 207, 273, 331, 334-335
in the narrative program for computer adoption, 34-37, 43, 45, 53, 66, 69, 89, 111, 159, 185, 200, 249, 251, 271-274, 285, 324, 352n.4, 363n.2, 367n.4, 372, 375
and state control, 12, 43, 85, 111, 154
Dorgelès, Roland, 14
Downing, Hazel, 201, 379
Douglas, C., 389
Dramaturgy, 23, 28-29, 381
Dreyfus, Hubert L., 255, 257, 382
Dreyfus, Stuart E., 255, 257, 382
Duhamel, Georges, 14
Dumont, René, 322, 366n.1, 382
Duncan, Starkey, 267, 382
Durkheim, Émile, 11, 365n.3, 382

E

East African Computer News, ix, xi, 39, 58, 63, 65, 66-67, 79, 82, 83, 182, 211, 221, 228, 229, 340, 341, 350n.9, 361n.1, 361n.2, 381-382
Eco, Umberto, 100, 253, 258, 368n.7, 383
École Nationale Supérieure de Statistique et Économie Appliquée, 132
Economy. *See* Ivory Coast; Kenya
Education, 48, 52, 77, 118-128, 379, 382, 394
and accreditation, 7, 18, 216
budget for, 124-125, 150, 154, 205
ceiling of, 198-199, 277
colonial, 206
and computers, 21, 24, 48, 49, 70-71, 84, 92, 96-97, 117-126, 153, 198, 205, 206, 213-216, 217, 218, 226-227, 246, 254-255,

278, 286, 303, 308, 322, 326, 334, 343-344, 346, 354n.2, 357n.3, 358n.7, 361n.1, 361n.2, 364n.10, 373, 388, 389, 392
crisis in, 209
and educators, 118, 121, 206, 226-227, 256, 259, 260, 261, 262, 264, 265, 266-267, 363-364n.6, 371
and employment, 214-216, 218, 225-226, 242, 243, 380, 394
international, 118
and learning, 7, 61, 126, 206, 220, 227, 245-246, 249, 255-256, 259, 260, 262, 274-277, 308, 313, 316, 329, 367n.4, 382
local, 118, 152-153
philosophies of, 118-120, 126, 205-210, 227, 255, 257
planning of, 212-213, 219
policies for, 218
postgraduate, 7, 115, 205, 216, 222-223
standards of, 79, 97, 111, 115-116, 206, 215, 219, 221, 229, 235, 244, 334, 361n.1, 361n.2, 361-362n.7, 362n.10, 362n.11
and students, 255-258, 277-278
and universities, 109, 115-116, 119-124, 204, 205, 216, 218, 221-222, 226-227, 235, 383
vocational, 212, 214, 220, 223, 228, 362n.9
See also Schools
Ekman, Paul, 263, 383
Eldon, Michael, 355n.4, 383
Ellul, Jacques, 79-80, 383
Employment, 44
and careers, 80, 219, 238-240, 242-243, 345-347
and education (*see* Computers; Education)
possibilities for, 44, 49, 214-217, 223, 225, 230-231, 234, 237-238, 243, 292-293, 306, 308, 346, 350n.4, 350n.6, 351n.11, 361n.5, 362n.10, 362n.11, 362n.12, 384, 386
statistics for, 45, 48, 244, 381
See also Computers; Education; Schools
Engineers, 19, 24, 39, 43, 48, 80, 93-95, 108, 112, 115, 116-125, 129-130, 132-140, 142-147, 149-154, 168, 210-216, 290, 300, 308-309, 343-344, 347, 361n.4, 373, 374, 377
and division of labor, 93, 276, 291-292
training of, 43, 115-121, 125, 142-146, 149-154, 205, 211-212, 215, 218, 222-223, 361n.4, 373, 374
women as, 94-95, 146, 292-293
Enterprises, 55, 57, 71, 73-75, 101, 111, 121, 134, 141-151, 222, 274, 318, 361-362n.7
government, 13, 44, 51, 109, 111, 308, 318, 324
and informal economy, 45, 349n.3, 353n.10, 380, 385
local 13, 121, 170, 228, 324, 353n.9
private, 5, 21-22, 44, 47, 48, 62, 67, 109-110, 115-117, 128, 133-135, 138, 141-143, 151, 157, 160, 170, 185, 189, 202, 212, 215-216, 225, 228, 289-290, 298, 308, 317, 324, 343, 345, 357n.3, 358n.7, 359n.4, 360n.5, 361n.2, 361n.4, 361-362n.7, 365n.6, 373, 374, 394
size of, 44
small and medium-sized, 21, 44-47, 111, 124, 128-130, 134, 137, 170, 190, 238, 243, 289-290, 324, 334, 349n.3, 353n.9, 355n.4, 384

Entitlement
to read, 7, 184-185, 200, 275
and work organization, 7, 10, 19, 29, 60, 101, 275, 277, 289, 293, 310, 317, 346, 350n.8, 359n.3, 360n.6, 360n.10, 376
Entrepreneurs, 48, 91, 93, 110, 128, 170, 220, 243, 373
Equatorial Guinea, 143
Equipment, 19, 45, 69, 112, 123, 128, 133, 142, 148-151, 161, 201, 221, 312, 361-362n.7
national brand, 152
outmoded, outdated, or incompatible, 6, 17, 32, 35, 52, 74, 159, 173, 177, 274, 290, 301, 305, 309, 312, 318, 335, 361-361n.7
sale of, 288, 290, 300, 359n.4, 360n.5
up-to-date and compatible, 17, 32, 67, 109, 112, 128, 142, 161, 294-295, 303
and vocabulary, 52, 72
ESII (École Supérieure d'Ingénieurs Informaticiens), 120-121, 128, 373
Ethnography, vii, 9-10, 12, 49-50, 72, 160-161, 198, 250, 293, 322, 355n.8, 368n.9, 380, 381, 386, 393
Eurosoft, 142
Evolués, 14-15, 24, 373
Exchange, 287
contractual, 279, 286, 289-306, 311-312, 365n.5, 365n.6, 365n.7, 365-366n.10, 373, 376
economic, 286, 339-340, 388
foreign, 81, 291, 300, 368-369n.11
monetary, 73, 81, 219, 291
in narrative, 286-288
semiotic, 279
social, 287, 363-364n.6
speech systems as (*see* Discourse)
symbolic, 219, 286-288, 379, 388
verbal, 252, 253, 257, 259, 260, 262, 263, 264-265, 310-311, 364n.7, 366-367n.2, 367n.3
See also Contract; Discourse; Semiotics
Expression, 253, 273
by computer, 266
continuum of, 253, 257, 278
cultural, 313, 375
indexical (*see* Indexical Expressions)
and perception, 253, 257, 273, 278
terminal, 253
verbal, 310-311, 363-364n.6
See also Communication; Signs, terminal
Eyerman, Ron, 77, 383

F

Fabbri, Paolo, x, 368n.9, 387
Fabian, Johannes, 331, 393
Fazal, N.G. (Mrs.), ix, 230-235
Fernandez, James 312, 383
Fisher, Renée, 354n.2, 383
Flexner, Stuart Berg, 383
Flynn, Pierce J., xi, 363n.1, 386
Folktales, 3-4, 28, 337-341, 368n.11
Foster, Philip J., 220, 383
Foucault, Michel, 325, 383
Fragile computer contract. *See* Contract, fragile computer
France
computerization of, 93, 110, 136, 205, 213, 389
economy of, 110, 350n.7, 361n.6
influences on Ivory Coast, 43, 47, 93, 107-109, 133, 137, 145-146, 149, 152, 215, 291, 294, 299, 341, 361n.6
social class in, 93, 380
Friedland, Mary, ix, 226, 362n.8, 362n.9, 383
Friesen, W.V., 263, 383

Future-orientation, 12, 37, 40, 98-100, 102, 110, 116, 136, 142, 240, 243, 315, 321, 324, 326, 333-335, 343, 347, 352n.5
See also Analepsis

G

Gabon, 11, 92, 143, 357n.5, 371, 373
Gandoulou, Justin-Daniel, 15, 383
Garfinkel, Harold, 7, 35, 51, 184. 278, 287, 314, 330, 360n.6, 368n.9, 383, 384, 388
GCSC (Government Computer Services Center of Kenya), 86, 92, 203, 373
Gender
and the computer profession, 83, 94-96, 146, 292-293
and division of labor, 15, 24, 94-95, 107, 139, 146, 183, 292-293
and feminization of clerical work force, 24, 349-350n.4, 351n.11, 392
and mobility aspirations, 24, 94-95, 293
in office work, 24, 94-95, 107, 183, 190, 386, 389
Genette, Gérard, 310, 365n.9, 384
Génie logiciel (software whiz), 121, 373
Gestures, 8, 256, 258, 264, 265, 266-267, 268, 277-278, 364n.7
backchannel, 267, 364n.7
demonstrative, 261, 263, 265, 267, 269-270, 271, 278, 363n.4
of effort, 258, 273-274
illustrative, 269
mimetic, 363n.4
and ritual (*see* Ritual) as signs, 363n.4, 364n.7
See also Ritual and gestures
Ghana, 220, 382
Giddens, Anthony, 27, 351n.1, 352n.3, 384
Gildin, Hilail, 364n.1, 384
Gill, Colin, 213, 384
GIMI (Groupement Interprofessionnelle pour le Management et l'Informatique), 22, 129, 215-216, 355n.1, 361n.6, 373
Ginguay, Michel, 354n.2, 384
Gintis, Herbert, 17, 20, 220, 380
Giri, Jacques, 322, 344, 351n.13, 366n.1, 384
Goffman, Erving, 54-55, 384
Goldthorpe, John, 14, 384
Goodwin, Charles, 363-364n.6, 384
Gouldner, Alvin, 12, 30, 57-58, 77, 326, 367n.5, 384
Goupil Computers, 133, 146
Graffins College (Kenya) ix, 230, 235-237
Grand Bassam (Ivory Coast), 326, 328
Greimas, Algirdas Julien, x, 4, 28-29, 245, 254, 278, 283, 284, 287, 315, 323, 349n.1, 352n.4, 363n.2, 363n.4, 364n.2, 365n.10, 366n.2, 367n.4, 384, 385
Gusfield, Joseph R., 349n.2, 385

H

Habermas, Jürgen, 12, 326, 367n.5, 385
Habitus, 27, 30, 34, 72, 310, 351n.1, 373
Haferkamp, Hans, 27, 385
Halbherr, Peter, 77, 385
Hamilton, Marcia, 219, 385
Hand-calculation and processing, 19-20, 31, 73-74, 87, 92, 158, 166, 168, 171, 187, 191, 206, 305, 316
Harper, Malcolm, 44, 186, 385
Heim, Michael, 53, 275, 385

Hénault, Anne, 349n.1, 364n.4, 385
Hewlett-Packard Computers, 148, 150
Honeywell Computers, 148, 149
Hotel industry, 109
 and banking, 109, 134
 use of computers in, 91, 109, 130, 134-135, 204, 219
Houphouët-Boigny, Félix (President of Ivory Coast), 113
Hume, David, 283, 364n.1, 385

I

IAI (Institut Africain d'Informatique), 94, 122, 211, 373
Iancano, Suzanne, 185, 366n.12, 386
IBM (International Business Machines), 45, 65, 68, 86-87, 108-109, 133, 138-139, 142, 146, 148-149, 154, 155, 158, 161-166
 in France, 77, 384
 in Ivory Coast, 10, 108-109, 131, 133, 136, 138-139, 146, 148-149, 291-292, 358n.8
 in Kenya, 35, 45, 65, 154, 161-166, 174-175, 209, 229, 243, 268, 269-270, 306, 356n.6, 358n.8, 364n.8
 ranking of, 146, 148, 358n.8, 358n.9
ICDC (Industrial and Commercial Development Corporation of Kenya), 170, 204, 359n.4, 373
ICL Computers, 37, 46-47, 66, 148, 158, 208, 221, 239-240
Ideology, 26, 28, 96, 173, 218, 277, 366n.12, 382, 388
 and social change, 214, 318
 See also Information
Iimu (plural, Aimu), 337, 373
Ikiara, G.K., 353n.10, 385
Ikonicoff, Moises, 388
Indexical expressions, 261, 263, 268, 313-314, 328, 363-364n.6, 379
 and events, 246
India, 116, 345
Industry, 38, 41, 43, 48, 115, 117, 157, 212, 343-344, 356n.9, 359n.1, 359n.4, 361n.1, 361n.5, 361-362n.1, 373, 374
Inforafric, 151
Informaticien (computer expert) 373
 See also Computers, experts with
Information, 80, 113, 117-118, 325
 democratization of, 79, 97-98, 214-215, 243, 317-319, 325
 dissemination of, 215, 243, 325
 era of, 43
 explosion of, 325, 334, 387
 and ideology, 97, 214-215, 317-319, 344, 366n.12, 391
 processing of, 118, 127, 221, 223, 243, 255, 278, 325, 363-364n.6, 373
 rhetoric of, 116
 serial, 325
 See also Knowledge
Informatics, 40, 80, 91, 108, 111, 116-117, 120-127, 213, 217, 367n.5, 373, 382, 387
Informatique (information processing), 108, 373
Informatique de gestion (business computer systems), 373
Ingle, Marcus D., 219, 385
INSET (Institut National Supérieur de l'Enseignement Technique), 6, 94, 116, 118-123, 211, 214, 222, 225, 308, 334, 361n.4, 373, 374, 375
INSTAT (Institut National pour la Sous-traitance et l'Assistance Technique, France), 132, 374
Institute of Computer Science, University of Nairobi, xii, 178, 181, 207, 211, 221-222
 curriculum of, 221-222, 308, 374, 384

See also University of Nairobi
Insurance companies, 17, 44, 66, 174-176, 189, 199, 203, 219, 238, 306, 324, 349n.4, 360n.9
Interaction
embodied, 273, 275
human-machine, 7, 20, 26, 27-30, 34, 36, 51-54, 58, 61-62, 63, 64, 71, 74-76, 101, 117, 242-246, 249-250, 255-257, 259, 260, 261, 273-275, 277, 322, 323, 335, 352n.2, 352n.4, 354n.3, 356n.12, 363n.2, 363n.5, 363-364n.6, 392
mechanization of, 263, 352n.3
nonverbal, 249, 256-257, 261, 262, 263, 269, 270, 271, 277-278, 314, 363n.4, 383
social, 10, 71, 75-76, 93, 117, 184, 224, 249-250, 274-275, 310, 314, 323, 356n.12, 363n.5, 372, 381
and social distance, 275-276
Inventory, 74-75, 191, 275
by computer, 24, 27, 45-46, 47, 68, 73, 75, 101, 130, 136, 163, 187, 201, 251, 267, 269, 270, 304, 324, 329
and invoices, 36, 267, 268, 269, 270
of stock, 24, 45-47, 68-70, 73, 75, 123, 130, 134, 137, 141, 163, 191-192, 251, 267, 269, 275, 324, 364n.8, 372, 376
Investment, 42-43, 49
foreign, 12-13, 42-43, 49, 109, 152, 159, 243
local, 17, 109, 243, 353n.10, 359n.4
strategies of, 44, 159, 195
ISI (Institut Supérieur d'Informatique), 94, 120-121, 128, 211, 374, 375
IUT (Institut Universitaire de Technologie), 211, 292, 374
Ivanov, Vjaceslav, 100, 351n.14, 356n.10, 386, 394
Ivory Coast, 103, 107-123
computer policies of, 5, 11, 21, 41-49, 52, 55, 57, 75, 83, 92, 94, 95-96, 98, 113, 118, 120-122, 129, 160, 202, 205-206, 212-215, 217, 279, 292-299, 303, 309, 342-346, 351n.10, 353-354n.1, 357n.1, 357n.2, 357n.4, 358-359n.10, 359n.2, 365n.4, 365n.5, 365n.6, 376, 382
computer world of, 41, 45-46, 55, 76, 114-115, 127-130, 137-138, 141, 143, 147-150, 152-154, 202, 217, 344
development plans of, 43, 49, 52, 114-116, 143-154
economy of, 41, 47, 294, 334, 342, 357n.5, 358-359n.10, 365n.4, 371, 380
education in, 48, 92, 94, 95, 98, 113-116, 147, 205-206, 210, 215, 217, 279, 332, 343, 346, 353-354n.1
employment in, 44, 109, 144-145, 215, 217, 351n.10, 351n.11, 386
informatics plans of, xi, xii, 11, 19, 21, 39-40, 43, 48, 52, 110-130, 150-154, 210, 215, 294-297, 306, 326-327, 352n.5, 353-354n.1, 355n.1, 357n.2, 357n.3, 357n.4, 358n.7, 359n.2, 365n.5, 365n.6, 372, 374, 390
Ministry of Finance, 131
Ministry of Justice, 117, 132
Ministry of Planning, 132
Ministry of Transportation, 132
National University of, 308, 334, 351n.10, 382
sales of computers in, 113, 143, 289-290, 293-294, 353n.7, 358n.8, 358n.9, 358-359n.10, 365n.4
See also Computer; Education; Employment; National Informatics Plans (Ivory Coast)

J

Jacobs, Jerry, 263, 391
Japan, 12, 46, 164
Jedlicki, Claudio, 388
Jeune Afrique Économie, 146, 148, 151, 358n.8, 358n.9, 391, 393
Jones, Gareth, 20, 24, 82, 187, 198, 275, 351n.11, 360n.9, 381
Jones, Michael Owen, 9, 386
Joshi, Heather, 42, 386
Journées Africaines d'Informatique, 107-109, 386
Jules-Rosette, Bennetta, 12, 95, 191, 331, 349n.3, 353n.8, 356n.7, 356n.8, 360n.7, 386

K

Kamba
  carvers, 69, 78, 191-198
  stories of, 337-340, 389
Kanter, Rosabeth Moss, 16, 386
Kareithi, Maxwell, 67, 83, 354n.4, 355n.3, 386
Kaypro Computers, 74-75, 90, 178-179, 191-192, 356n.6
Kenya, 22, 45, 52, 55, 57, 71-73, 75-76, 130, 135, 157-204, 390
  colonial experience of, 190
  computer associations in, 83, 211, 229, 355n.1
  computer policies of 40, 41-45, 83-96, 158-160, 198, 202, 207, 216, 217, 279, 294, 298, 300-301, 308, 316-318, 342, 350n.7, 352-353n.6, 355n.5, 356n.6, 361n.3, 362n.12
  computer world of, 40, 41, 76, 128, 202, 209, 217, 243
  development plans of, 85-86, 89-90, 153, 157, 346, 359n.1, 361n.3, 390
  economy of, 18, 37, 41, 79, 157, 359n.1, 381
  education in (public and private), 7, 52, 84, 92, 97, 160, 205-206, 214, 216-218, 220, 226-229, 230-237, 249-279, 346, 361n.1, 361n.2, 361n.3, 362n.12, 381
  employment in, 17-18, 44, 351n.11, 360, 362n.12, 381
  financial managers in, 86-87
  independence of 90, 157, 170
  religion in, 96
  sales of computers in, 6, 84, 158-159, 289, 292, 294, 353n.7, 358n.8, 358n.9, 360n.5, 361n.3, 390
  Treasury of, 86, 203, 316
  See also Computers; Education; Employment; Vendors
Kenya Commercial Bank, 203
Kenya Craft Cooperative Union, 194, 196
Kenya Industrial Estates, 45
Kenya Polytechnic Institute, ix, 175, 207-208, 216, 223-225, 226, 238, 242, 246, 361n.2, 374
Kenyan shilling, 178, 196, 232, 360n.7, 374
Keypunchers, 10, 24, 32, 82-83, 95, 171, 187, 292, 372, 376, 382
KIE (Kenya Institute of Education), 223, 374
Kling, Rob, 185, 366n.12, 386
KNEC (Kenya National Examinations Council), 203, 207, 231, 374
Knorr-Cetina, Karin D., 368n.9, 386
Knowledge, 14, 57, 60, 388, 389, 390
  abstract, 58, 275, 311
  cultural, 254, 256-258, 278, 314, 329-330, 392
  embodied, 246, 258, 266
  organization of, 66
  popularization and dissemination of 3, 5, 27, 52, 66, 72, 201, 325, 336, 355n.1

and skill, 254, 257
and technology, 12, 21, 27, 58, 60, 61, 66, 70-71, 79, 200, 208, 220, 242, 244-246, 257, 273-274, 275-276, 303, 315, 324-325, 336, 344, 355n.1, 355n.2
transfer of, 27, 61, 206, 336
Korea, 48, 116, 345
KPCU (Kenya Planters Cooperative Union), 64-65, 203
Kragby, Raymond, x, 123-128
Kruchten, Philippe, 354n.2, 383
KTDA (Kenya Tea Development Authority), 65, 203
Kumar, Krishna, 48-49, 386

L

Labor
division of, 11, 51, 93, 172, 177, 293, 365n.3, 375, 382
and learning, 244-246, 249, 273, 275-277
and organization of the workplace, 379, 380
and productivity, 3, 7, 18, 21, 44, 54, 62, 68, 116, 249, 273, 278, 303, 323, 341, 344
Labor force, 41, 83, 96, 112, 117, 198, 213, 219, 349-350n.4, 351n.11, 361n.5, 361-362n.7
internal, 24, 108, 198
Lall, Sanjaya, 48, 387
Landowski, Eric, 287, 387
Langdon, Steven W., 13, 359n.4, 387
Language: computer, 30, 52, 57, 58, 72, 87-89, 100-101, 116, 124, 250, 251, 252, 257-258, 266, 271, 277, 293, 301, 313, 316, 328-329, 354n.2, 384
creolization of, 315, 366n.11, 382
and dictionaries, 52, 383, 384, 385, 390
everyday, 52, 56-57, 336
figurative, 54, 56-57, 68, 258, 261, 331
functions of, 72, 315, 392
metaphorical, 56, 61, 62, 67, 71, 315, 324, 332, 336
metonymic, 313
natural, 100, 314
oral, 206, 250, 276, 310, 393
and syntax, 329, 366n.11, 366-367n.2
See also Data; Discourse
Lanzarotti, Mario, 388
Larsen, Judith, 75, 391
Last train of the twentieth century, 107, 152, 304, 324, 333, 352n.5, 374
Latour, Bruno, 368n.9, 387
Lecallo, Denis, 45, 387
Lemarchand, René, 14, 387
Le Mon, R., 205, 207, 213, 389
Leonard, David K., 18, 19, 86, 87, 88, 316, 355n.5, 390
Lesson plans: for computer instruction, 126-127, 250, 251, 255-256, 261, 262, 263, 264, 280-281, 329, 363n.6
Levin, Henry M., 213, 361n.5, 391
Lewis, Suzanne Grant, xi, 21, 22, 52, 83, 96, 342, 350n.5, 353n.7, 368n.12, 387
Liberia, 311
Literacy
computer, 205, 209, 212, 228, 230, 255, 361-362n.7, 389
cultural, 254, 257-258, 278, 389
universal, 205
See also Language; Knowledge
Little, Kenneth, 25, 80, 387
Livington, Eric, 368n.9, 384, 388
Logic, 171
algorithmic, 206, 210, 253-255, 312
computer, 138, 254

commonsense, 255, 313
formal, 311-312, 366-367n.2, 371
linear, 257
situational, 311-312
syllogistic, 311-312
Lubell, Harold, 42, 386
Luke, Timothy, 25, 387
Lussato, Bruno, 53, 354n.3, 387
Lynch, Michael, 368n.9, 384, 388
Lyotard, Jean François, 11, 200, 326, 336, 367n.3, 367n.5, 368n.10, 388

M

MacCannell, Dean, x, 325, 388
MacCannell, Juliet Flower, x, 325, 388
MacGaffey, Wyatt, 356n.7, 388
Machalek, Richard, 78, 392
Maffesoli, Michel, 76, 388
Magic
traditional, 337-341, 345
and technology, 4, 28, 337-341, 345
Main test. *See* Tests
MALD (Ministry of Agriculture and Livestock Development of Kenya), 65, 88-91, 159, 203, 374
Malawi, 96
Mali, 11, 109, 342, 357n.5, 371
Mami wata, 331-332, 375
Mannheim, Karl, 3, 37, 102, 388
Mapping
cognitive, 31, 253-254, 257, 277, 313
gestural, 266, 277
relational, 266
Market, 118, 123, 128, 141
for computers, 17, 52, 84, 109, 112, 138, 141-146, 297, 302, 308, 310, 318, 237, 358n.8, 360n.5, 365n.6, 376
for employment, 226, 234, 236-237, 243, 277, 362n.10, 391
international, 109, 342
labor, 24, 220, 361n.5
local, 108, 128, 294-295, 302, 307, 310, 339, 342, 349n.3, 353n.10
postindustrial, 17, 214
for software, 52, 140, 141-142, 341, 342
sanctions of, 295, 297, 301, 306
Markus, M.L., 18, 388
Marshalls East Africa, 46-47, 203
Masini, Jean, 13, 42, 388
Mastery, 34, 37, 53, 69, 71, 111, 121, 375
of computers in the workplace, 35, 37, 53, 55, 66-71, 73, 84, 89, 98, 102, 111, 158, 173, 200, 219, 244, 272-274, 276-277, 323-324, 346, 352n.3, 364n.8, 366n.12
embodied, 55, 258
and learning, 69, 71, 121, 207, 226, 249, 254, 258, 272-273, 276-277, 323, 334, 346
in the narrative program, 41, 53, 58, 102, 249, 258, 276, 277, 285, 363n.2, 375, 376
Mathematics, 121, 212, 335
and computer training, 221-225, 228, 255
Mauritania, 133-134
Mauss, Marcel, 286, 388
Mazuri, Ali A., 41-42, 98-99, 343, 352n.6, 354n.3, 356n.9, 388
Mbiti, John, xii, 337, 338, 339, 340, 368n.11, 389
Medicine
modern, 314, 344
traditional African, 338-339, 341, 345
Mehan, Hugh B., x, 212, 255-256, 389

Memory
computer, 10, 33, 149, 184, 164, 374
and orality, 206, 276
Menzies, Heather, 158, 217, 389
MFP (Ministry of Finance and Planning of Kenya), 90-91, 203, 356n.6, 375
MIAG (Maîtrise d'Informatique Appliquée à la Gestion), 122, 374, 375
Microcomputers, 18, 20, 21, 42, 52, 58, 124-125, 134, 142, 151, 353n.7
cost of, 80, 160, 298, 301, 303, 305, 358n.9
use of, 19, 32, 45, 46-47, 52, 62, 72, 74, 78, 83, 86, 87-91, 98, 111, 142, 158, 179, 189, 226, 232, 237, 256-257, 301, 304, 355n.5, 356n.6, 368-369n.11, 374, 375, 380, 385, 387, 390, 392, 394
See also Computerization; Computers
Minc, Alain, 6, 110, 205, 214, 326, 357n.4, 389
MOA (Ministry of Agriculture of Kenya), 18, 33, 65, 85, 86, 88, 316, 355n.5, 374, 375
Mobility
in firms, 133
horizontal, 83, 324
problems of, 217, 220, 276, 293, 309
vertical, 15, 24, 37, 80, 83, 95, 99, 133, 171, 174, 185, 198, 201, 217, 243, 293, 309, 313, 324
Modeling systems, 26, 100-103, 313-314, 356n.10, 394
computer, 250, 254, 261, 277, 313-314
cultural, 102, 312
primary, 100, 277, 392
secondary, 100-102, 277
tertiary, 314
Models, 8, 17, 26, 28, 102, 312, 324, 326-328, 334, 346, 360n.10
computer, 26, 68, 327, 351n.14, 352n.2
of the future, 326
Modernity, vii, 13, 51, 158, 322, 323, 331, 342, 379
broken promise of, 13-16, 24, 347
Modernization, 23, 42, 343, 350n.6, 388
Moi, Daniel Arap (President of Kenya), 85-86, 157
Mombasa (Kenya), 162, 164, 349n.3
Monitoring
instructional 261-263, 363-364n.6
reflexive, 27, 68, 97, 263, 323, 352n.3, 360n.8, 363n.6
Mouly, Jean, 42, 386
Mozart (as computer entry), 257-258, 259, 278, 362-363n.1, 386
Mukerji, Chandra, 99, 389
Mushioki, Willie, viii

N

Nairobi (Kenya), 73-74, 130, 160-167, 174, 249, 334
City Council of, 38, 203
commercial district of, 123, 128, 174-176, 235
computer world of, 126, 128, 167, 174, 239, 342
cooperatives in, 191
industrial area of, 161, 191
retail traders in, 167-170
See also Kenya
Nairobi Handicraft Cooperative, 33, 74, 142-147, 191-198, 202
management committee of, 31, 192-197, 304-305
See also Cooperatives, handicraft
Narrative
complications of, 31, 68, 285, 308, 325, 337, 367n.4

contract type, 286-287, 365n.9
cultural, 3, 4-5, 10, 26, 27-28, 29, 51, 67, 102, 337, 349n.2, 351n.1, 368n.11, 376
deconstruction of, 325, 367n.4
format of, 257
and models, 101-102, 254
postmodern, 30, 335-341, 367n.3, 367n.5, 368n.10
program, 10, 249-250
program for computer adoption, 4-5, 17, 27, 28, 29, 30-40, 43, 45, 52, 65, 75-76, 101, 102, 112, 159, 200, 249, 251, 254, 271-273, 278-279, 284-285, 295, 299-300, 313, 317-318, 322, 325, 331, 336-337
and social practices, 27, 200, 249, 254, 278, 336, 345
structure of, 5, 27-28, 101-102, 112, 254, 271, 283, 287, 325, 336
technological, 65, 92-93
and text, 101, 249-250, 251, 254
transformation of, 254, 325

National Informatics Plans (Ivory Coast), xi-xii, 19, 21, 39, 40, 43, 48, 49, 52, 109, 111-130, 151-154, 210-211, 215, 295-298, 303, 306, 308, 326
See also Ivory Coast

Nationalization, 43, 48-49, 118-121, 152, 153
of business, 48-49, 343, 353n.10
of computer elite, 118-121, 222, 343-344, 346
of computer industry, 51, 206, 215, 343-344
See also Class, elite

NCR Computers (National Cash Register), 10, 142, 148-149, 155, 171-174, 181-182, 203, 208, 381
ranking of, 148, 358n.8

Negroponte, Nicholas, 205

New mechanical solidarity. *See* Solidarity, new mechanical

New technicians of the sacred, 5, 26, 57, 72, 75-85, 93-94, 98-99, 107, 122, 202, 216, 266, 347, 355n.2, 364n.9, 375, 391

N'Guessan, Seri-Bi, x, 118-119, 120-121, 122

Niger, 57, 109, 133

Nigeria, 151

Nora, Simon, 6, 110, 205, 214, 326, 357n.4, 389

Nora-Minc Report, 6, 110, 205, 214, 326, 357n.4, 389
See also Computerization of society; France

Nordin, John P., ix, 315, 389

Norman, Colin, 17, 18, 389

Nzioka, Gideon, 63, 389

O

OAU (Organization of African Unity), 351n.13, 367

Objects, 75, 328
anthropomorphic, 53, 56, 75-76, 354n.6
cultural, 3, 52, 78, 312, 330, 354n.6, 356n.10, 379
inanimate, 4, 8, 53, 363n.5
sacred, 28, 72, 78, 266, 337, 354n.7
and subjects, 29, 34, 39, 51, 53, 71, 323, 337, 340, 349n.1, 363n.2, 364n.2, 366-367n.2, 372
tools as, 71, 75, 258, 266, 314, 323, 332, 339, 342, 346, 363n.5, 368n.7, 368n.8, 380
valued, 37, 51, 61, 65, 77, 200, 249, 268, 273, 304, 310, 317, 323, 326-327, 336, 360n.10, 364n.9, 366-367n.2

Oceanography, 99

OCM (Office Central de la Mécano-

graphie), 111, 113, 132, 143, 371, 375, 376
OECD (Organization for Economic Cooperation and Development), viii, xi, 146, 213, 350n.7, 389
Olivetti Computers, 148, 158, 203, 222
ranking of, 148
Onduleur (surge suppressor), 143-144, 373
Ong, Walter, 206, 349, 389
Ordinateur (Computer or data processing system), 109, 147, 375
See also Computers; *Système d'exploitation*
Osborne Computers, 90-91, 203, 356n.6
Oyomno, Mr., ix, 223, 225

P

Papagiannis, George J., 205, 207, 213, 228, 389
Papert, Seymour, 205, 206, 207, 228, 255, 257, 389
Parastatal organizations, 38, 45, 62, 131, 157, 175, 306
Parsons, Talcott, 284, 336, 337, 344, 365n.3, 389-390
Passeron, Jean Claude, 220, 380
Peirce, Charles Sanders, 314, 328, 390
Performance, 28-29, 76
computer, 53, 76, 278
in interaction, 28, 53, 55, 364n.7
job, 55, 276, 362n.12
in narrative, 366-367n.2
oral, 127, 206, 276, 364n.10, 389, 393
public, 28-29
Perrolle, Judith A., 18, 110, 185, 209, 212, 243, 390
PDG (President Director General), 108, 147, 375
Pichette, Serge, 97, 390
Pinckney, Thomas C., 18, 19, 86, 87, 88, 316, 355n.5, 390
Polanyi, Michael, 84, 336, 368n.10, 390
Policy, 3
computer, 3, 10, 11, 13, 30, 43, 56, 109-111, 128, 133, 146, 151-154, 201-202, 205-206, 212-214, 244, 287, 292, 294, 299, 303-304, 322, 342, 345-346, 352n.3, 357n.2, 357n.3, 363n.2
government, 4, 19, 25, 49, 103, 109, 112-113, 133, 148, 151-154, 160, 214-219, 288, 294, 299-303, 318, 322, 325, 343, 345-346, 352n.3, 355n.4, 366n.1, 388
public, 30, 80, 250, 278, 323, 349n.2, 394
Postmodernity, 11-12, 17, 26, 30-32, 84, 153-154, 278, 300, 304, 317, 322-323, 326, 332-333, 341, 345, 347, 367n.3, 375, 376
and art, 328
dream of, 13, 16, 26, 313, 323, 341, 347
and narrative, 326
signs of, 8, 322-329, 333, 343, 344-345
simulation of, 16-17, 26, 31-32, 247, 313, 317, 343, 351n.14
See also Simulation
Pralong, S., 20, 394
Price Waterhouse, 47, 155, 159, 204, 300-302, 390
Prince, Gerald, 286, 390
Professionalization, 21-24, 77-78, 82, 83-84, 94-95, 115-123, 125-154, 185, 214-218, 225
standards of, 214-218, 302, 350n.5
See also Computer Professionals
Professionals, xi, 14, 19, 24, 30, 53, 69, 71, 87, 198, 373, 376, 383
See also Computer Professionals

Progiciel de gestion (business computer software), 109, 136, 376
Programmers, 19, 24, 32, 34, 36, 58-60, 66, 69-73, 80, 88-89, 99, 108, 115, 135-136, 138, 142, 146, 171-172, 176, 182, 188-189, 207, 210, 213, 217-219, 218, 220, 225, 236, 237-243, 245, 249, 253, 255-256, 276-277, 291-292, 305, 313, 372, 375, 376
senior, 16, 20-21, 25, 35, 60, 68-69, 71-73, 99, 112, 168-169, 188, 238-240, 274, 309, 315, 375, 376
Programming, 137-140, 150, 291
applications of, 221, 275, 298, 305, 310, 316
skills of, 70, 169, 189, 213, 221, 231, 254, 256, 276-277, 313, 316, 376
See also Computers, programs for
Progressive computer contract. *See* Contract, progressive computer
Pupitreur, 147, 376
See also Staff, data entry

Q

Qualifying test. *See* Tests, qualifying in narrative

R

Radnor, Michael, 3, 393
Recoding. *See* Codes
Recordkeeping, 34, 39, 65, 123, 183, 201, 360n.6
and computers, 10, 35, 65, 123, 157, 178, 183, 194, 201, 221, 251, 278, 329, 340, 359n.1, 360n.6
Recruitment
on the job, 94, 95, 161, 217
professional, 77, 94, 95, 161, 175
and religious sects, 85, 95-96
in schools, 208, 213, 225, 226, 231, 234, 238, 292
Redundancy
in computer training, 225
in narrative, 285
of work tasks, 285
Referential illusion, 16, 32, 153, 244, 315, 331
Regulation, 42, 111, 117, 129
ambivalence and results of, 4, 6, 13, 22, 39, 42, 160, 207, 217, 289, 300, 308, 309, 318, 326, 342-343, 346
of computer imports, 40, 43, 45, 109, 142-146, 148, 153, 159, 288, 295, 300-302, 307, 309, 334, 342-343, 355n.4, 368n.11
of computers, 6, 45, 113, 159, 212, 297-301, 307-308, 309, 310, 318, 326, 342, 344-345, 352-353n.6, 357n.3, 365n.5, 365n.6, 367n.5
of customs, 6, 17, 45, 91-92, 109, 117, 131, 159, 291-292, 301, 309, 357n.4
of distribution of computers, 17, 109-110, 137, 142-146, 148-153, 289, 295
of employment, 215
of foreign exchange, 81, 216-217, 300, 358-359n.10
See also Control, government
Religion
and believers, 14, 78-80 313-314
and discourse (*see* Discourse)
Christianity as, 368n.11
and ideology, 96-97, 313
and modeling systems, 90-100, 313-314
and parables, 101
and sects, 77-79, 96-97, 99, 102, 321, 356n.7, 356n.8, 386
and technology, 99-101, 356n.7
traditional African, 34, 331-332, 337-341, 368n.11, 373, 375

See also Discourse; Ritual; Symbols
Reputation, 15, 73
of computer elite, 15, 30, 171, 174, 218
in the workplace, 74, 171, 174
Retail traders, 91, 170-174, 197-198, 219, 273-274, 302, 329
and sales, 162
Revolution
administrative, 17-18
computer, 40, 97, 107, 116, 121, 151-153, 341-343, 387
industrial, 18
Rhetorical strategies. *See* Discourse and rhetorical strategies
Ritual
and behavior, 8, 28, 255, 266, 381
and ceremonies, 29, 206, 344, 368n.6, 393
and codes of empowerment, 8
and gestures, 257, 266, 267, 269-270, 276, 341, 363n.4
of initiation, 78, 125, 206, 245, 276, 332-333, 341, 368n.6
and representations, 102
and symbols, 34, 101-102, 368n.6, 393
Rodney, Walter, 321, 390
Rogers, Everett, M., 75, 391
ROSTA (Regional Office for Science and Technology for Africa), 221
Roszak, Theodore, 53, 205, 254, 255, 352n.2, 354n.3, 391
Rothenberg, Jerome, 77, 391
Rousseau, Jean-Jacques, 364n.1, 391
Rule, James, 18, 379
Rumberger, Russell W., 213, 361n.5, 391
Rumelhart, David E., 252, 363n.6, 391

S

Sacks, Harvey, 314, 383
SADMIL (Ivory Coast), 155
Salaman, Graeme, 76, 356n.11, 391
Sapeurs, 15, 376
Sartre, Jean-Paul, 330, 368n.8, 391
Savané, Ibrahim Sy, 358n.9, 391
Schechner, Richard, 53, 391
Schegloff, Emanuel A., 9, 54, 349n.2, 391
Schools, 41, 48, 249, 322, 380
in California, 41, 255-256
for computer training, ix, 24, 43, 52, 58, 70, 98, 111-112, 118-128, 209, 212-220, 221-237, 255-256, 357n.5, 361n.1, 361n.2, 363n.2
curricula of, 118-119, 126, 128, 205, 208, 213, 217, 218, 221-225, 226-227, 239, 255, 303, 322
in Ivory Coast, 43, 111-112, 115-117, 118-128, 214, 217, 218, 343
in Kenya, 43, 70, 79, 92, 208-209, 213-214, 216-218, 226-237, 249-250, 361n.1, 361n.2
leavers of, 15, 126, 208, 217, 229, 362n.10
secondary, 7, 15, 70, 115, 125, 189, 205-207, 213, 214, 217, 218, 226-229, 238-239, 243, 321, 334, 361n.1
secretarial, 207, 229, 235
and universities, 41, 109, 115-116, 118-123, 128, 150, 205-207, 221-225, 235
See also Education
Schuster, Ilsa, 349n.4, 391
Schutz, Alfred, 54, 363n.3, 391
Schwartz, Howard, 263, 391
Scott, Robert J.P., viii, xii, 207, 213, 222, 227, 229, 350n.7, 353n.7, 362n.12, 392
Scribner, Sylvia, 311-312, 381
Sebeok, Thomas A., x, 100, 277, 314, 351n.14, 356n.10, 392
Secularization, 77, 98

of education, 98
Semiotics, 8-9, 29, 100
and analysis, 4, 8-9, 30, 273, 297, 337, 363n.2, 365-366n.10
and concept of contract, 8, 28-29, 30, 112-113, 283-288, 295, 297
of invention, 253-256, 258, 362-363n.1, 368n.8
of narrative, 4, 8-9, 200, 382, 383, 387
and processes of signification, 4, 9, 27, 29, 274, 277-278, 312, 322, 329, 363n.4, 364n.7
and semiosis, 328, 368n.7
theory in, 8, 100-103, 277-278, 383, 385, 388
See also Signs; Sociosemiotics; Symbols; Text
Sénégal, 109, 133, 142, 205, 357n.5, 371
SGI (Secrétariat Général à l'Informatique), 108, 113, 129, 143-144, 344, 357n.2, 371, 375, 376, 392
Sheya, Mohammed, S., 21, 52, 353n.7, 387
SIEPAT-Ordinateur, x, 123-128, 136, 334
Signs
analeptic, 8, 19, 324, 333-335, 376
computers as, 16, 52, 72, 294, 312, 313-317, 324, 328-333, 345-347
cultural, 7, 98
functions of, 31, 252
and interpretants, 328-333, 368n.7
of media, 328-329
and models, 100-103, 250, 261, 312-315, 317, 324, 326, 376, 380
nonverbal, 314, 363n.4, 364n.7
of progress, 31, 74-75, 153-154
referential, 16, 31, 312-315, 326-333, 367n.3, 368n.8
reflexive, 323, 325
religious, 102
representational, 8, 16, 31, 250, 317-319, 324-325, 328, 343, 376
and signifiers, 324, 329-333
structure of, 328
and symbols, 3, 52, 60, 72-73, 75, 100-103, 294, 318, 326, 386
technological, 328-333
terminal, 8, 17, 58, 252-253, 278, 319, 321-347, 352n.4, 366-367n.2, 367n.3, 367n.4, 375, 376
totalizing, 8, 324, 330-335, 376
visual, 228-333
See also Semiotics; Sociosemiotics; Symbols; Text
Simmel, Georg, 54, 392
Simulation, 16-17, 26, 31, 33-34, 53, 247, 254, 313, 326-327, 352n.2, 354n.3, 360n.10, 376, 379
of apparel, 326-327
and models, 312-317, 326-327, 343, 351n.14, 352n.2, 376
See also Postmodernity
Sinclair Computers, 222, 226
Smelser, Neil, x, 84, 392
Snow, David, 78, 392
Social drama, 28-29, 376
Société Général de Banque, 47, 134
Sociosemiotics, 9, 250, 279, 283-284, 287, 312, 313, 322, 346, 365-366n.10
See also Semiotics
Söderqvist, Thomas, 383
Solidarity, 55
mechanical, 11
new mechanical, 7, 8, 11, 26, 34, 54, 55, 64, 73, 219, 244, 263, 323, 375
organic, 11
SSII (Sociétés de Services et d'Ingénerie en Informatique), 6, 22, 109, 128-151, 153, 210, 216, 297-299, 309, 358n.6, 358-359n.10, 376

Sperber, Dan, 58, 392
Staff, 18, 125, 202, 350n.5, 354n.5, 377
  clerical (*see* Workers, clerical)
  computer, 92, 109, 159, 169, 171, 182, 185, 188, 198, 201, 202, 288, 306, 359n.3, 376, 386
  data-entry, 95, 98, 159, 168-169, 172, 175, 176, 182, 185, 188, 198, 200-201, 217, 291-292, 306-308, 372, 376
  operations, 66, 80, 95, 213, 386
Starehe Boys Centre (Nairobi), ix, 203, 207, 217, 226-228, 237-244, 256-268, 276, 334-335, 362n.8, 362n.9, 362n.11
Storytelling, 206, 286, 368n.11
Stover, William James, 354n.3, 392
Stratification, 16, 82, 198
  economic, 256
  and employment, 83, 214, 217
  social, 23, 220, 256
  and social inequality, 30, 220
Strom, Sharon Hartman, 22, 392
Structuration, 27, 351n.1, 383
Structure
  business, 181
  narrative, 27, 28
  organizational, 27, 56-57, 74-75, 118, 146, 147, 180
  semiotic, 100, 283, 327
  social, 27, 220, 385, 389
  See also Narrative, structure of
Success
  computer-related, 6, 16, 30, 33-34, 39, 51-76, 90, 91, 101, 159, 168, 182-183, 184, 217, 243, 257, 273, 277, 300, 307, 309, 318, 331 337
  in relation to computer failures, 37, 54-56, 61-62, 63, 67, 76, 101, 250, 257, 286, 300, 307
Suchman, Lucy A., 56, 392
Sudan, 51, 72, 159, 342, 354n.7, 358n.9, 382
Sudnow, David, 252, 393
Survey polls, 117, 329, 358n.8, 358n.9, 360n.5, 387
Svensson, Lennart G., 383
SYDAM (Système de Dédouanement Automatique de Marchandises), 149
Syllogisms. *See* Logic, syllogistic
Symbols
  of change, 12, 13, 26, 28, 31, 75, 294, 318, 386
  computers as, 12, 26, 75, 171, 266, 275, 294, 299, 304, 307, 315, 316, 318, 321, 323, 346, 360n.10
  condensation and unification of, 330
  of modernity, 3, 12-13, 347, 360n.10
  of postmodernity, 12, 30, 72, 315, 317
  of prestige, 196
  referential, 368n.6
  religious, 34, 72, 77
  and signs, 3, 31, 52, 100-102, 294, 325, 386
  and society, 327
  See also Postmodernity; Semiotics; Signs
Syncretism, 318, 331
Système d'exploitation (computer operating system), 147, 376
  See also Computers; Systems, operating
Systems, 61-63
  computer, 16-17, 20-21, 35, 37, 57, 62, 63, 64, 68-69, 72, 74, 108, 131, 136, 142-145, 161-164, 168, 172, 177-183, 185-186, 190, 195-196, 200, 219-224, 232-233, 242-246, 275-276, 285, 298-300, 305-306, 309, 314. 316-317, 364n.8, 365n.8, 371, 376, 380

design of, 242, 252, 373
failure of, 334-335
for fiscal reporting, 85, 117, 123, 132, 136, 157, 178
inappropriate, 309
of justification, 102
multi-user, 24, 73-74, 78, 180, 182, 232, 287
on-line, 32, 46, 64, 69, 136, 150, 164, 167, 239, 360n.9
opacity of, 275-276, 323
operating, 72, 117, 124-126, 133, 147, 173-176, 181, 186-188, 275
religious, 100-103
of representation, 334-335
semiotic, 99-103
Systems analyst, 20, 25, 56-57, 118, 129, 134, 138, 147, 184, 188-189, 198, 208, 213, 216, 218, 221, 223, 274, 277, 286, 293, 354n.5, 354n.6, 371, 372, 376
SYTRANSPAC, 149, 357n.4. *See also* TRANSPAC
Szombati-Fabian, Ilona, 331, 393

T

Tabflow (Kenya), 203
Tactile grounding, 266-268, 277-278
and learning, 206, 246, 265
Taiwan, 48, 345
Tannen, Deborah, 349n.2, 393
Tanzania, 21, 22, 44, 52, 57, 80, 83
computer associations in, 22-23, 80, 83, 96, 98, 218, 342, 350n.5, 350-351n.9, 371
computer installations in, 353n.7, 354n.3
computer policy of 22, 301, 342
computer profession in, 96
small enterprises in, 44
See also COMPAT
Technicians, 57, 72, 119, 292, 332
See also New technicians of the sacred
Technology, 8, 33, 48, 152-154, 157, 201
accountability of, 35, 36, 56, 58, 71-73, 139, 199, 201, 243-244, 278, 303, 324
advances in, 116-117, 307, 317, 321, 323-325, 343, 367n.5, 376
appropriate, 85, 91, 110-111, 116, 158, 166, 180, 184
and competition, 91
cost of 79, 298, 301, 303, 307, 309
dissemination of, 45, 52, 80, 85, 159, 224, 243, 287, 288, 291-293, 297, 299, 310, 315-316, 322, 332, 340-344, 351n.12, 353n.7, 353-354n.1, 358n.9, 387
failures of, 34, 37, 40, 55-56, 75, 92, 101, 160, 317, 334
fear of, 101, 304-305, 329-334, 337, 343
ghettoization of, 159
high, 48, 76, 98, 153-154, 212, 244, 331, 391
impact on social change, vii, 3, 10, 20, 23, 40, 48, 64, 110-111, 117, 159, 215, 244, 289, 316, 332, 344, 356n.12, 359n.3, 368-369n.11, 379, 390, 392, 394
industrial, 43, 332
innovation of, 25-26, 27, 83, 97, 100-103, 110, 191, 215, 285, 287, 297, 305, 317-319, 334-335, 343-345, 352n.4, 356n.12, 367n.5, 392
and nature, 336-340
new, 3-4, 9, 12, 20, 21, 25-26, 29, 47-53, 61, 72, 76, 80, 83, 92-93, 98, 103, 110-111, 117, 151, 160, 180, 200, 212, 225, 229, 242-243, 250-279, 285, 286, 298, 310, 312, 316-319, 325-326, 333-334, 341-347, 349n.2, 352n.6, 354n.3, 357n.1, 360n.10, 361n.5, 364n.8, 364n.9, 364n.10, 365-

366n.10. 366n.11, 366n.12, 367n.3, 367n.4, 368n.11, 371, 372, 373, 375, 376, 381, 391
power of, 16, 27-28, 30, 39, 54-57, 58, 72-74, 76, 101, 184-185, 220-221, 244-245, 254, 273, 278, 311, 315-316, 321, 324, 326, 336-341, 355n.8
progress of, 109, 325, 334, 343
and science, 117, 254, 303-304, 334-336, 343-344, 367n.3, 367n.5, 368n.9, 368n.10, 383, 384, 387, 388, 389
and social practices, 17, 30, 69, 250-279, 287-288, 297, 310-311, 314, 324, 326, 333, 336, 345, 368n.8, 368n.9, 375
standardization of, 159
successes of, 34, 37, 54, 58, 120, 160, 170, 184, 285, 343
and technicalization, 98
transfer of, 75, 185, 297, 345, 368-369n.11, 390, 393
vicissitudes of, 324
See also Computers; Knowledge; Success, computer-related
Telecommunications, 4, 38, 64, 108-111, 113-114, 149-150, 206, 224, 239-240, 295, 299, 326, 338, 352n.2, 357n.1, 357n.3, 357n.4, 367n.5
Telematics, 110-111, 113
Terminal utterance, 278, 323, 366-367n.2, 367n.4
Tests, 285
aptitude, 208, 231, 235-236, 238, 362n.12
entrance, 231, 235, 362n.12
main in narrative, 34, 285, 336-337, 367n.4, 374
qualifying in narrative, 3, 285, 295, 317, 337, 367n.4, 376
See also Credentials; Signs, terminal
Text, 10, 250, 287
biblical, 101
computer, 101, 184, 200-201, 250-252, 256, 258, 259, 260, 261, 262, 263, 264, 265, 269-270, 275, 277-278, 314, 360n.6, 363n.3, 372
format of, 29, 251
and genres, 310, 365n.9
oral, 206-207, 252, 349n.2
and paratext, 310-312, 318, 365n.9
rules for, 254, 365n.9
semiotic analysis of, 29, 66, 101, 184, 275, 349n.1, 349n.2
social implications of 287, 312, 365-366n.10
technical 30, 285
technological, 294
written, 66, 250, 251, 254, 283, 365n.7, 393
Time
clock, 252, 253, 259, 260, 261, 262, 263, 264, 265, 269-270, 363n.3
computer, 250-253, 259, 260, 261, 262, 264, 265, 269-270, 363n.3
digital, 252
linear, 252, 363n.3
lived, 252, 256, 363n.3
and narrative, 252, 284, 382
real, 252, 271
Toffler, Alvin, 53, 393
Togo, 134, 145
Toporov, V.N., 100, 351n.14, 356n. 10, 394
Totalization. *See* Signs, totalizing
Tourism, 157, 219, 326
Tourist art, 191, 326, 386
Toyota Dealership, 161-167, 203, 238, 272
See also Automobile dealerships
Trance
and absorption, 252
computer, 252, 277, 352n.3, 354n.4

and disengagement, 267-268
TRANSPAC, 357n.4
See also SYTRANSPAC
Travailleurs
nonproductifs, 147
productifs, 147
See also Workers
Tupper, Patricio, 358n.8, 393
Turkle, Sherry, 53, 219, 245, 252, 315, 354n.3, 393
Turner, Glyn (teacher), ix, 226
Turner, Victor, 28-29, 330, 368n.6, 393
Tyler, Stephen A., 355n.8, 393
Typewriters, 141
in relation to computers, 24, 74, 141, 171, 189-190

U

UNESCO, 25, 221, 351n.12, 393
Uganda, 57
Ungar, Sanford J., 357n.5, 359n.1, 393
Unions, 21, 64-65, 111, 196, 215, 218, 361n.6
and syndicates, 22, 215, 361n.6
UNISYS Computers, 138, 143, 146, 148, 149
ranking of, 148, 358n.8, 358n.9
United States, 132
banking and finance in, 17, 149
in comparison with African nations, 12, 14, 17, 44, 46, 217, 228-229, 350n.6, 350n.7
computer training in, 361n.5
and computers, 17-18, 46, 75, 217, 219-221, 350n.7, 361n.5
and employment, 44, 350n.6
historical figures in, 258, 260
University of Nairobi, viii, 18, 203, 216, 221-223, 227, 361n.1, 361n.2
Chiromo campus of, 221, 374
See also Institute of Computer Science, University of Nairobi
Urry, John, 14, 379
USAID (United States Agency for International Development), 86, 91
Valet, Blagué, x, 123-128, 206, 334
Vansina, Jan, 206, 393
VAX, 151
Vendors, 6, 17, 21-22, 24, 45, 52, 57, 71-72, 83, 109-113, 129, 133, 138-139, 142, 146-153, 299-300, 316, 373
competition among, 17, 109, 112, 146, 148, 150-154, 158, 288, 290-294, 308, 309, 339, 351n.10, 358n.7, 358n.8, 358n.10, 376
complaints of and about, 17, 293-295, 301-302, 305-306
of computer hardware, 6, 70, 81, 83-85, 91, 96, 288, 291, 298-300, 305-306, 309, 316, 351n.10, 358n.8, 358-359n.10, 365n.8, 371, 372
of computer software, 189, 288-289, 298, 358n.7, 371, 372
lobby of, 21, 84, 97, 129, 302, 361n.3, 371, 373
role in computer training, 17, 25, 52, 57, 70, 79, 146, 148, 209, 230-231, 239, 307, 316, 361n.3, 365n.8
transactions of, 287-290, 299-300, 305-306, 309, 372
Vidal, G., 358n.9, 382
Videotape
analysis of, 7, 161, 250, 252, 256, 259, 260, 262, 263, 264, 265, 269-270, 312, 346, 362-363n.1, 364n.7
games on, 252
impact of, 352n.2
VISICALC, 87-88, 251

W

Wad, Atul, 3, 393
Wallace, John B., 57, 354n.5, 393
Wang Computers, 65, 74, 86, 146, 148, 150, 155, 158, 173, 203-204, 233
Warhol, Andy, 327
Warren, Bill, 49, 394
Weber, Max, 77, 220, 363n.5, 394
Weekes-Vagliani, Winifred, viii, 20, 394
Wells, Louis T., 48, 394
Wescott, Clay G., 85, 86, 87, 88, 90, 92, 318, 356n.6, 394
Westernization, 51, 98, 278, 345, 368n.11
White, Stephen K., 25, 387
Williamson, N., 205, 207, 213, 389
Wilson, David (teacher), ix, 226, 238, 362n.8, 362n.9
Wilson, Ernest J., 157, 394
Woolgar, Steve, 368n.9, 387
Word processing, 24, 36-37, 53, 70, 74, 118, 125, 176-178, 183, 188-189, 229-233, 235, 352n.3, 372, 379, 385
Word processors, 36-37, 95, 98, 171, 188-189, 200, 220-221, 229, 233
Work
- impersonalization and mechanization of, 36, 275, 352n.3, 356n.11, 360n.9
- organization of, 52-53, 58, 147, 187-191, 200-201, 211-213, 360n.10, 375, 391
- specialization of, 11, 135, 137, 216-217, 220, 274-275, 354n.5, 360n.9, 360n.10
- technology-centered, 200, 354n.4

Workers, 23, 53, 55, 57, 66-69, 70-74, 111-112, 129-130, 147, 153-154, 193, 200-202, 224, 225, 244, 306, 345-347, 349n.3, 373, 381
- autonomy of, 137-138, 187-188, 202, 215, 219, 240, 244, 275, 276
- blue collar, 351n.11
- categories of, 24, 42-44, 122-125, 134-135, 147, 152-153, 188, 216-217, 224-225
- clerical, 14, 18, 23, 24, 54, 57, 64, 74, 82-83, 87-90, 98, 134, 158, 164, 169, 179, 186, 188, 197-201, 240, 350n.5, 350n.6, 350n.7, 351n.11, 352n.3, 373, 377, 379, 392
- hierarchy of, 58, 59, 60, 71-72, 147, 216-217, 271-278, 293
- and internal labor market, 24, 41-42, 96, 108-112, 115-116, 122, 152-153, 293, 350n.6
- as managers, 17, 18-20, 22, 42, 43, 44, 54, 57, 60-63, 68, 71-73, 75, 78, 91-93, 130, 135, 139, 152, 158-160, 165, 168-169, 184-187, 190-191, 200-202, 209, 215, 218, 222, 242-246, 275, 288, 292-293, 306-307, 313, 324, 344, 347, 354n.5, 354n.6, 356n.11, 361n.2
- and mobility, 24, 41-42, 98, 122-123, 306
- nonproductive, 83, 134, 377
- productive, 55, 83, 87, 134, 202, 377
- semiskilled, 67
- stratification of, 147, 214, 293
- supervision of, 145, 200-201, 216, 220, 238-239, 244-246, 255-256, 309, 352n.3, 356n.4, 359n.3
- unskilled, 153, 213
- white collar, 14, 15, 20, 26, 350n.6, 351n.11, 381

World Bank, 25, 86, 125, 131, 355n.5, 384
Wray, Brian, ix, 228-229, 353n.9, 362n.9
Wynn, Eleanor Herasimchuk, 57, 394

X

Xerox Computers, 203

Y

Yamoussoukro (Ivory Coast), x, 118, 373, 374

Z

Zaire (former Belgian Congo), 14, 96, 387, 393
Zaliznjak, A.A., 100, 310, 351n.14, 356n.10, 394
Zambia, 96, 321, 368n.6, 386
Zimbabwe, 151
Zuboff, Shoshana, 62, 245, 246, 249, 275, 276, 364n.10, 394

Bei Fragen zur Produktsicherheit wenden Sie sich bitte an:
If you have any questions regarding product safety, please contact:

Walter de Gruyter GmbH
Genthiner Straße 13
10785 Berlin
productsafety@degruyterbrill.com